Introduction to Proof in Abstract Mathematics

Andrew Wohlgemuth
University of Maine

Dover Publications, Inc.
Mineola, New York

Copyright

Copyright © 1990, 2011 by Andrew Wohlgemuth
All rights reserved.

Bibliographical Note

This Dover edition, first published in 2011, is an unabridged republication of the work originally published in 1990 by Saunders College Publishing, Philadelphia. The author has provided a new Introduction for this edition.

Library of Congress Cataloging-in-Publication Data

Wohlgemuth, Andrew.
   Introduction to proof in abstract mathematics / Andrew Wohlgemuth. — Dover ed.
     p. cm.
   Originally published: Philadelphia : Saunders College Pub., c1990.
   Includes index.
   ISBN-13: 978-0-486-47854-8 (pbk.)
   ISBN-10: 0-486-47854-8 (pbk.)
    1. Proof theory. I. Title.

QA9.54.W64 2011
511.3'6—dc22

20010043415

Manufactured in the United States by Courier Corporation
47854804   2018
www.doverpublications.com

Introduction to the Dover Edition

The most effective thing that I have been able to do over the years for students learning to do proofs has been to make things more explicit. The ultimate end of a process of making things explicit is, of course, a formal system—which this text contains. Some people have thought that it makes proof too easy. My own view is that it does not trivialize anything important; it merely exposes the truly trivial for what it is. It shrinks it to its proper size, rather than allowing it to be an insurmountable hurdle for the average student.

The system in this text is based on a number of formal inference rules that model what a mathematician would do naturally to prove certain sorts of statements. Although the rules resemble those of formal logic, they were developed solely to help students struggling with proof—without any input from formal logic. The rules make explicit the logic used implicitly by mathematicians. After experience is gained, the explicit use of the formal rules is replaced by implicit reference. Thus, in our bottom-up approach, the explicit precedes the implicit. The initial, formal step-by-step format (which allows for the explicit reference to the rules) is replaced by a narrative format—where only critical things need to be mentioned. Thus the student is led up to the sort of narrative proofs traditionally found in textbooks. At every stage in the process, the student is always aware of what is and what is not a proof—and has specific guidance in the form of a "step discovery procedure" that leads to a proof outline. The inference rules, and the general method, have been used in two of my texts (available online) intended for students different from the intended readers of this text:

(1) *Outlining Proofs in Calculus* has been used as a supplement in a third-semester calculus course, to take the mystery out of proofs that a student will have seen in the calculus sequence.

(2) *Deductive Mathematics—An Introduction to Proof and Discovery for Mathematics Education* has been used in courses for elementary education majors and mathematics specialists.

Andrew Wohlgemuth
August, 2010

Preface

This text is for a course with the primary purpose of teaching students to do mathematical proofs. Proof is taught "syntactically". A student with the ability to write computer programs can learn to do straightforward proofs using the method of this text. Our approach leads to proofs of routine problems and, more importantly, to the identification of exactly what is needed in proofs requiring creative insights.

The first aim of the text is to convey the *idea* of proof in such a way that the student will know what constitutes an acceptable proof. This is accomplished with the use of very strict inference rules that define the precise syntax for an argument. A proof is a sequence of steps that follow from previous steps in ways specifically allowed by the inference rules. In Chapter 1 these rules are introduced as the mathematical material is developed. When the material gets to the point where strict adherence to the rules makes proofs long and tedious, certain legitimate shortcuts, or abbreviations, are introduced. In Chapters 2 and 3 the process of proof abbreviation continues as the mathematics becomes more complex. The development of the idea of proof starts with proofs made up of numbered steps with explicit inference rules and ends with paragraph proofs. An acceptable proof at any stage in this process is by definition an argument that one could, if necessary, rewrite in terms of a previous stage, that is, without the conventions and shortcuts.

The second aim of the text is to develop the students' ability to *do* proofs. First, a distinction is drawn between formal mathematical statements and statements made in ordinary English. The former statements make up what is called our *language*. It is these language statements that appear as steps in proofs and for which precise rules of inference are given. Language statements are printed in boldface type. Statements in ordinary English are considered to be in our *metalanguage* and may contain language statements. The primary distinction between metalanguage and language is that, in the former, interpretation of statements (based on context, education, and so forth) is essential. The workings of language, on the other hand, are designed to be mechanical and independent of any interpretation. The language/metalanguage distinction serves to clarify and smooth the transition from beginning, formal proofs to proofs in narrative style. It begins to atrophy naturally in Chapter 3. Presently almost all mathematics students will have had experience in computer programming, in which they have become comfortable operating on at least two language levels—in, say, the operating system and a programming language. The language/metalanguage distinction takes advantage of this.

Language statements are categorized by their form: for example, "**if** . . ., **then** . . ." or ". . . **and** . . .". For each form, two inference rules are given, one for proving and one for using statements of that form. The inference rules are designed to do two jobs at once. First, they form the basis for training in the logic of the arguments generally used in mathematics. Second, they serve to guide the development of a proof. Previous steps needed to establish a given step in a proof are dictated by the inference rule for proving statements having the form of the given step. Inference rules and theorems are introduced in a sequence that ensures that early in the course there will be only one possible logical proof of a theorem. The discovery of this proof is accomplished by a routine process fitting the typical student's previous orientation toward mathematics and thus easing the transition from computational to deductive mathematics. Henry Kissinger put the matter simply:

> The absence of alternatives clears the mind marvelously.

Theorems in the text are given in metalanguage and contain language statements. The first task in developing a proof is to decide which language statements are to be assumed for the sake of proof (the hypotheses) and which one is to be proved (the conclusion). Although there is no routine procedure for doing this, students have no trouble with it.

Our approach is a compromise between the formal, which is precise but unwieldy, and the informal, which may, especially for beginning students, be ambiguous. A completely formal approach would involve stating theorems in our language. This approach would necessitate providing too many definitions and rules of inference before presenting the first theorem for proof by students. Our approach enables students to build on each proof idea as it is introduced. Understanding of standard informal mathematical style, which we have called metalanguage, is conditioned by a gradual transition from a formal foundation. Our gradual transition contrasts with the traditional juxtaposition of the formal and informal in which an introductory section from classical logic is followed abruptly by narrative style involving language that has not been so precisely defined.

Our syntactical method has worked well in practice. A much greater percentage of our students can give correct proofs to theorems that are new to them since the method was adopted. The value of a student's ability to function in an ill-defined environment has been replaced by the value of doing a lot of hard work. Students no longer become lost in an environment with which they cannot cope or in which their only hope lies in memorization.

Chapters 1, 2, and 3 cover material generally considered to be core material. In Chapters 1 and 2, illustrative examples use computational properties of the real numbers, which are now introduced very early in the school curriculum. These properties are given in Appendix 1 and can be either used implicitly throughout or introduced at some stage deemed appropriate by the instructor.

Preface vii

Chapters 4, 5, and 6 are written in increasingly informal style. Except for the treatment of free variables in Chapter 4, these chapters are logically independent and material from them can be used at the discretion of the instructor. Chapter 4 contains introductory material on sequences and continuous functions of a real variable. Chapter 5 contains material on the cardinality of familiar sets, and Chapter 6 is an introduction to an axiomatically defined algebraic structure in the form of some beginning group theory. These chapters illustrate how the proof techniques developed in Chapters 1 through 3 apply to material in abstract algebra and advanced calculus.

The only way to learn how to do proofs is by proving theorems. The text proofs provided in Chapters 1, 2, and 3 serve mainly to (1) illustrate the use of inference rules, (2) demonstrate some basic idea on the nature of proof or some specific technique, or (3) exemplify the rules of the game for doing the proofs given as exercises. This is a departure from standard mathematical exposition in which the student is a spectator to the main development and many computational examples, and easy results are "left" for exercises. Thus, in standard exposition, the organization and presentation of definitions and theorems have the goal of facilitating proofs given in the text or illustrating mathematical concepts. In our text, the definitions, theorems, and rules of inference—and the sequence in which they are presented—have the goal of organizing the theorems *to be proved by the student*. The text is therefore a compromise between text-free teaching methods, in which organization sets up student proofs of theorems but in which there are no illustrative proofs or proof methods, and standard exposition, in which organization sets up proofs done by the author and student exercises are secondary.

Andrew Wohlgemuth
Orono, Maine

Suggestions for Using This Text

Chapters 1 through 3 can be used as a text for a sophomore-level one-semester course prerequisite for full courses in abstract algebra and advanced calculus. In our course at Maine, the core consists of students' proving those numbered theorems whose proofs are listed as exercises. A sample syllabus is given on page 347. The entire text can be used for a two-semester course. "Additional Proof Ideas" at the ends of some sections present proof in various traditional ways that supplement our basic approach. This material may be included as taste or emphasis dictates but is not necessary for a basic course. Problems on this material are identified as "Supplementary Problems" in pertinent sections. The text's precise proof syntax, which enables students to recognize a valid proof, also makes it possible to use undergraduates as graders (good students who have previously taken the course, for example). Thus the text's approach to proof makes possible a teaching environment that provides quick feedback on many proofs, even in large classes.

Many students arrive in upper-level courses with no clear idea of just what constitutes a proof. Time is spent dealing not only with new mathematics and significant problems, but with the idea of proof. Some reviewers have suggested that our text could be used by students independently to supplement upper-level texts. Chapter 1 and selected portions of Chapters 2 and 3 could be used to replace the introductory sections on proofs and logic of advanced texts.

The practice exercises are given as self-tests for understanding of the inference rules. Answers to practice exercises are given in the text. Solutions to other problems are given in the Solutions Manual. A few problems, identified as such, will be very challenging for beginning students. The setup and initial progress in a proof attempt are possible using our routine procedure—without hints. Hints are more appropriately given, on an individual basis, after the student has had time to get stuck. The real joy in solving mathematical problems comes not from filling in details after being given a hint, but from thinking of creative steps oneself. Premature or unneeded hints, infamous for taking the fun out of driving a car, can also take the fun out of mathematics.

Acknowledgments

Thanks are due to many people for their contributions to the text. Professor Chip Snyder patiently listened to ideas in their formative stages. His wisdom as a teacher and understanding as a mathematician were invaluable. The reviewers, Professors Charles Biles, Orin Chein, Joel Haack, and Gregory Passty, are responsible for significant material added to the original manuscript as well as for stylistic improvements. Professors Robert Franzosa and Chip Snyder have made helpful suggestions based on their use of the first three chapters of the text in our course at Maine. Robert Stern, Senior Mathematics Editor at Saunders, exercised congenial professional control of the process of text development. Mary Patton, the Project Editor, effectively blended the needs of a different kind of text with presentable style.

Contents

Chapter 1 **Sets and Rules of Inference** 1
    1.1 Definitions 1
    1.2 Proving *For All* Statements 8
    1.3 Using *For All* and *Or* Statements 20
    1.4 Using and Proving *Or* Statements 30
    1.5 *And* Statements 39
    1.6 Using Theorems 47
    1.7 Implications 53
    1.8 Proof by Contradiction 65
    1.9 *Iff* 78
    1.10 *There Exists* Statements 84
    1.11 Negations 94
    1.12 Index sets 100

Chapter 2 **Functions** 109
    2.1 Functions and Sets 109
    2.2 Composition 121
    2.3 One-to-One Functions 129
    2.4 *Onto* Functions 139
    2.5 Inverses 147
    2.6 Bijections 155
    2.7 Infinite Sets 159
    2.8 Products, Pairs, and Definitions 166

Chapter 3 **Relations, Operations, and the Integers** 173
    3.1 Induction 173
    3.2 Equivalence Relations 186
    3.3 Equivalence Classes 192
    3.4 Well-Defined Operations 199
    3.5 Groups and Rings 205

Contents

              3.6   Homomorphisms and Closed Subsets of $\mathbb{Z}$ 212
              3.7   Well-Defined Functions   215
              3.8   Ideals of $\mathbb{Z}$ 218
              3.9   Primes   220
              3.10 Partially Ordered Sets   222

Chapter 4   **Proofs in Analysis**   231
              4.1   Sequences   231
              4.2   Functions of a Real Variable   238
              4.3   Continuity   243
              4.4   An Axiom for Sets of Reals   247
              4.5   Some Convergence Conditions for Sequences   253
              4.6   Continuous Functions on Closed Intervals   257
              4.7   Topology of $\mathbb{R}$ 261

Chapter 5   **Cardinality**   265
              5.1   Cantor's Theorem   265
              5.2   Cardinalities of Sets of Numbers   270

Chapter 6   **Groups**   277
              6.1   Subgroups   277
              6.2   Examples   282
              6.3   Subgroups and Cosets   287
              6.4   Normal Subgroups and Factor Groups   292
              6.5   Fundamental Theorems of Group Theory   297

Appendix 1   **Properties of Number Systems**   303

Appendix 2   **Truth Tables**   309

Appendix 3   **Inference Rules**   315

Appendix 4   **Definitions**   319

Appendix 5   **Theorems**   331

Appendix 6   **A Sample Syllabus**   347

                      Answers to Practice Exercises   349

                      Index   363

Chapter 1

Sets and Rules of Inference

1.1 DEFINITIONS

We begin this section with a review of some notation and informal ideas about sets now common in the school curriculum. A *set* is a collection of things viewed as a whole. The things in a set are called *elements* or *members* of the set. The expression "$x \in A$" means that x is a member of set A and is read "x is an element of A" or "x is a member of A". The expression "$x \notin A$" means that x is not a member of A. We will assume that all sets are formed of elements from some universal set U under consideration. The set of real numbers will be denoted by $\mathbf{R}$, integers by $\mathbf{Z}$, and positive integers by $\mathbf{N}$.

We will not give a formal definition of set. It will be an undefined term—a starting point in terms of which we will define other things. While the idea of a set is undefined, we will assume that any particular set may be defined by giving a *rule* for deciding which elements of the universal set are in the particular set and which are not.

For example, if $\mathbf{R}$ is our universal set, then the set of all real numbers between 0 and 1 is written $\{x \in \mathbf{R} \mid 0 < x < 1\}$. This is read "the set of all x in $\mathbf{R}$ such that zero is less than x and x is less than one". In the expression "$\{x \in \mathbf{R} \mid 0 < x < 1\}$", the symbol "$x$" is used as a *local variable* or a *dummy variable*. That is, x has no meaning outside the expression and any other letter (except $\mathbf{R}$, of course) would do as well. Thus $\{x \in \mathbf{R} \mid 0 < x < 1\}$ and $\{y \in \mathbf{R} \mid 0 < y < 1\}$ have exactly the same meaning. Sometimes the rule for deciding set membership is not explicitly stated but is nevertheless obvious. For example, a set may be given by listing its elements, such as $\mathbf{N}_3 = \{1, 2, 3\}$. In particular, $\mathbf{N} = \{1, 2, 3, 4, \ldots\}$ and $\mathbf{Z} = \{0, 1, -1, 2, -2, \ldots\}$. The set $\{1, 2, 3, \ldots, k\}$ will be denoted by $\mathbf{N}_k$.

A *statement* is a mathematical expression that is either *true* or *false*. For example, $2 \in \{x \in \mathbf{R} \mid x < 5\}$ (true), $3^2 + 4^2 = 5^2$ (true), and $b\sqrt{2} = a$ for some

1

$a, b \in \mathbb{Z}$ (false) are statements. The rule $0 < x < 1$ used to define the set $A = \{x \in \mathbb{R} \mid 0 < x < 1\}$ is an example of an *open statement*; that is, the truth of $0 < x < 1$ depends on the value substituted for x. For example, if $x = \frac{1}{2}$, then $0 < x < 1$ is true, and if $x = 2$, then $0 < x < 1$ is false. The open statement $0 < x < 1$ is considered to be a *property* that real numbers may or may not have—the defining property of the set A above. Thus if a is some real number that makes $0 < a < 1$ true, then a satisfies the defining property of the set A and is therefore a member of A. If $0 < a < 1$ is false, then $a \notin A$. In our definitions of some particular set, say, A, we will always have in mind some universal set of elements that are allowed to be substituted in the open statement that gives the defining property for A.

In the definition $A = \{x \in \mathbb{R} \mid 0 < x < 1\}$, the universal set is $\mathbb{R}$. When this universal set is either clear or not important to our discussion, we will sometimes not mention it explicitly. For example, A above can be written $A = \{x \mid 0 < x < 1\}$. In order to avoid logic contradictions, it is important to know what sort of thing can be substituted for the variable in an open statement. This is the role of the universal set.

The set with no elements is called the *empty set* and is denoted by $\emptyset$. One reason for considering an empty collection as a set is to provide consistency in notation. For example, we want $\{a \in \mathbb{R} \mid a^2 = -1\}$ to be a set as well as $\{a \in \mathbb{R} \mid a^2 = 2\}$.

A formal mathematical proof consists of a numbered sequence of statements. Each statement must follow logically from previous statements (or steps) in the sequence. In our proofs, beside each statement we will give in parentheses: (1) the numbers of the previous steps from which the given statement follows, (2) a semicolon, and (3) the reason that the given statement follows logically from the indicated previous steps. For example, in the hypothetical proof steps below, the justification (1, 3; Theorem 14) for Step 4 indicates that Statement 4 is a logical consequence of the statements in Steps 1 and 3 when Theorem 14 is applied:

1. ...
2. ...
3. ...
4. $x \in B$ (1, 3; Theorem 14)

In the set $D = \{x \in \mathbb{R} \mid x < 1\}$, the open statement $x < 1$ gives the property that defines the set. Thus, if $y \in D$, then y must have the property defining D, namely, $y < 1$. Conversely, if $a < 1$, that is, if a satisfies the defining property, then we must have $a \in D$. This shows the two ways the definition of a particular set can be used as a reason for proof steps.

Example 1:
Define $C = \{x \in \mathbb{R} \mid x < 2\}$. The definition of C tells us in both cases below why Step 2 follows from Step 1.

1.1 Definitions

    1. $a \in C$
    2. $a < 2$ (1; def. C)
    1. $b < 2$
    2. $b \in C$ (1; def. C)

This example illustrates our first inference rule:

Inference Rule *Set definition rule: If an element is in a set, we may infer that it satisfies the defining property. Conversely, if it satisfies the defining property, we may infer that it is in the set.*

Note in Example 1 that it is understood, but not stated, that all elements are in the universal set $\mathbb{R}$. If we know that $a \in C$, then by the definition of C we know both $a < 2$ and $a \in \mathbb{R}$, but the latter is not stated.

In this text we will have certain standard phrases and very precise ways of dealing with these phrases. They will be part of our formal proof language (for brevity's sake called our *language*) and will be printed in boldface type. Thus the letters **A, B, X, ... a, b, x, ...**, which stand for numbers, sets, or elements in sets, will be part of our language. Statements about numbers (such as $x^2 + 2 < 3$) or set membership (such as $a \in X$ or $5 \notin A$) will also be part of our language. Sentences like the one you are now reading, the purpose of which is to communicate ideas in ordinary English, are part of our informal language (called *metalanguage*[†]) and printed in ordinary type. Metalanguage statements may contain symbols or statements of our language, but the reverse is not possible. Thus we may write, "For all real numbers x, if $x \in C$, then $x < 2$." This sentence is in our metalanguage but contains the statements $x \in C$ and $x < 2$, which are part of our language.

As we progress, we will enlarge our stock of language statements. These are the statements used in proof steps and for which very precise rules of inference are given. Our theorems, on the other hand, will be given in metalanguage and will contain language statements. For example, consider the following statement:

Theorem (example) *If $X = \{x \in \mathbb{R} \mid x < 1\}$ and $a \in X$, then $a < 2$.*

In developing a proof of this theorem, our first task will be to decide, from the metalanguage statement, which language statements we may assume for the sake of proof (these are called *hypotheses*) and which statement we are to show

---

[†] The prefix *meta*(mĕt´ ȧ) in mathematics and logic means "higher" or "transcending". Thus language and metalanguage are on different levels—analogous to computer science, where we can operate on more than one level in different systems and languages. The primary distinction between metalanguage and language is that, in the former, interpretation of statements (based on context, education, and so on) is essential. The workings of language, on the other hand, are designed to be mechanical and independent of any interpretation.

(this is called the *conclusion*). The hypotheses and conclusion will form the first part of our proof:

Proof:

Assume: 1. $X = \{x \in \mathbb{R} \mid x < 1\}$
              2. $a \in X$
Show: $a < 2$

The statements of the hypotheses and conclusion must be part of our language. The conclusion will always be the last step in the proof. The hypotheses may be used to draw inferences or written down as steps at any stage of the proof. We could, for example, continue with the development of the proof of the sample theorem as follows:

1. $a \in X$ (hyp. 2)
2. $a < 1$ (1, hyp. 1; def. X)
       .
       .

(last step)   k. $a < 2$

From the facts that $a \in X$ (Step 1) and that X is defined to be $\{x \in \mathbb{R} \mid x < 1\}$ (hypothesis), we may infer that $a < 1$; that is, a has the property defining the set X. The formal justification for this is "definition of X" since, by our rule of inference, this is how we use the definition. To complete the proof, we need to show that $a < 2$. This follows from the step $a < 1$, the fact that $1 < 2$, and the "transitivity" of the order relation "$<$" on the real numbers.

Appendix 1 gives the algebraic and order properties of the familiar number systems. These are the properties derived from the axioms of arithmetic stressed in school mathematics. It is these properties that let us solve equations and inequalities and perform the ordinary computations and algebraic manipulations with which you are familiar from previous courses—and which allow us to conclude $a < 2$ from $a < 1$ above. As we develop our mathematical and logical ideas, we will occasionally use examples from the sets $\mathbb{R}$, $\mathbb{Z}$, and $\mathbb{N}$. For these examples (only), we will allow changes in proof steps according to the following inference rule:

Inference Rule
     Computation rule: Steps in a proof that follow from other steps by the familiar techniques of algebra applied to $\mathbb{R}$, $\mathbb{Z}$, and $\mathbb{N}$ are allowed. Justification for such steps is given, for example, as "property of the real numbers" (abbreviated "prop. $\mathbb{R}$"). These properties are found in Appendix 1.

A complete proof of our sample theorem is then given by:

1.1 Definitions

Proof:

> Assume: 1. $X = \{x \in \mathbf{R} \mid x < 1\}$
> 2. $a \in X$
>
> Show: $a < 2$
>
> 1. $a \in X$ (hyp. 2)
> 2. $a < 1$ (1, hyp. 1; def. X)
> 3. $1 < 2$ (prop. **R**)
> 4. $a < 2$ (2, 3; prop. **R**) ∎

Proofs are differentiated from other parts of the text by shading, and the symbol ∎ signals the end of a proof.

Step 4 follows from Steps 2 and 3 by the transitive property of $<$. Since we will not define the real numbers, it is not possible to prove either the transitive property or the fact $1 < 2$. Students who have not yet fully understood what a proof *is* generally find proofs of such things hopelessly unenlightening. Only a mathematician could feel that the idea of proof was more fundamental than the fact $1 < 2$, which would therefore require proof.

In our proofs we will usually combine adjacent steps whose justification is "property of **R** (or **Z** or **N**)". Thus we may omit Step 3 above and go right to Step 4 (now renumbered Step 3). The new steps would be:

> 1. $a \in X$ (hyp. 2)
> 2. $a < 1$ (1, hyp. 1; def. X)
> 3. $a < 2$ (2; prop. **R**)

In doing proofs, we will assume that the hypotheses of the theorem being proved are *true* (for the sake of argument). We will also assume all previously proved theorems are true. Thus a proof for us will be a sequence of statements (that is, steps), the truth of each of which follows by our rules of inference from the truth of previous steps, hypotheses, and theorems.

Example 2:
Consider the statements

(a) For sets A, B and a real number x, if $x \in A$, then $x \in B$.

(b) For real numbers x, y, if $x \not< y$ and $x \neq y$, then $y < x$.

Chapter 1 Sets and Rules of Inference

The hypotheses for (a) are

$$A, B \text{ sets}$$
$$x \text{ real}$$
$$x \in A$$

We consider the symbols "A", "B", and "x" defined for the sake of a proof so that we may use them in the proof without definition in the proof itself. We will include as hypotheses an identification of all such symbols used in the proof and the language statements we may take as true for the sake of argument.

Hypotheses and conclusion for (a) are:

Assume: A, B sets
           x real
           $x \in A$
Show: $x \in B$

Hypotheses and conclusion for (b) are:

Assume: $x, y \in \mathbb{R}$
    1. $x \not< y$
    2. $x \neq y$
Show: $y < x$

Exercises 1.1

Practice Exercises

1. Define $A = \{z \in \mathbb{R} \mid z \neq 1\}$.
   (a) Suppose we are given Step 4 and the reason for Step 5 in some proof:

   4. $b \in A$
   5.   (4; definition of A)

   What *must* Step 5 be?
   (b) Suppose we are given the following:

   4.
   5. $b \in A$ (4; definition of A)

   What *must* Step 4 be?

2. Suppose we are given the following:

   5. $x \in B$
   6. $x \leq 7$ (5; def. B)

1.1 Definitions

       7. $x \leq 8$ (6; prop. **R**)
       8. $x \in C$ (7; def. C)

What *must* the definitions of sets B and C be?

3. Suppose we are given the following:

       5. $1 \leq a \leq 3$
       6. $a \in X$ (5; def. X)
       7. $b \in X$
       8.   (7; def. X)

What must the definition of X be? What must Step 8 be?

4. Each of the following metalanguage statements contains pieces that are language statements. For each metalanguage statement, decide which language statements we may assume in a proof and which we need to show. Label them appropriately.

   (a) For all real numbers x, y, z, if $x < y$ and $y < z$, then $x < z$.
   (b) Let A and B be sets. Suppose $x \in A$. Prove $x \in B$.
   (c) Let $C = \{y \in Z \mid y \neq 0\}$. If $a = 1$, then $a \in C$.
   (d) If $x \leq y \leq z$, then $y \leq 2$.
   (e) For sets X and Y: if $x \in X$, then $x \in Y$.

5. What metalanguage statements might be interpreted by the following hypotheses and conclusions?

   (a)

       Assume: $x, y \in \mathbf{R}$
                       $x > 0$
                       $y > 0$
       Show: $xy > 0$

   (b)

       Assume: A, B sets
                       $x \notin A$
       Show: $x \in B$

   (c)

       Assume: $x, y \in \mathbf{R}$
                       $0 < x < y$
       Show: $x^2 < y^2$

1.2 PROVING *FOR ALL* STATEMENTS

A set *A* is called a *subset* of a set *B* if every element of *A* is an element of *B*. For example, {1, 2, 3} is a subset of {0, 1, 2, 3, 4, 5}. If $A = \{1, 2\}$ and $B = \{1, 2\}$, then *A* is a subset of *B* since each element of *A* (namely, 1 and 2) is an element of *B*. The statement "every element of *A* is an element of *B*" will be put into formal language as **for all *x* such that** $x \in A: x \in B$ or, equivalently, $x \in B$ **for all** $x \in A$. (The colon can be read aloud as a comma [a pause] or by the words "we have".) This gives us our first formal definition.

Definition   For sets *A and B, A is a subset of B (written* $A \subseteq B$*) provided that for all x such that* $x \in A: x \in B$.

Our next rule of inference will allow us to use definitions other than the definition of a set as justification for proof steps. The preceding definition tells us that the *relationship* ($A \subseteq B$) between *A* and *B* holds provided a *condition* (**for all *x* such that** $x \in A: x \in B$) is true. In using this definition, we may infer that the condition is true if the relationship $A \subseteq B$ has been established in a previous step. Conversely, if the condition has been established, the defined relationship may be inferred.

Example 1:

1. $A \subseteq B$
2. **for all** *x* **such that** $x \in A: x \in B$ (1; def. $\subseteq$)

Example 2:

1. **for all** *x* **such that** $x \in A: x \in B$
2. $A \subseteq B$ (1; def. $\subseteq$)

The definition of set containment can be used as justification for inferring Step 2 from Step 1 in both examples above.

Suppose we had the step $A \subseteq B$ in a proof. It is customary in informal mathematics to use the justification "def. $\subseteq$" to say more than the defining condition **for all *x* such that** $x \in A: x \in B$. In fact, the validity of **for all *x* such that** $x \in A: x \in B$ is combined with other previous steps to draw further inferences—all with the justification "def. $\subseteq$". The precise way in which we will be allowed to do this is given in Section 1.4. For now, however, we will limit ourselves to the one-step inference illustrated by Examples 1 and 2—which is given in the following general rule of inference:

1.2 Proving *For All* Statements

Inference Rule

Definition rule: Suppose some relationship has been defined. If the relationship holds (in some proof step or hypothesis), then the defining condition (only) may be inferred. Conversely, if the defining condition is true, then the relationship may be inferred.

The next example illustrates a form we will use for exercises. In this type of exercise, you are to fill in the steps or reasons marked with an asterisk (*). In general, an asterisk plays the role of a "wild card" and thus need not represent the same thing each time it occurs. If there is more than one solution, any *one* will do.

Example 3:

1. $S \subseteq T$
2. * (1; def. $\subseteq$)

Solution:

1. $S \subseteq T$
2. for all x such that $x \in S$: $x \in T$ (1; def. $\subseteq$)

In this example we know that Step 2 must come from Step 1 as a result of our using the definition of $\subseteq$. By our definition rule, Step 2 must give the defining property.

Example 4:

1. *
2. for all y such that $y \in S$: $y \in T$ (1; def. $\subseteq$)

Solution:

1. $S \subseteq T$
2. for all y such that $y \in S$: $y \in T$ (1; def. $\subseteq$)

Example 5:

1. *
2. $X \subseteq Y$ (1; def. $\subseteq$)

Solution:

1. for all t such that $t \in X$: $t \in Y$
2. $X \subseteq Y$ (1; def. $\subseteq$)

Example 6:
1. **for all** x **such that** $x \in A$: $x \in B$
2. $A \subseteq B$ (1; *)

Solution:
1. **for all** x **such that** $x \in A$: $x \in B$
2. $A \subseteq B$ (1; def. $\subseteq$)

Since open statements are neither true nor false, they cannot be used as proof steps. Proof steps are true statements, the truth of which follows from previous steps and theorems (all true) by our formal rules of inference. There are two ways the statement $x \geq 0$, for example, could appear in a proof: (1) when x has been previously defined, and (2) when the statement $x \geq 0$ is *quantified*, as in **for all natural numbers** x: $x \geq 0$. The meaning of this last statement is that every natural number satisfies the property given by the open statement $x \geq 0$.

The symbol "x" in the open statement $x \geq 0$ is called a *free* variable. When we put the quantifier **for all natural numbers** x: in front of $x \geq 0$, then x is called a *bound* variable (or *local* variable), and the *scope* of x is limited to the statement itself: that is, x has no meaning outside the statement. We will shortly see how to deal with such quantified statements in proofs.

Similarly, if the set B has been previously defined in a proof but x has not, then x is free in the statement $x \in B$, which is an open statement. **For all** x **such that** $x \in A$: $x \in B$ is not an open statement, however, and x is bound in this statement. The x is frequently called a *dummy* variable in this case, as it is when used to define a set. Note that the statement **for all** x **such that** $x \in A$: $x \in B$ tells us something about A and B (that $A \subseteq B$) but nothing about x. We don't come away from the statement knowing anything about x.

We will use capital script letters to denote language statements. For example, we might use $\mathcal{P}$ to represent the statement $x \in C$. If we wish to emphasize that x is a free variable in $\mathcal{P}$, we write $\mathcal{P}(x)$ instead of just $\mathcal{P}$.

If $\mathcal{P}(x)$ represents $x \in A$ and $\mathcal{Q}(x)$ represents $x \in B$, then the statement **for all** x **such that** $x \in A$: $x \in B$ is represented by **for all** x **such that** $\mathcal{P}(x)$: $\mathcal{Q}(x)$. Note that x is free in $\mathcal{P}(x)$ and in $\mathcal{Q}(x)$ but not in **for all** x **such that** $\mathcal{P}(x)$: $\mathcal{Q}(x)$. The meaning of the statement **for all** x **such that** $\mathcal{P}(x)$: $\mathcal{Q}(x)$ is that if x is any element for which $\mathcal{P}(x)$ is true, then $\mathcal{Q}(x)$ is true for that x. To prove this language statement, we therefore can select an arbitrary x for which $\mathcal{P}(x)$ is true and then show that $\mathcal{Q}(x)$ is a true statement about this x. We now make this a formal rule of inference that tells us how to prove such **for all** statements.

Inference Rule   *Proving **for all** statements: In order to prove a statement of the form **for all** x such that $\mathcal{P}(x)$: $\mathcal{Q}(x)$, assume that x is an arbitrarily chosen (general) element such that $\mathcal{P}(x)$ is true. Then establish that $\mathcal{Q}(x)$ is true.*

1.2 Proving *For All* Statements

In a proof, we will indent steps in which an assumption is made and continue the indentation for all steps that depend on this assumption. Also, if a new symbol is defined whose sole purpose is to prove some statement, we will indent the steps containing this symbol. For example, employing the **for all** rule would dictate steps such as:

$i-2.$ $\cdots$
$i-1.$ $\cdots$
    $i.$ Let x be chosen arbitrarily such that $\mathscr{P}(x)$.
      $\cdot$
      $\cdot$
    $k.$ $\mathcal{Q}(x)$ (   )
$k+1.$ **for all** x such that $\mathscr{P}(x)$: $\mathcal{Q}(x)$ (i—k; rule for proving **for all** statements)

This proof format is considered part of the inference rule.

It is not necessary to give a justification for steps, such as step i, in which an assumption is made or that serve to define a new symbol. Steps i through k will be written so as to establish the truth of $\mathcal{Q}(x)$. These steps are all indented since they are based on the assumption that x has been chosen to make $\mathscr{P}(x)$ true. The x has no meaning outside these indented steps, and the *scope* of x is the set of steps i through k. Indeed, the variable we use in Step $k+1$ may be different from the variable in the indented steps. For example,

    $i.$ Let t be chosen arbitrarily such that $\mathscr{P}(t)$.
      $\cdot$
      $\cdot$
    $k.$ $\mathcal{Q}(t)$ (   )
$k+1.$ **for all** x such that $\mathscr{P}(x)$: $\mathcal{Q}(x)$ (i—k; rule for proving **for all** statements)

In Statement $k+1$, x is a local (bound) variable that has no meaning outside this statement. In Steps i through k, t is considered a fixed quantity that has the same meaning for each step. The truth of Step $k+1$ depends not on the assumption that $\mathscr{P}(t)$ is true for the chosen t (in fact, this is meaningless) but only on the fact that $\mathcal{Q}(t)$ is true whenever $\mathscr{P}(t)$ is true. Since Step $k+1$ does not depend on the assumption valid for Steps i through k, we need not indent it. If we think of indenting as "pushing" one level deeper, then the rule for proving **for all** statements allows us to "pop" one level back up.

It is not legitimate to use a proof step as a reason for subsequent steps after we have popped up from the indentation level of the step. The only exceptions to this are given by specific rules of inference, such as the rule for proving **for all** statements. The indented Steps i through k above are like a subroutine in a structured computer program, the only purpose of which is to prove Step $k+1$. The variable t is defined only in this block of Steps i through k. After we pop back up in Step $k+1$, the truth of Steps i through k ceases to exist for us.

We define $Q(x)$ **for all** x **such that** $\mathcal{P}(x)$ to mean exactly the same thing as **for all** x **such that** $\mathcal{P}(x)$: $Q(x)$.

We will use standard abbreviations for many of our language statements. For example,

$$\text{for all } x \text{ such that } x \in A: x \in B$$

will be shortened to

$$\text{for all } x \in A: x \in B$$

The following language statements all say the same thing:

$$\text{for all } x \text{ such that } x \in A: x \in B$$

$$\text{for all } x \in A: x \in B$$

$$x \in B \text{ for all } x \text{ such that } x \in A$$

$$x \in B \text{ for all } x \in A$$

It is common in mathematical logic to use abbreviations in statements. "**For all**" is abbreviated "∀". For example, "**for all** $x \in A: x \in B$" can be written "∀ $x \in A: x \in B$". Since our development is intended to lead to customary informal paragraph proofs, we won't use such abbreviations in statements. In order to fit justifications for proof steps on a single line, however, it will be necessary to have abbreviations for the logical inference rules. Hence the phrase "rule for proving **for all** statements" will be abbreviated "pr. ∀".

The following examples illustrate the use of the rule for proving **for all** statements.

Example 7:

    1. Let $x \in C$ be arbitrary.
    2. $\cdots$
    3. $x \in D$
    4. *   (1—3; pr. ∀)

Solution:

    1. Let $x \in C$ be arbitrary.
    2. $\cdots$
    3. $x \in D$
    4. for all $x \in C: x \in D$ (1—3; pr. ∀)

1.2 Proving *For All* Statements

Example 8:

1. *
2. · · ·
3. · · ·
4. *
5. for all $x < 4$: $x \in A$ (1—4; pr. ∀)

Solution:

1. Let x be arbitrary such that $x < 4$.
2. · · ·
3. · · ·
4. $x \in A$
5. for all $x < 4$: $x \in A$ (1—4; pr. ∀)

The meaning of our **for all** statement can be clarified by specifying what it means for such a statement to be false.

For All Negation

In order that the statement $Q(x)$ for all x such that $\mathcal{P}(x)$ be false, it is required only to have some particular value x_0 for which $\mathcal{P}(x_0)$ is true but $Q(x_0)$ is false.

In particular, the statement **for all** $x \in A$: $x \in B$ is false only if there is some x_0 such that $x_0 \in A$ is true but $x_0 \in B$ is false. If $\varnothing$ is the empty set, the statement $x \in \varnothing$ can never be true. Therefore the statement **for all** $x \in \varnothing$: $x \in A$ must be *true* for any set A. Thus the empty set is a subset of every set A—by definition.

Note that we don't prove that the empty set is a subset of some set A by using our proof format for showing **for all** $x \in \varnothing$: $x \in A$. There are two ways that the statement **for all** x **such that** $\mathcal{P}(x)$: $Q(x)$ could be true. The first is that there are no values of x that make $\mathcal{P}(x)$ true. In this case, **for all** x **such that** $\mathcal{P}(x)$: $Q(x)$ is said to be vacuously true. The other possibility is that there are values of x that make $\mathcal{P}(x)$ true but these x also make $Q(x)$ true. It is this possibility that our rule covers. Thus a complete *proof* that **for all** x **such that** $\mathcal{P}(x)$: $Q(x)$ is true would run along these lines: "If there are no values of x such that $\mathcal{P}(x)$ is true, then the statement is vacuously true. If there are values of x that make $\mathcal{P}(x)$ true, then select one arbitrarily and show that $Q(x)$ is true for this x. The statement is therefore true in either case." Mathematicians don't bother to mention trivialities like vacuously true statements when doing proofs; in this way our inference rule models customary informal practice.

We will develop a routine method for discovering proof steps. The first step in this method is to determine the hypotheses and conclusion from the statement of a theorem. The next step depends on the form of the "top-level" language statement of the conclusion. This top-level statement may contain pieces that are themselves language statements, but is itself not contained in any larger statement.

Example 9:
The top-level statement in

(a) **for all** x such that $x \in A$: $x \in B$

is a **for all** statement of the form **for all** x **such that** $\mathcal{P}(x)$: $\mathcal{Q}(x)$. It contains the language statements $x \in A$ and $x \in B$ but is itself not contained in a larger statement.

The top-level statement in

(b) $\{x \in \mathbb{R} \mid x < 1\} \subseteq \{x \in \mathbb{R} \mid x < 2\}$

is an assertion of set containment of the form $C \subseteq D$. It contains the language statements $x < 1$, $x < 2$, and $x \in \mathbb{R}$.

Our next example illustrates our computation rule of inference and our routine procedure for discovering proof steps.

Example 10:
Define $D = \{x \in \mathbb{R} \mid x < 1\}$ and $C = \{x \in \mathbb{R} \mid x < 2\}$.
Prove that $D \subseteq C$.

To start our proof, we identify the hypotheses and conclusion. The identification of these statements is part of the proof and therefore included in the shaded text.

Proof:
  Assume: $D = \{x \in \mathbb{R} \mid x < 1\}$
           $C = \{x \in \mathbb{R} \mid x < 2\}$
  Show: $D \subseteq C$

Our procedure dictates that at this stage in the proof development you should ignore the hypotheses and focus solely on what it means for the conclusion to be true. This will always be seen by the form of the top-level statement of the conclusion. The statement $D \subseteq C$ asserts that one set is a subset of another set. We have to consider at this point what it means for this to be a true statement. Since this is a mathematics text, the meaning is found in our formal definition of "subset". Excessive genius is not needed to come to this conclusion. Nothing but the *definition* can give us the meaning. It is frequently helpful to put the conclusion on scrap paper and then analyze what it means by definition.

1.2 Proving *For All* Statements

Scrap Paper

Show: $D \subseteq C$ ← This becomes the last line of the proof.

That is, show: $x \in C$ **for all** $x \in D$ ← By definition, this must be the next-to-the-last line.

The **for all** rule tells us how to prove the statement in the scrap-paper box above. The first steps we write down are

---

1. Let $x \in D$ be arbitrary.
   .
   .
k. $x \in C$
k+1.   for all $x \in D$: $x \in C$ (1—k; pr. ∀)
k+2. $D \subseteq C$ (k+1; def. ⊆)

By working backward from the conclusion, we have been led to determine the first step in the proof as well as the last few steps. After writing these steps down, our job is to make the connection from Step 1 to Step k. Since x was picked arbitrarily in the set D, the only thing we could possibly know about x is the property it must have by virtue of being in D. That is, Step 2 follows from Step 1 by using the definition of D. There is no other choice. Similarly, to show Step k, $x \in C$, we *must* use the definition of C. This gives us Steps 2 and $k-1$, which we now add to our developing proof:

---

1. Let $x \in D$ be arbitrary.
2. $x < 1$ (1; def. D)
   .
   .
k−1. $x < 2$
k. $x \in C$ (k−1; def. C)
k+1.   for all $x \in D$: $x \in C$ (1—k; pr. ∀)
k+2. $D \subseteq C$ (k+1; def. ⊆)

It is important to realize that so far all the steps and reasons were dictated by necessity. There remains only to make the connection between Steps 2 and

$k-1$, and this connection is clear: $x < 2$ follows from $x < 1$ by a property of the real numbers. Even though this connection is obvious, it is the only part of the proof that can in any way be considered creative. Everything else is inevitable. Step $k-1$ is now seen to be Step 3, and the complete proof is given by:

Proof:

Assume: $D = \{x \in \mathbf{R} \mid x < 1\}$
$C = \{x \in \mathbf{R} \mid x < 2\}$
Show: $D \subseteq C$

1. Let $x \in D$ be arbitrary.
2. $x < 1$ (1; def. D)
3. $x < 2$ (2; prop. **R**)
4. $x \in C$ (3; def. C)
5. for all $x \in D$: $x \in C$ (1—4; pr. $\forall$)
6. $D \subseteq C$ (5; def. $\subseteq$)

The language statement $x \in C$ **for all** $x \in D$ has the same meaning as "if x is an arbitrarily chosen element of D, then $x \in C$". In Section 2.3 we will model, in our language, such metalanguage "if . . ., then . . ." statements involving arbitrarily chosen elements. For now, proof steps involving an arbitrary choice will be used to prove **for all** statements according to our rule.

We define the statement in Step 1 to be equivalent to any one of the following:

Let $x \in D$ be arbitrary.

Let x be an arbitrarily chosen element of D.

Let x be chosen arbitrarily such that $x \in D$.

Let x be a completely general element of D.

Step 1 serves the purpose of defining the symbol "x", not making an assumption about a previously defined symbol. This definition remains in force for Steps 1 through 4, the scope of variable x.

Let us summarize the logic behind our definition of set containment and the inference rule used to show containment. Suppose we wish to show $A \subseteq B$ for some sets A and B. Possibly $A = \emptyset$. In this case, $A \subseteq B$ since the empty set is a subset of every set. If A is not empty, then we must establish that every element of A is in B. Thus our inference rule leads us to choose arbitrarily an element of A, call it something, and then prove it is in B. In proving, say,

1.2 Proving *For All* Statements   **17**

$A \subseteq B$, we don't bother to mention that A may be empty. We *assume* there is at least one element in A because if there is not, then we are done.

The material in this section is of course directed toward understanding how definitions and inference rules work in a proof—not merely toward understanding what it means for one set to be contained in another.

Additional Proof Ideas

A more formal approach to proofs would dictate that, in justifying each step, (1) only previously steps, the hypotheses, definitions, and previously proven theorems can be referred to before the colon, and (2) only rules of inference can be referred to after the colon. A proof of Example 9 using this scheme would be:

Proof:

Assume: 1. $D = \{x \in \mathbb{R} \mid x < 1\}$
2. $C = \{x \in \mathbb{R} \mid x < 2\}$
Show: $D \subseteq C$

1. Let $x \in D$ be arbitrary.
2. $x < 1$ (1, hyp. 1; rule for using definitions)
3. $x < 2$ (2; computation rule: prop. **R**)
4. $x \in C$ (3, hyp. 2; rule for using definitions)
5. for all $x \in D: x \in C$ (1—4; pr. ∀)
6. $D \subseteq C$ (5, def. ⊆; rule for using definitions)

Compare the proof above with the steps below, which show our accepted proof style given in the text:

1. Let $x \in D$ be arbitrary.
2. $x < 1$ (1; def. D)
3. $x < 2$ (2; prop. **R**)
4. $x \in C$ (3; def. C)
5. for all $x \in D: x \in C$ (1—4; pr. ∀)
6. $D \subseteq C$ (5; def. ⊆)

In both proofs we apply the rule for using the definition of a set to Step 1 in order to get Step 2, but in our less formal, accepted style, we think of applying the definition of D to Step 1 in order to get Step 2. The definition rule is implicit in this. Thus definitions—and, later, theorems—can be used after the colon when we think of applying these to proven steps in order to get new steps. Our accepted style is a contraction of the more formal style and is guided by common paragraph style. No one ever says, in a paragraph proof, that $\mathcal{P}$ follows from $\mathcal{Q}$ by some rule of inference. People do say that $\mathcal{P}$ follows from $\mathcal{Q}$ by some definition.

Exercises 1.2

Practice Exercises

1. In the proof steps below, what must Step 5 be?

    (a)   1. Let x be arbitrary such that $x < 1$.
           2. $\cdots$
           3. $\cdots$
           4. $x \notin C$
           5. *   (1—4; pr. $\forall$)

    (b)   1. $X = \{x \in Z \mid x < 0\}$
           2. $\cdots$
           3. $\cdots$
           4. $y \in X$
           5. *   (1, 4; def. X)

    (c)   1. $\cdots$
           2. $\cdots$
           3. $\cdots$
           4. $S \subseteq T$
           5. *   (4; def. $\subseteq$)

    (d)   1. $\cdots$
           2. $\cdots$
           3. $\cdots$
           4. for all $t \in X: t \in Y$
           5. *   (4; def. $\subseteq$)

    (e)   1. Let $t \in B$ for an arbitrary t.
           •
           •

1.2 Proving *For All* Statements

            5. *
            6.   for all $t \in B$: $t \in X$ (1—5; pr. $\forall$)

   (f) 1. $t \in N$
          2. ...
          3. ...
          4. $t \in A$
          5. *
          6. ...
          7. $p \in S$
          8. for all p such that $p \notin T$: $p \in S$ (5—7; pr. $\forall$)

2. In each case below, give the justification for Step 5.

   (a) 1. Let $x \in A$ be arbitrary.
          2. ...
          3. ...
          4. $x \in B$
          5. for all $x \in A$: $x \in B$ (*)

   (b) 1. ...
          2. for all $t \in A$: $t \in B$
          3. $x \in A$
          4. $x \in B$
          5. $A \subseteq B$ (*)

   (c) 1. ...
          2. ...
          3. ...
          4. $A \subseteq B$
          5. for all $t \in A$: $t \in B$ (*)

3. Give all the proof steps you can (with appropriate gaps) that lead to showing $S \subseteq T$.

Straightforward Problems

4. Define $A = \{z \in Z \mid z \neq 1\}$ and $B = \{z \in Z \mid z \geq 2\}$.
Prove $B \subseteq A$.

5. Define $X = \{x \in \mathbb{R} \mid 2x > x\}$ and $Y = \{x \in \mathbb{R} \mid x > 0\}$.
Prove $X \subseteq Y$.

1.3 USING *FOR ALL* AND *OR* STATEMENTS

Statements of the form

$$\text{for all } x \text{ such that } \mathcal{P}(x) \colon \mathcal{Q}(x)$$

can be generalized to more than one variable. For example,

$$\text{for all } x, y \text{ such that } \mathcal{P}(x, y) \colon \mathcal{Q}(x, y)$$

We will consider our inference rule for proving **for all** statements to apply in this more general setting. A format for proving the statement above would be

    i. Let x, y be arbitrary such that $\mathcal{P}(x, y)$.
       ·
       ·
    k. $\mathcal{Q}(x, y)$
  k+1. for all x, y such that $\mathcal{P}(x, y)\colon \mathcal{Q}(x, y)$ $(i$—$k;\ \text{pr. } \forall)$

Each of the following is an example of a **for all** phrase that quantifies statements in our language:

    for all $x \in \mathbf{R}$:

    for all **real numbers** x:

    for all **sets** C:

    for all **sets** A, B:

    for all **integers** x, y:

    for all $x, y \in A$:

    for all $x \in A, y, z \in B$:

    for all x, y, z such that $x \in A, y \in B, z \in B$:

The rule for proving **for all** statements will allow us to prove statements quantified with phrases like those listed above.

Example 1:

    i. Let $x \in A, y, z \in B$ be arbitrary.
       ·
       ·
    k. $\mathcal{Q}(x, y, z)$
  k+1. for all $x \in A, y, z \in B\colon \mathcal{Q}(x, y, z)$ $(i$—$k;\ \text{pr. } \forall)$

1.3 Using *For All* and *Or* Statements

Example 2:

    i. Let A and B be arbitrary sets.

    ·

    ·

    k. $A \cap B \subseteq B$

k+1. **for all sets A, B: $A \cap B \subseteq B$**   (i—k; pr. $\forall$)

We now give, for a single variable, a rule for *using*, as a reason, a **for all** statement that we know is true.

Inference Rule

Using for all statements: If we know that the statement for all x such that $\mathscr{P}(x)$: $\mathcal{Q}(x)$ is true and if we have $\mathscr{P}(t)$ as a step in a proof for any variable t, then we may write $\mathcal{Q}(t)$ as a step in the proof.

The rule for using a **for all** statement will be abbreviated "us. $\forall$" in proof step justifications. A proof format for employing this inference rule would be:

    i. $\mathscr{P}(t)$

    ·

    ·

    j. **for all x such that $\mathscr{P}(x)$: $\mathcal{Q}(x)$**

    k. $\mathcal{Q}(t)$ (i, j; us. $\forall$)

Example 3:

1. $x \in A$
2. **for all $t \in A$: $t \in B$**
3. *   (1, 2; us. $\forall$)

Solution:

1. $x \in A$
2. **for all $t \in A$: $t \in B$**
3. $x \in B$ (1, 2; us. $\forall$)

Example 4:

1. **for all $t < 1$: $t \in S$**
2. $y < 1$
3. *   (1, 2; us. $\forall$)

Chapter 1   Sets and Rules of Inference

Solution:
1. for all $t < 1$: $t \in S$
2. $y < 1$
3. $y \in S$ (1, 2; us. ∀)

Example 5:
1. for all sets A such that $A \subseteq B$: $A \subseteq C$
2. *
3. $X \subseteq C$ (1, 2; us. ∀)

Solution:
1. for all sets A such that $A \subseteq B$: $A \subseteq C$
2. $X \subseteq B$
3. $X \subseteq C$ (1, 2; us. ∀)

This rule will also be considered to apply to more than one variable.

Example 6:
1. for all x, y such that $|x| < |y|$: $x^2 < y^2$
2. $|a| < |b|$
3. *   (1, 2; us. ∀)

Solution:
1. for all x, y such that $|x| < |y|$: $x^2 < y^2$
2. $|a| < |b|$
3. $a^2 < b^2$ (1, 2; us. ∀)

Theorem 1.3.1   Suppose A, B, and C are sets. If $A \subseteq B$ and $B \subseteq C$, then $A \subseteq C$.

In order to prove this theorem, we first decide what we are given and what we need to show:

Proof:
Assume: A, B, C sets
               1. $A \subseteq B$
               2. $B \subseteq C$
Show: $A \subseteq C$

1.3 Using *For All* and *Or* Statements

Now consider only the conclusion and what it means:

Scrap Paper

Show: $\quad A \subseteq C$ ⎫
⎬ By definition of $\subseteq$, this means
That is, show: for all $x \in A$: $x \in C$ ⎭

This gives us the first and last three lines of our proof:

1. Let $x \in A$ be arbitrary.
   .
   .
k. $\quad x \in C$
k+1. for all $x \in A$: $x \in C$ $\qquad$ (1—k; pr. $\forall$)
k+2. $A \subseteq C$ $\qquad$ (k+1; def. $\subseteq$)

The lines we have written come from focusing on the conclusion and on what it means for it to be true—by definition. These lines give us the form of the proof. It is now possible to connect Step 1 and Step k (that is, to fill in the remaining lines of the proof) by using the hypotheses. Step 1 tells us that $x \in A$, and the hypothesis $A \subseteq B$ involves the set A. It will clearly be productive to use this hypothesis. We may infer from $A \subseteq B$ only what is given by the definition of set containment (since at this stage we have only the definition and no previous theorems involving $\subseteq$). This gives us Step 2:

1. Let $x \in A$ be arbitrary.
2. **for all $t \in A$: $t \in B$** $\qquad$ (hyp. 1; def. $\subseteq$)

We have used "t" as a local variable in Statement 2 since the symbol "x" is already in use from Step 1. We now apply Statement 2 to the information in Step 1 by the rule for using **for all**:

1. Let $x \in A$ be arbitrary.
2. **for all $t \in A$: $t \in B$** $\qquad$ (hyp. 1; def. $\subseteq$)
3. $x \in B$ $\qquad$ (1, 2; us. $\forall$)

Using the hypothesis that $B \subseteq C$, we can now complete the proof:

Proof:

Assume: A, B, C sets
    1. $A \subseteq B$
    2. $B \subseteq C$
Show: $A \subseteq C$

1. Let $x \in A$ be arbitrary.
2. for all $t \in A$: $t \in B$ (hyp. 1; def. $\subseteq$)
3. $x \in B$ (1, 2; us. $\forall$)
4. for all $t \in B$: $t \in C$ (hyp. 2; def. $\subseteq$)
5. $x \in C$ (3, 4; us. $\forall$)
6. for all $x \in A$: $x \in C$ (1—5; pr. $\forall$)
7. $A \subseteq C$ (6; def. $\subseteq$)

Notice that we are free to use "t" again as a local variable in Step 4 since its scope as a local variable in Step 2 does not go beyond Step 2. The scope of the variable x, however, goes from Step 1 through Step 5. Also, we don't write the hypotheses down again in the proof steps but rather *use* the hypotheses to deal with the variables and steps obtained by considering the meaning of the conclusion.

Definition   *Given sets A and B, the **union** of A and B, written $A \cup B$, is defined by $A \cup B = \{x \mid x \in A \text{ or } x \in B\}$.*

We take the word "**or**" in this definition, as always in mathematics, in the inclusive sense, that is, x is in $A \cup B$ if it is in A or in B or in both. If we are going to use this definition, we need to know how to use and prove **or** statements.

Inference Rule   *Using **or** statements (preliminary version): If we know that $\mathcal{P}$ or $\mathcal{Q}$ is true and if we can show that $\mathcal{R}$ is true assuming $\mathcal{P}$ and also that $\mathcal{R}$ is true assuming $\mathcal{Q}$, then we may infer $\mathcal{R}$ is true.*

This rule is abbreviated "us. **or**". A proof scheme for using **or** statements in proofs would look like this:

    $i.$ $\mathcal{P}$ or $\mathcal{Q}$
Case 1 $i+1.$ Assume $\mathcal{P}$ is true.
    $i+2.$ $\cdots$

1.3 Using *For All* and *Or* Statements

$\quad\quad\quad\quad\quad\bullet$
$\quad\quad\quad\quad\quad\bullet$
$\quad\quad\quad\quad j.\quad \mathcal{R}$ is true.
$\quad$Case 2$\quad j+1.\quad$Assume $\mathcal{Q}$ is true.
$\quad\quad\quad\quad j+2.\quad \cdots$
$\quad\quad\quad\quad\quad\bullet$
$\quad\quad\quad\quad\quad\bullet$
$\quad\quad\quad\quad k.\quad \mathcal{R}$ is true.
$\quad k+1.\quad \mathcal{R}$ is true. $\quad\quad\quad\quad\quad\quad\quad\quad (i\text{---}k;\ \text{us. }\mathbf{or})$

Think for a moment to see why this inference scheme does logically establish $\mathcal{R}$ if we know that $\mathcal{P}$ or $\mathcal{Q}$ is true but can't know which is true.

The case labels for Statements $i+1$ and $j+1$ above serve to separate pieces of the proof that must remain distinct. Even though Steps $i+1$ through k are all at the same indentation level, it is not legitimate to use any statement from Case 1 ($i+1\text{---}j$) in establishing a statement from Case 2 ($j+1\text{---}k$). In order to invoke the rule for using **or** statements, it is necessary to have the line we seek to establish (like Line $k+1$) as the last line in each case.

The rule for using **or** statements is needed to establish the following theorem:

Theorem 1.3.2$\quad$ For sets A, B, and C, if $A \subseteq C$ and $B \subseteq C$, then $A \cup B \subseteq C$.

Proof:
$\quad$*Assume:*$\quad\quad A, B, C$ sets
$\quad\quad\quad\quad\quad\quad\quad$ 1. $A \subseteq C$
$\quad\quad\quad\quad\quad\quad\quad$ 2. $B \subseteq C$
$\quad$*Show:*$\quad\quad\quad A \cup B \subseteq C$

Scrap Paper
$\quad$*Show:*$\quad A \cup B \subseteq C \quad\leftarrow$ At the top level, this is a statement that
$\quad\quad\quad\quad\quad\quad\quad\quad\quad\quad\quad\quad$ one set is contained in another.
$\quad\quad\quad\quad\quad\quad\quad\quad$ By definition of $\subseteq$, we have
$\quad\quad\quad\quad\quad\quad\quad\quad\quad\quad\quad\quad\downarrow$
$\quad$*That is, show:*$\quad$ for all $x \in A \cup B$: $x \in C$

When we have introduced a local variable (here x), it is time to bail out of this scrap-paper analysis and set up the first steps in the proof:

1. Let $x \in A \cup B$ be arbitrary.
   .
   .
   k. $x \in C$
   k+1. **for all** $x \in A \cup B: x \in C$ (1—k; pr. ∀)
   k+2. $A \cup B \subseteq C$ (k+1; def. ⊆)

We are led naturally to these steps by considering first the top-level form of the conclusion and then the top-level form of the step preceding the conclusion. Attempts to make statements based on the definition of union before these steps are set up will be disastrous. Set union is the new idea introduced in the theorem, and it is natural to be thinking about it, but discipline in proceeding according to the routine backward procedure will be rewarded.

Let us proceed. We now have from Step 1 that x is an arbitrarily chosen but henceforth fixed element of the set $A \cup B$. What can we say about x? Only one thing—that which we know by the *definition* of this set. *Now* is the time to use the definition of union. We get:

1. Let $x \in A \cup B$ be arbitrary.
2. $x \in A$ **or** $x \in B$ (1; def. ∪)

Just as we must use the definition of union to argue from Step 1, we must use the rule for using **or** to argue from Step 2. We seek step k. $x \in C$:

1. Let $x \in A \cup B$.
2. $x \in A$ **or** $x \in B$ (1; def. ∪)
Case 1   3. Assume $x \in A$.
   .
   .
   j. $x \in C$
Case 2   j+1. Assume $x \in B$.
   .
   .
   k−1. $x \in C$
   k. $x \in C$ (2, 3—k−1; us. **or**)
   k+1. **for all** $x \in A \cup B: x \in C$ (1—k; pr. ∀)
   k+2. $A \cup B \subseteq C$ (k+1; def. ⊆)

1.3 Using *For All* and *Or* Statements

The entire proof so far has been dictated by our grammar, definitions, and rules of inference. We had no choice but to get the steps above. Now is the time in a proof when generally a little ingenuity is called for—when we need to make the connections. In our case, we need to go from 3. $x \in A$ to j. $x \in C$ and from $j+1$. $x \in B$ to $k-1$. $x \in C$. Now that our task is clearly defined and we can carry the backward analysis no further, it is at last time to consider the hypotheses. We need to get $x \in C$ from $x \in A$. But $A \subseteq C$ by hypothesis. This means that the only intermediate step comes from the definition of set containment. We use the hypothesis $B \subseteq C$ in a similar way. We now have a complete proof:

Proof:

   Assume: A, B, C sets
                  1. $A \subseteq C$
                  2. $B \subseteq C$
   Show: $A \cup B \subseteq C$

          1. Let $x \in A \cup B$ be arbitrary.
          2. $x \in A$ or $x \in B$ (1; def. $\cup$)
Case 1   3. Assume $x \in A$.
          4. for all $t \in A$: $t \in C$ (hyp. 1; def. $\subseteq$)
          5. $x \in C$ (3, 4; us. $\forall$)
Case 2   6. Assume $x \in B$.
          7. for all $t \in B$: $t \in C$ (hyp. 2; def. $\subseteq$)
          8. $x \in C$ (6, 7; us. $\forall$)
          9. $x \in C$ (2—8; us. **or**)
        10. for all $x \in A \cup B$: $x \in C$ (1—9; pr. $\forall$)
        11. $A \cup B \subseteq C$ (10; def. $\subseteq$) ∎

Summary of Our Routine Approach to Determining Steps in a Proof

After determining the hypotheses and conclusion, focus on what it means for the conclusion to be true. This will be seen by the top-level form of the conclusion and by considering either (1) what this means by definition or (2) which inference rule is needed to show the appropriate kind of statement. Ignore the hypotheses at first. It is generally best to continue developing steps from the bottom up, adding steps at the top as they are dictated by the inference rules. Further steps that follow inevitably from those already written are added, both from the top down and from the bottom up—narrowing the gap between the top and bottom steps. When this straightforward process can be carried no further, it is time to consider the hypotheses for help in bridging the remaining gap or gaps in the proof. As you get better at doing proofs, you will develop a sense of where they need to go and will start to write them from the top down. Don't rush it.

Exercises 1.3

Practice Exercises

1. In the proof fragment (a) below, Step 3 is to follow from Steps 1 and 2 and the rule for using **for all**. There is only one statement that can replace the asterisk to make logical proof steps that follow our rules of inference. The same is true for Steps 4 and 7. Fill in Steps 3, 4, and 7 in such a way as to make logical proof steps. In the other proof fragments, similarly fill in the proof steps marked by an asterisk. In cases where more than one logical replacement is possible, any one will do.

    (a)   1.   for all $t \in A$: $t \in B$
              2. $x \in A$
              3.   *   (1, 2; us. $\forall$)
              4.   *
              5. $y \in B$ (1, 4; us. $\forall$)
              6. $t \in A$
              7.   *   (1, 6; us. $\forall$)

    (b)   1.   *
              2.   · · ·
              3.   *
              4.   for all $x \in Y$: $x \in Z$ (1—3; pr. $\forall$)
              5.   *
              6. $a \in Z$ (4, 5; us. $\forall$)

    (c)   1.   *
       Case 1   2.   Assume $x \in A$.
                    3.   · · ·
                    4.   *
       Case 2   5.   Assume $x \in B$.
                    6.   · · ·
                    7.   *
                    8. $a \in B$ (1—7; us. **or**)

1.3 Using *For All* and *Or* Statements   29

(d)   1. $a \in A$ or $b \in A$
Case 1   2.   Assume $a \in A$.
       3.   ...
       4. $b \in C$
Case 2   5.   *
       6.   ...
       7.   *
       8.   *   (1—7; us. **or**)

Note: In part (a) above, the scope of the local variable t of Step 1 does not go beyond this step. We are therefore free to use t in Step 6 as a global variable. Although this use is legitimate, it is to be discouraged. As a general rule, avoid using the same symbol for both a local and a global variable. One exception to this rule is that, in proving a statement (such as **for all**) containing a local variable, we frequently use the same symbol in the previous steps used as justification.

2. (a)   1.   *
       2.   ...
       3.   *
       4.   **for all** $x \in A, y \in B$: $x^2 + 1 < y$ (1—3; pr. $\forall$)

(b)   1. $x \in S$
       2.   *
       3. $x \in T$ (1, 2; us. $\forall$)

(c)   1.   *
       2. $x \in S$ or $x \in T$ (1; def. $\cup$)

(d)   1. $s \in M \cup N$
       2.   *   (1; def. $\cup$)

(e)   1. $x \in S$ or $x \in T$
       2.   *   (1; def. $\cup$)

(f)   1.   *
       2.   ...
       3.   *
       4.   **for all** sets A, B: $A \subseteq A \cup B$ (1—3; pr. $\forall$)

(g)   1.   Let A and B be arbitrary sets.
       2.   ...
       3. $Q(A, B)$
       4.   *   (1—3; pr. $\forall$)

(h) 1. $a \in B \cup C$
2. *   (1; def. $\cup$)
Case 1   3. *
4. ...
5. *
Case 2   6. *
7. ...
8. *
9. $a \in X$ (2—8; us. **or**)

(i) 1. $z < 1$ or $x > 5$
Case 1   2. *
3. ...
4. $x \in A$
Case 2   5. *
6. ...
7. *
8. *   (1—7; us. **or**)

(j) 1. Let $x \in A$ be arbitrary.
    .
    .
k. $x \in B \cup C$
k+1. *   (1—k; *)

Straightforward Problems

3. Define $X = \{x \in R \mid x > 1\}$ and $Y = \{x \in R \mid x > 10\}$.
   Prove **for all** $a \in X \cup Y: a \neq 0$.

4. Assume $X = \{x \in R \mid x > 1\}$ and $Y \subseteq X$.
   Prove **for all** $x \in Y: x > 1$.

1.4 USING AND PROVING *OR* STATEMENTS

Consider again the following parts of the proof of Theorem 1.3.2:

Assume: A, B, C sets
1. $A \subseteq C$
2. $B \subseteq C$

1.4 Using and Proving *Or* Statements

```
Case 1   3.  Assume x ∈ A.
         4.  for all t ∈ A: t ∈ C      (hyp. 1; def. ⊆)
         5.  x ∈ C                     (3, 4; us. ∀)
```

In order to keep our proofs from becoming too lengthy and tedious, we will use certain shortcuts. For example, in Step 4 above we have the defining property for set containment ($A \subseteq C$). This property follows from Hypothesis 1 by our definition rule. Instead of writing **for all** $t \in A$: $t \in C$ as a proof step, we want to identify this in our minds with the hypothesis $A \subseteq C$, which *means* **for all** $t \in A$: $t \in C$. In order to use the hypothesis $A \subseteq C$, therefore, we use the **for all** statement. We apply the **for all** statement to Step 3, $x \in A$, to conclude $x \in C$—but we think of this as applying the definition of set containment to $x \in A$ to conclude $x \in C$. The **for all** statement becomes implicit, and our new contracted proof steps are:

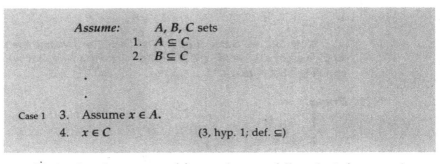

```
         Assume:    A, B, C sets
                1.  A ⊆ C
                2.  B ⊆ C
         .
         .
Case 1   3.  Assume x ∈ A.
         4.  x ∈ C                     (3, hyp. 1; def. ⊆)
```

Notice that the new proof format does not follow the inference rules we have so far. According to our definition rule, the *only* thing we are allowed to infer from $A \subseteq C$ and the definition of set containment is the statement **for all** $t \in A$: $t \in C$—not the statement $x \in C$. We now want to extend our definition rule so that it allows this two-step process: first obtain the defining property and then, without writing down the defining property as a proof step, apply the property to the existing proof steps to infer new steps.

Inference Rule
Extended definition rule: When the statement $\mathscr{P}$ is the defining property for some definition, it is permissible to either use $\mathscr{P}$ or prove $\mathscr{P}$ (according to appropriate rules) without writing $\mathscr{P}$ itself as a step. For justification for the step inferred, give the definition and not the rule for using or proving $\mathscr{P}$.

Example 1:
Consider the proof fragment above for Theorem 1.3.2. The statement **for all** $x \in A$: $x \in C$ is the defining property for $A \subseteq C$. It is permissible to use **for all**

$x \in A: x \in C$ without writing it down as a proof step. For the justification for Step 4, give "def. $\subseteq$" and not "us. $\forall$".

For a while we will use the notation "def.^{2}" for this extended definition rule in step justification. The former, one-step definition rule will be distinguished by writing "def.^{1}". The notation "def." will stand for either of the above, and ultimately we will use only the ambiguous "def.".

Example 2:

    1. $a \in M$
    2. $M \subseteq N$
    3. *   (1, 2; def.^{2} $\subseteq$)

Solution:

    1. $a \in M$
    2. $M \subseteq N$
    3. $a \in N$ (1, 2; def.^{2} $\subseteq$)

Note that we think of $M \subseteq N$ in Step 2 as *meaning* **for all** $t \in M: t \in N$. Applying **for all** $t \in M: t \in N$ to $a \in M$ gives $a \in N$, but we think of this as applying $M \subseteq N$ to $a \in M$.

Example 3:

    1. $a \in M$
    2. $M \subseteq N$
    3. *   (2; def.^{1} $\subseteq$)

Solution:

    1. $a \in M$
    2. $M \subseteq N$
    3. **for all** $t \in M: t \in N$ (2; def.^{1} $\subseteq$)

The effect of using the extended definition rule is to focus our attention on the mathematical objects we're talking about and not on the logical form of statements about them. Ultimately our logic will be entirely implicit. We will use the extended definition rule when using $\subseteq$ in the proof of Theorem 1.4.1.

The language word "**or**" is called a *connective* since, given two language statements $\mathscr{P}$ and $\mathscr{Q}$, we can construct the new statement $\mathscr{P}$ **or** $\mathscr{Q}$. The meaning

1.4 Using and Proving *Or* Statements

of the connective "**or**" is completely determined when we have rules for both using and proving **or** statements. We get to the second of these now.

Inference Rule

Proving or statements: If $\mathcal{P}$ has been established as a line in a proof, then $\mathcal{P}$ or $\mathcal{Q}$ may be written as a line. (Symmetrically, we may write $\mathcal{P}$ or $\mathcal{Q}$ if we have established $\mathcal{Q}$.) This rule is abbreviated "pr. or".

A proof format for using this rule would be:

i. $\mathcal{P}$
  .
  .
  .
j. $\mathcal{P}$ or $\mathcal{Q}$†   (i; pr. or)

This inference rule is used in the following proof.

Theorem 1.4.1

For sets A, B, and C: if $A \subseteq B$ or $A \subseteq C$, then $A \subseteq B \cup C$.

Proof:
  Assume: A, B, C sets
           $A \subseteq B$ or $A \subseteq C$
  Show: $A \subseteq B \cup C$

Notice that we have only one fact on set containment in the hypothesis, not two as we had in Theorem 1.3.2 with the "and" statement. We don't know $A \subseteq B$ is a fact, and we don't know $A \subseteq C$ is a fact. We know that one or the other is a fact but not which one.

Scrap Paper
  Show: $A \subseteq B \cup C$
                                                  ⎞ By definition of $\subseteq$
  That is, show: for all $x \in A$: $x \in B \cup C$ ⎠

---

† From the rule for proving **or**, $\mathcal{Q}$ or $\mathcal{P}$ could just as well have been written as Step j. The idea of *equivalence* is defined in Section 1.7. Theorem 1.7.3L asserts that $\mathcal{P}$ or $\mathcal{Q}$ is equivalent to $\mathcal{Q}$ or $\mathcal{P}$.

34   Chapter 1   Sets and Rules of Inference

      1.   Let $x \in A$ be arbitrary.
            ·
            ·
      k. $x \in B \cup C$
$k+1$. **for all** $x \in A: x \in B \cup C$ (1—k; pr. $\forall$)

Our scrap-paper analysis has given us Steps 1, k, and $k+1$. We ought to consider next what we need to know about x in order that it be shown in $B \cup C$. It is clear that x must satisfy the condition defining $B \cup C$. This gives us step $k-1$:

      1.   Let $x \in A$ be arbitrary.
            ·
            ·
$k-1$. $x \in B$ **or** $x \in C$
    k. $x \in B \in C$ ($k-1$; def.^{1} $\cup$)
$k+1$. **for all** $x \in A: x \in B \cup C$ (1—k; pr. $\forall$)

We are now at the point of needing to prove an **or** statement (Step $k-1$). Since we know the truth of an **or** statement (hypothesis), we proceed by cases:

          1.   Let $x \in A$ be arbitrary.
          2. $A \subseteq B$ **or** $A \subseteq C$ (hyp.)
Case 1   3.   Assume $A \subseteq B$.
          4. $x \in B$ (1, 3; def.^{2} $\subseteq$)
          5. $x \in B$ **or** $x \in C$ (4; pr. **or**)
Case 2   6.   Assume $A \subseteq C$.
          7. $x \in C$ (1, 6; def.^{2} $\subseteq$)
          8. $x \in B$ **or** $x \in C$ (7; pr. **or**)
          9. $x \in B$ **or** $x \in C$ (2—8; us. **or**)
        10. $x \in B \cup C$ (9; def.^{1} $\cup$)
        11. **for all** $x \in A: x \in B \cup C$ (1—10; pr. $\forall$)
        12. $A \subseteq B \cup C$ (11; def.^{1} $\subseteq$)

1.4 Using and Proving *Or* Statements

In this proof it was useful to write the hypothesis as Step 2 in order to later invoke the rule for using **or**. Notice the advantageous use of retreat. In Step 4 we have $x \in B$, and from this we infer the weaker condition $x \in B$ **or** $x \in C$, that is, we seem to give up some of the force of what we know about x. But it is this same weak condition, $x \in B$ **or** $x \in C$, that follows from Step 7, $x \in C$, and thus holds in both Case 1 and Case 2. Thus we are able to invoke the rule for using an **or** statement—in this case to prove another **or** statement.

There is another way in which we will be allowed to use **or** statements. This involves the idea of the negation of a statement. If $\mathcal{P}$ is any statement, then the *negation* of $\mathcal{P}$ is written "$\sim\mathcal{P}$" (read "not $\mathcal{P}$") and is the assertion "the statement $\mathcal{P}$ is false". Consequently, if $\mathcal{P}$ really is false, then $\sim\mathcal{P}$ is true, and if $\mathcal{P}$ is true, then $\sim\mathcal{P}$ is false. If, in a proof, we have established that both $\mathcal{P}$ and $\sim\mathcal{P}$ are valid at the same time, we say that we have reached a contradiction—which we denote by the symbol #. For steps we might have:

1. · · ·
2. $\mathcal{P}$
3. · · ·
4. $\sim\mathcal{P}$ # Step 2   (reason 4 is true)
5. · · ·

We will also be allowed to use, without explicit mention, the fact that for a statement $\mathcal{P}$, $\sim(\sim\mathcal{P})$ means the same as $\mathcal{P}$. Thus $\sim\mathcal{P}$ is the negation of $\mathcal{P}$, and $\mathcal{P}$ is the negation of $\sim\mathcal{P}$. So our steps leading to a contradiction might look like this:

1. · · ·
2. $\sim\mathcal{P}$
3. · · ·
4. $\mathcal{P}$ # Step 2   (reason 4 is true)
5. · · ·

The other way of using an **or** statement involves obtaining a contradiction:

Inference Rule   *Using **or** statements (preliminary version, second part): If we know that $\mathcal{P}$ or $\mathcal{Q}$ is true and can establish a contradiction assuming $\mathcal{P}$ is true, then we know that $\mathcal{Q}$ is true.*

A format for using this rule would be:

   i. $\mathcal{P}$ or $\mathcal{Q}$
   i+1.   Assume $\mathcal{P}$.

Chapter 1 Sets and Rules of Inference

$\qquad \vdots$

$\quad\quad\quad\quad$ j. (Contradict some previous step here.)
$\quad$ j+1. $\mathscr{Q}$ $\qquad\qquad\qquad\qquad\qquad\qquad\qquad$ (i—j; us. **or**)

Example 4:
$\quad$ 1. $x < 2$
$\quad$ 2. $\mathscr{P}$ or $\mathscr{Q}$
$\quad\quad$ 3. Assume $\mathscr{P}$.
$\quad\quad\quad \vdots$
$\quad\quad$ j. $x > 3$
$\quad$ j+1. $x \geq 2$ # Step 1 $\qquad$ (*)
$\quad$ j+2. * $\qquad\qquad\qquad\qquad$ (*; us. **or**)

Solution:
$\quad$ 1. $x < 2$
$\quad$ 2. $\mathscr{P}$ or $\mathscr{Q}$
$\quad\quad$ 3. Assume $\mathscr{P}$.
$\quad\quad\quad \vdots$
$\quad\quad$ j. $x > 3$
$\quad$ j+1. $x \geq 2$ # Step 1 $\qquad$ (j; prop. **R**)
$\quad$ j+2. $\mathscr{Q}$ $\qquad\qquad\qquad\qquad$ (2—j+1; us. **or**)

For future reference we now put the two parts of our inference rule together to get the final version. In this version we suppose that we know the truth of an **or** statement involving k constituent statements:

$$\mathscr{P}_1 \text{ or } \mathscr{P}_2 \text{ or } \mathscr{P}_3 \text{ or } \cdots \text{ or } \mathscr{P}_k$$

Each of these k statements corresponds to a case in a proof, giving us Case 1, Case 2, ..., Case k.

Inference Rule

Using **or** statements: If we know $\mathscr{P}_1$ or $\mathscr{P}_2$ or $\cdots$ or $\mathscr{P}_k$ is true and if we prove $\mathscr{R}$ is true in all cases that do not lead to a contradiction, then we infer that $\mathscr{R}$ is true. If all cases lead to a contradiction, then we infer the negation of the most recently assumed statement.

1.4 Using and Proving *Or* Statements

As an example of a proof format involving this rule, we have:

    1.   ...
    2. $\mathcal{P}_1$ or $\mathcal{P}_2$ or $\mathcal{P}_3$

Case 1   3.   Assume $\mathcal{P}_1$ is true.
              .
              .
          5.   (Contradict some previous step.)

Case 2   6.   Assume $\mathcal{P}_2$ is true.
          7.   ...
          8. $\mathcal{R}$ is true.

Case 3   9.   Assume $\mathcal{P}_3$ is true.
         10.   ...
         11. $\mathcal{R}$ is true.
         12. $\mathcal{R}$ is true.   (2—11; us. **or**)

A natural example of the use of this rule where all cases do lead to a contradiction is given in Section 1.8. In Section 1.7 we give a theorem that almost always proves more useful for proving **or** statements than the rule presented in this section. The reason we have given the rule as stated here is that it embodies the meaning of **or**: to prove $\mathcal{P}$ **or** $\mathcal{Q}$, prove $\mathcal{P}$ or prove $\mathcal{Q}$.

The following extension of our rule for proving **or** statements to statements involving more than two pieces is clear:

Inference Rule   *Proving or statements: We may write $\mathcal{P}_1$ or $\mathcal{P}_2$ or $\cdots$ or $\mathcal{P}_k$ if we have established any one of $\mathcal{P}_1$ through $\mathcal{P}_k$.*

Example 5:

    1.   *
    2. $x \in A \cup B$ (1; def.¹ $\cup$)

Solution:

    1. $x \in A$ or $x \in B$
    2. $x \in A \cup B$ (1; def.¹ $\cup$)

Example 6:

    1.   *
    2. $x \in A \cup B$ (1; def.² $\cup$)

Solution:

1. $x \in A$
2. $x \in A \cup B$ (1; def.² ∪)

Example 6 illustrates the use of the extended definition rule: By the definition of ∪, $x \in A \cup B$ means $x \in A$ **or** $x \in B$. If we apply the rule for proving **or**, we need either $x \in A$ or $x \in B$ as a preceding step. Either of these statements would be satisfactory as Step 1. The intermediate step $x \in A$ **or** $x \in B$ in going from $x \in A$ to $x \in A \cup B$ is not written down.

Exercises 1.4

Practice Exercises

1. Use the extended definition rule to fill in the proof steps marked by an asterisk (*). Then fill in step "$2\frac{1}{2}$", the step that has not been written but has been applied to get Step 3. There may be more than one correct response, and if so, any one will do.

    (a) 1. $A \subseteq B$
        2. $x \in A$
        3. *   (1, 2; def.² ⊆)

    (b) 1. $X \subseteq Y$
        2. *
        3. $t \in Y$ (1, 2; def.² ⊆)

    (c) 1. Let $x \in A$ be arbitrary.
        2. $x \in B$
        3. *   (1, 2; def.² ⊆)

    (d) 1. $\cdots$
        2. $x \in B$
        3. *   (2; def.² ∪)

    (e) 1. $\cdots$
        2. *
        3. $t \in X \cup Y$ (2; def.² ∪)

2. In the proof fragments below, fill in the proof steps marked with an asterisk (*). There may be more than one correct response, and if so, any one will do.

(a) 1. $x \in A \cup B$
    2. *   (1; def.^{1} $\cup$)

(b) 1. *
    2. $x \in A \cup B$ (1; def.^{1} $\cup$)

(c) 1. **for all** $x \in A$: $x \in B$
    2. *   (1; def.^{1} *)

(d) 1. $x < 10$
    2. $x \in A \cup B$
    3. $x \in A$ **or** $x \in B$ (2; *)

Case 1   4. Assume $x \in A$.
            .
            .

            j. $x \geq 10$ # Step 1
Case 2   j+1. Assume $x \in B$.
            .
            .

            k. $x = 5$
k+1. *   (3—k; *)

Straightforward Problems

3. Let A and B be sets. Prove $A \subseteq A \cup B$.
4. Let $A \subseteq C$ and $B \subseteq D$. Prove $A \cup B \subseteq C \cup D$.

Supplementary Problem

5. Assume **for all** $x \in A$: $\mathcal{P}(x)$ and **for all** $x \in B$: $\mathcal{P}(x)$. Prove **for all** $x \in A \cup B$: $\mathcal{P}(x)$.

1.5 *AND* STATEMENTS

Definition   *For any sets A and B, the **intersection** of A and B is the set $A \cap B$ defined by $A \cap B = \{x \mid x \in A$ **and** $x \in B\}$.*

To deal with **and** statements in a proof, we have two rules:

Chapter 1 Sets and Rules of Inference

Inference Rule

Using **and** statements: If $\mathcal{P}$ **and** $\mathcal{Q}$ is a step in a proof, then $\mathcal{P}$ can be written as a step and $\mathcal{Q}$ can be written as a step. Abbreviation: "us. &".

Format:

    i. $\mathcal{P}$ **and** $\mathcal{Q}$
    j. $\mathcal{P}$ (i; us. &)
    k. $\mathcal{Q}$ (i; us. &)

Example 1:

1. $x < 1$ **and** $x \in A$
2. *   (1; us. &)

Solution:

1. $x < 1$ **and** $x \in A$
2. $x < 1$ (or 2. $x \in A$)   (1; us. &)

Inference Rule

Proving **and** statements: In order to show $\mathcal{P}$ **and** $\mathcal{Q}$ in a proof, show $\mathcal{P}$ and also show $\mathcal{Q}$. Abbreviation: "pr. &".

The proof format for using this rule for proving **and** is:

    i. $\mathcal{P}$
        .
        .
    j. $\mathcal{Q}$
    k. $\mathcal{P}$ **and** $\mathcal{Q}$ (i, j; pr. &)

Example 2:

1. *
2. *
3. $x \in M$ **and** $x \in N$ (1, 2; pr. &)

Solution:

1. $x \in M$
2. $x \in N$
3. $x \in M$ **and** $x \in N$ (1, 2; pr. &)

1.5 *And* Statements

We will not have occasion to use rules for more general **and** statements but will assume for the sake of symmetry with our **or** statement that our rules apply as in the following examples:

Example 3:

1. *
2. *
3. *
4. $\mathcal{P}$ and $\mathcal{Q}$ and $\mathcal{R}$ (1, 2, 3; pr. &)

Solution:

1. $\mathcal{P}$
2. $\mathcal{Q}$
3. $\mathcal{R}$
4. $\mathcal{P}$ and $\mathcal{Q}$ and $\mathcal{R}$ (1, 2, 3; pr. &)

Example 4:

1. $\mathcal{P}$ and $\mathcal{Q}$ and $\mathcal{R}$
2. *   (1; us. &)

Solution:

1. $\mathcal{P}$ and $\mathcal{Q}$ and $\mathcal{R}$
2. $\mathcal{Q}$ (or 2. $\mathcal{P}$ or 2. $\mathcal{R}$)   (1; us. &)

Theorem 1.5.1   *For sets A, B, and C, if $A \subseteq B$ and $A \subseteq C$, then $A \subseteq B \cap C$.*

    Proof: Exercise 6.

Definition   *A set A is **equal** to a set B (written $A = B$) provided that $A \subseteq B$ and $B \subseteq A$.*

Example 5:

1. $A = B$
2. *   (1; def.[1] =)
3. *   (2; us. &)
4. *   (2; us. &)

Chapter 1 Sets and Rules of Inference

Solution:

1. $A = B$
2. $A \subseteq B$ and $B \subseteq A$ (1; def.^{1} =)
3. $A \subseteq B$ (2; us. &)
4. $B \subseteq A$ (2; us. &)

Example 6:

1. $A = B$
2. *   (1; def.^{2} =)
3. *   (1; def.^{2} =)

Solution:

1. $A = B$
2. $A \subseteq B$ (1; def.^{2} =)
3. $B \subseteq A$ (1; def.^{2} =)

Example 7:

1. *
2. *
3. $A = B$ (1, 2; def.^{2} =)

Solution:

1. $A \subseteq B$
2. $B \subseteq A$
3. $A = B$ (1, 2; def.^{2} =)

Our first theorem involving the idea of set equality (Theorem 1.5.2) is a triviality. Nevertheless it is very important to see exactly why it is a triviality. All theorems must be provable—even trivial ones. There is no such thing as a trivial theorem that can't be proved.

Theorem 1.5.2   For sets A and B,

(a) $A \cap B = B \cap A$
(b) $A \cup B = B \cup A$

1.5 *And* Statements

We will prove part (a) and leave (b) as Exercise 7. In proving Theorem 1.5.2b, you may use all the inference rules in this section—not only those preceding the statement of the theorem. This applies in general for the entire text.

Proof:
Assume: A, B sets
Show: $A \cap B = B \cap A$

Scrap Paper
That is, show: $A \cap B \subseteq B \cap A$ By definition of $=$
and $B \cap A \subseteq A \cap B$

That is, show: 1. $A \cap B \subseteq B \cap A$ By definition of **and**
2. $B \cap A \subseteq A \cap B$

Again we have taken the conclusion and reworded it by definition. (Note that the definition of intersection is not involved at this top grammatical level.) This analysis gives us the first and last lines of a section of our proof. First:

1. Let $x \in A \cap B$ be arbitrary.
   .
   .
   .
k. $x \in B \cap A$
k+1. for all $x \in A \cap B$: $x \in B \cap A$ (1—k; pr. ∀)
k+2. $A \cap B \subseteq B \cap A$ (k+1; def. ⊆)

Step k+2 is one of the statements in the second set of conclusions we had on scrap paper. We now need to establish the other statement: $B \cap A \subseteq A \cap B$. Notice, however, that this second statement is just the first statement with the roles of A and B reversed. Since "A" and "B" are just names for arbitrarily given sets, we will have proved the *content* in the second statement, $B \cap A \subseteq A \cap B$, when we have proved Steps 1 through k+2. We will henceforth be allowed to write:

k+3. $B \cap A \subseteq A \cap B$ (1—k+2; symmetry)

We have the following general rule for using symmetry. (Following our discussion of Theorem 1.5.2, which illustrates the first part of the rule, we will give an example illustrating the second part.)

Inference Rule

*Using symmetry: If $\mathscr{P}(A_1, B_1, C_1, \ldots)$ is any statement that has been proved for arbitrary (that is, completely general) $A_1, B_1, C_1, \ldots$ in the hypotheses, and if $A_2, B_2, C_2, \ldots$ is any rearrangement of $A_1, B_1, C_1, \ldots$, then $\mathscr{P}(A_2, B_2, C_2, \ldots)$ is true. The foregoing also applies to variables inside a **for all** statement; that is, if **for all** $A_1, B_1, C_1, \ldots$: $\mathscr{P}(A_1, B_1, C_1, \ldots)$ is true, then **for all** $A_1, B_1, C_1, \ldots$: $\mathscr{P}(A_2, B_2, C_2, \ldots)$ is true.*

For example, if $A_1 = A$ and $B_1 = B$ in Step $k+2$ above, then $A_2 = B$ and $B_2 = A$ in Step $k+3$. Thus our proof takes the form:

    1. Let $x \in A \cap B$ be arbitrary.
       .
    k. $x \in B \cap A$
    $k+1$.   for all $x \in A \cap B$: $x \in B \cap A$ (1—k; pr. $\forall$)
    $k+2$. $A \cap B \subseteq B \cap A$ ($k+1$; def. $\subseteq$)
    $k+3$. $B \cap A \subseteq A \cap B$ (1—$k+2$; symmetry)
    $k+4$. $A \cap B \subseteq B \cap A$ and $B \cap A \subseteq A \cap B$ ($k+2, k+3$; pr. &)
    $k+5$. $A \cap B = B \cap A$ ($k+5$; def. =)

Note that in going from $k+2$ to $k+3$ using symmetry, we are doing this:

$$A \cap B \subseteq B \cap A$$
$$\downarrow \quad \downarrow \quad \downarrow \quad \downarrow$$
$$B \cap A \subseteq A \cap B$$

not this:

$$A \cap B \subseteq B \cap A$$
$$B \cap A \subseteq A \cap B$$

To this point there has been nothing to indicate why we referred to Theorem 1.5.2a as trivial. Indeed, the analysis we have undertaken so far does not look essentially different from our analysis of other proofs. The reason for our assertion of triviality is that mathematicians don't even think of the steps we have written so far. These steps have become automatic—unconscious. The only area of thought concerns connecting Steps 1 and k above, and doing this is very easy. Not only easy but inevitable, for in Step 1 the only fact we know about x is that it is in a certain set. And we know only *one* thing about this set, that which is

1.5 *And* Statements

given by its defining property, namely, the definition of intersection. Thus we get Steps 2 through 5, which complete our proof.

Proof:

1. Let $x \in A \cap B$ be arbitrary.
2. $x \in A$ and $x \in B$ (1; def.^{1} $\cap$)
3. $x \in A$ (2; us. &)
4. $x \in B$ (2; us. &)
5. $x \in B$ and $x \in A$ (3, 4; pr. &)
6. $x \in B \cap A$ (5; def.^{1} $\cap$)
7. for all $x \in A \cap B$: $x \in B \cap A$ (1—6; pr. $\forall$)
8. $A \cap B \subseteq B \cap A$ (7; def.^{1} $\subseteq$)
9. $B \cap A \subseteq A \cap B$ (1—8; symmetry)
10. $A \cap B \subseteq B \cap A$ and $B \cap A \subseteq A \cap B$ (8, 9; pr. &)
11. $A \cap B = B \cap A$ (10; def.^{1} =)

Let us see how our extended definition rule can shorten this proof. From Step 1 by the usual definition rule, we have the statement $x \in A$ and $x \in B$. This is the statement $\mathscr{P}$ of the extended definition rule. Instead of writing this down, we go immediately to Steps 3 and 4, which are obtained by *using* the **and** statement. The justification for Steps 3 and 4 is now "definition of $\cap$".

Continuing, we apply this thinking in reverse order. Instead of *proving* Step 5 by Steps 3 and 4, we consider that we have proved Step 6 by these steps—because Step 5 gives the defining property ($\mathscr{P}$ of the rule) for the set defined in Step 6.

Similarly, Step 7 can be omitted since it gives the defining property for Step 8—and 10 is omitted since it gives the defining property for 11.

The shortened proof is:

1. Let $x \in A \cap B$ be arbitrary.
2. $x \in A$ (1; def.^{2} $\cap$)
3. $x \in B$ (1; def.^{2} $\cap$)
4. $x \in B \cap A$ (2, 3; def.^{2} $\cap$)
5. $A \cap B \subseteq B \cap A$ (1—4; def.^{2} $\subseteq$)
6. $B \cap A \subseteq A \cap B$ (1—5; symmetry)
7. $A \cap B = B \cap A$ (5, 6; def.^{2} =)

This will always be an acceptable form for proofs. The defining properties may be included as proof steps any time doing so adds clarity to a proof.

Note that, in our thinking, the statement of Step 1, set membership, has been identified with the set's defining property. So $x \in A \cap B$ *means* $x \in A$ **and** $x \in B$. To use $x \in A \cap B$, therefore, we use the rule for using an **and** statement. Thus the extended definition rule allows us not only to contract what we write down but also to contract our thought processes—to ignore certain trivia in order to better focus on more fruitful things. Later, we will see other, shorter ways in which this proof can be written.

As an example of how symmetry works within a **for all** statement, consider the following:

Example 8:

1. **for all** $x, y \in A$: $x^2 + y < z$
2. **for all** $x, y \in A$: $y^2 + x < z$ (1; symmetry)

In Step 1, x and y are local (bound) variables, and this step would be exactly the same as **for all** $i, j \in A$: $i^2 + j < z$. The rule for symmetry, however, allows us to interchange the roles of x and y in the statement $x^2 + y < z$. Step 2 is true because Step 1 is true for completely arbitrary elements of A. By this we mean that x and y are in no way constrained. Thus, from the truth of **for all** $x, y \in A$: $x^2 + y < z$, we have the truth of **for all** $y, x \in A$: $x^2 + y < z$. Step 2 follows from this by changing local variables.

Additional Proof Ideas

The rule for using symmetry within a **for all** statement in fact follows from the rules for using and proving **for all** statements. In Example 8, for instance, consider the following alternate way of obtaining Step 4 (the old Step 2 in Example 8) from Step 1:

1. **for all** $x, y \in A$: $x^2 + y < z$ (scope of local variables x, y stops here)
2. Let $x, y \in A$ be arbitrary.   (now use x, y as new global variables)
3. $y^2 + x < z$ (1, 2; us. ∀)
4. **for all** $x, y \in A$: $y^2 + x < z$ (2, 3; pr. ∀)

The use of x and y as local variables in Step 1 and global variables in Steps 2 and 3 can be confusing, to say the least, and it would be very poor style to develop an actual proof this way. The proof fragment above does show why the symmetry rule works within a **for all** statement, however. It also shows why the variables to which we apply symmetry need to be completely general. This generality must also hold for the variables in the hypotheses (which act like constants in the proof) to which we apply symmetry.

Exercises 1.5

Practice Exercises

1. 1. **for all** $x, y \in \mathbb{R}: x + y = z$
   2. *   (1; symmetry)

2. Assume, in the following proof fragment, that A and B are hypothesized as arbitrary sets.

   k. $A \cap B \subseteq B$
   k+1. *   (k; symmetry)

3. 1. **for all** $a, b, c \in \mathbb{R}: (a + b) + c = a + (b + c)$
   2. *   (1; symmetry)
   (There are five possibilities for Step 2 different from Step 1.)

4. 1. **for all** $x, y \in \mathbb{R}: x^4 + y^2 \geq 0$
   2. *   (1; symmetry)

5. Give all the proof steps you can leading up to a proof of $S = T$, where S and T are sets.

Straightforward Problems

6. Prove Theorem 1.5.1.
7. Prove Theorem 1.5.2b.

1.6 USING THEOREMS

When we assert the equality of two mathematical expressions, we mean that the two expressions symbolize or name the same single mathematical object. For example, if x and y are real numbers, then to say that $x = y + 1$ is to say that the symbol "x" and the symbol "$y + 1$" both represent the same real number. The use of the following inference rule is familiar to all students of algebra.

Inference Rule *Substitution: Any name or representation of a mathematical object can be replaced by another name or representation for the same object. It is necessary to avoid using the same name for different objects.*

Chapter 1 Sets and Rules of Inference

This rule is illustrated in the following sample proof steps:

Example 1:
1. $A \cap B = C$
2. $A = D$
3. $D \cap B = C$ (1, 2; sub.)

Example 2:
1. $x^2 + 3 = x$
2. $x = y + 1$
3. $(y+1)^2 + 3 = y + 1$ (1, 2; sub.)

Now let us consider the use of theorems as justification for proof steps. In general, theorems are written in our metalanguage and applied to steps in a proof—which are in our language. In order to use a theorem as justification for proof steps, we translate the metalanguage of the theorem into a statement in our language. At this stage in our development, Theorem 1.5.2 is the only theorem for which we can do this. In the next section, we will introduce *implications* as language statements. This will permit us to translate all theorems thus far into formal language statements.

Consider, for example, a language statement equivalent to Theorem 1.5.2b:

for all sets A, B: $A \cup B = B \cup A$

In order to apply Theorem 1.5.2b, then, we apply the rule for using a **for all** statement. This could be applied to sample proof steps as follows:

Example 3:
1. $(A \cup B) \cup C \subseteq A \cup (B \cup C)$
2. **for all sets X, Y: $X \cup Y = Y \cup X$**   (Thm. 1.5.2b)
3. $(A \cup B) \cup C = C \cup (A \cup B)$ (2; us. ∀)
4. $C \cup (A \cup B) \subseteq A \cup (B \cup C)$ (1, 3; sub.)

The effect of the steps above is to replace the left-hand side of the $\subseteq$ statement in Step 1 with the left-hand side in Step 4. Thus

$$(A \cup B) \cup C$$
$$C \cup (A \cup B)$$

Applying Theorem 1.5.2b has the effect of commuting the set C with the set $A \cup B$. In our application of theorems to proof steps, we will contract the steps above to obtain:

1.6 Using Theorems

    1. $(A \cup B) \cup C \subseteq A \cup (B \cup C)$
    2'. $C \cup (A \cup B) \subseteq A \cup (B \cup C)$ (1; Thm. 1.5.2b)

Thus the truth of the theorem is *applied* to Step 1 to obtain Step 2'. This is summarized in our formal rule for using theorems:

Inference Rule   *Using theorems: In order to apply a theorem to steps in a proof, find a language statement $\mathscr{P}$ equivalent to the statement of the theorem. Then $\mathscr{P}$ may be written as a new proof step or used, by substitution, to change a proof step.*

Example 3 may be continued with Steps 5 and 6 to illustrate the process on the right-hand side of the $\subseteq$ statement:

    1. $(A \cup B) \cup C \subseteq A \cup (B \cup C)$
    2. **for all sets** X, Y: $X \cup Y = Y \cup X$ (Thm. 1.5.2b)
    3. $(A \cup B) \cup C = C \cup (A \cup B)$ (2; us. $\forall$)
    4. $C \cup (A \cup B) \subseteq A \cup (B \cup C)$ (1, 3; sub.)
    5. $A \cup (B \cup C) = (B \cup C) \cup A$ (2; us. $\forall$)
    6. $C \cup (A \cup B) \subseteq (B \cup C) \cup A$ (4, 5; sub.)

We can use our rule to contract Steps 5 and 6:

    1. $(A \cup B) \cup C \subseteq A \cup (B \cup C)$
    2'. $C \cup (A \cup B) \subseteq A \cup (B \cup C)$ (1; Thm. 1.5.2b)
    3'. $C \cup (A \cup B) \subseteq (B \cup C) \cup A$ (2; Thm. 1.5.2b)

In Steps 1 through 3', we operated first on the left and then on the right of the $\subseteq$. We will allow simultaneous application of the rule to both sides as follows:

    1. $(A \cup B) \cup C \subseteq A \cup (B \cup C)$
    2''. $C \cup (A \cup B) \subseteq (B \cup C) \cup A$ (1; Thm. 1.5.2b)

Example 4:

    1. $(A \cup B) \cup C \subseteq A \cup (B \cup C)$
    2. $C \cup (A \cup B) \subseteq (B \cup C) \cup A$ (1; Thm. 1.5.2b)
    3. $C \cup (B \cup A) \subseteq (C \cup B) \cup A$ (2; Thm. 1.5.2b)

Theorem 1.5.2b is used in the proof of the following theorem:

Theorem 1.6.1   For sets A, B, and C,

    (a) $A \cap (B \cap C) = (A \cap B) \cap C$
    (b) $A \cup (B \cup C) = (A \cup B) \cup C$

Chapter 1 Sets and Rules of Inference

The proof of part (a) is Exercise 2. We now prove part (b), but you should first attempt to prove (b) yourself, looking at our proof only as necessary to confirm your steps.

Proof:

Assume: A, B, C sets
Show: $(A \cup B) \cup C = A \cup (B \cup C)$

Scrap Paper

Show: $A \cup (B \cup C) = (A \cup B) \cup C$ ⎞
That is, show: $(A \cup B) \cup C \subseteq A \cup (B \cup C)$ and ⎠ def. $=$
$A \cup (B \cup C) \subseteq (A \cup B) \cup C$

First, the definitions of set containment and equality lead us to:

1. Let $x \in (A \cup B) \cup C$ be arbitrary.
   .
   .
k. $x \in A \cup (B \cup C)$
k+1. $(A \cup B) \cup C \subseteq A \cup (B \cup C)$ (1—k; def. $\subseteq$)
k+2. $C \cup (A \cup B) \subseteq (B \cup C) \cup A$ (k+1; Thm. 1.5.2b)
k+3. $C \cup (B \cup A) \subseteq (C \cup B) \cup A$ (k+2; Thm. 1.5.2b)
k+4. $A \cup (B \cup C) \subseteq (A \cup B) \cup C$ (k+3; symmetry)
k+5. $(A \cup B) \cup C = A \cup (B \cup C)$ (k+1, k+4; def. $=$)

Note that in the hypothesis of Theorem 1.6.1, sets A, B, and C are completely arbitrary, and so we may use the symmetry rule. In going from Step $k+3$ to Step $k+4$, every occurrence of C is replaced by A, every occurrence of B remains B, and every occurrence of A is replaced by C. Otherwise, all symbols remain the same. This is how symmetry is used. Pictorially:

$$C \cup (B \cup A) \subseteq (C \cup B) \cup A$$
$$\downarrow \quad \downarrow \quad \downarrow \quad \quad \downarrow \quad \downarrow \quad \downarrow$$
$$A \cup (B \cup C) \subseteq (A \cup B) \cup C$$

In practice, Steps $k+2$ and $k+3$ are combined; that is, a theorem can be used more than once in making a single step. Pictorially:

1.6 Using Theorems

$k+1.$ $(A \cup B) \cup C \subseteq A \cup (B \cup C)$
$k+2.$ $C \cup (A \cup B) \subseteq (B \cup C) \cup A$
$k+3.$ $C \cup (B \cup A) \subseteq (C \cup B) \cup A$

(with crossing substitution arrows between the statements)

becomes

$k+1.$ $(A \cup B) \cup C \subseteq A \cup (B \cup C)$
$k+2'.$ $C \cup (B \cup A) \subseteq (C \cup B) \cup A$

Next we use the definition of union to make the connection between Steps 1 and k:

	1.	Let $x \in (A \cup B) \cup C$ be arbitrary.	
	2.	$x \in (A \cup B)$ or $x \in C$	(1; def. $\cup$)
Case 1	3.	$x \in A \cup B$	
	4.	$x \in A$ or $x \in B$	(3; def. $\cup$)
Case 1a	5.	$x \in A$	
	6.	$x \in A$ or $x \in B \cup C$	(5; pr. **or**)
Case 1b	7.	$x \in B$	
	8.	$x \in B$ or $x \in C$	(7; pr. **or**)
	9.	$x \in B \cup C$	(8; def. $\cup$)
	10.	$x \in A$ or $x \in B \cup C$	(9; pr. **or**)
	11.	$x \in A$ or $x \in B \cup C$	(4—10; us. **or**)
	12.	$x \in A \cup (B \cup C)$	(11; def. $\cup$)
Case 2	13.	$x \in C$	
	14.	$x \in B$ or $x \in C$	(13; pr. **or**)
	15.	$x \in B \cup C$	(14; def. $\cup$)
	16.	$x \in A$ or $x \in B \cup C$	(15; pr. **or**)
	17.	$x \in A \cup (B \cup C)$	(16; def. $\cup$)
	18.	$x \in A \cup (B \cup C)$	(2—17; us. **or**)
	19.	$(A \cup B) \cup C \subseteq A \cup (B \cup C)$	(1—18; def. $\subseteq$)
	20.	$C \cup (B \cup A) \subseteq (C \cup B) \cup A$	(19; Thm. 1.5.2b)
	21.	$A \cup (B \cup C) \subseteq (A \cup B) \cup C$	(20; symmetry)
	22.	$(A \cup B) \cup C = A \cup (B \cup C)$	(19, 21; def. =)

If we used the extended definition rule in all possible places, our proof would be:

	1.	Let $x \in (A \cup B) \cup C$ be arbitrary.	
Case 1	2.	$x \in A \cup B$	
Case 1a	3.	$x \in A$	
	4.	$x \in A \cup (B \cup C)$	(3; def. $\cup$)
Case 1b	5.	$x \in B$	
	6.	$x \in B \cup C$	(5; def. $\cup$)
	7.	$x \in A \cup (B \cup C)$	(6; def. $\cup$)
	8.	$x \in A \cup (B \cup C)$	(2—7; by cases)
Case 2	9.	$x \in C$	
	10.	$x \in B \cup C$	(9; def. $\cup$)
	11.	$x \in A \cup (B \cup C)$	(10; def. $\cup$)
	12.	$x \in A \cup (B \cup C)$	(1—11; by cases)
	13.	$(A \cup B) \cup C \subseteq A \cup (B \cup C)$	(1—12; def. $\subseteq$)
	14.	$C \cup (B \cup A) \subseteq (C \cup B) \cup A$	(13; Thm. 1.5.2b)
	15.	$A \cup (B \cup C) \subseteq (A \cup B) \cup C$	(14; symmetry)
	16.	$A \cup (B \cup C) = (A \cup B) \cup C$	(13, 15; def. =)

The justification for Steps 8 and 12 involves using the **or** statement implicit in the definition of union in Steps 2 and 1, respectively. We will use the words "by cases" to replace "us. **or**" when there is no explicit **or** statement written as a step. Notice here that we have dropped the word "assume" for the assumption in Steps 2, 3, 5, and 9. It will be implicit that the first statement in a case will be an assumption.

Our primary purpose in proving Theorem 1.6.1 has been to illustrate the use of theorems and symmetry in proofs. In Section 1.9 we will see a shorter way in which this proof can be written.

Theorem 1.6.2 For sets A, B, and C,

(a) $A \cap (B \cup C) = (A \cap B) \cup (A \cap C)$

(b) $A \cup (B \cap C) = (A \cup B) \cap (A \cup C)$

Proof: Exercises 3 and 4.

Exercises 1.6

Practice Exercise

1.
   1. $X \subseteq Y$
   2. $Y \subseteq Z$
   3. $X \subseteq Z$ (1, 2; *)

Straightforward Problems

2. Prove Theorem 1.6.1a.
3. Prove Theorem 1.6.2a, using:

   Assume: A, B, C sets
   Show: $A \cap (B \cup C) = (A \cap B) \cup (A \cap C)$

4. Prove Theorem 1.6.2b.

Supplementary Problem

5. Prove Theorem 1.6.2a again, but this time represent the metalanguage of the theorem with no hypotheses and the conclusion:

   Show: For all sets A, B, C: $A \cap (B \cup C) = (A \cap B) \cup (A \cap C)$

   Note that the conclusion is a language statement equivalent to the metalanguage statement of the theorem.

1.7 IMPLICATIONS

If $\mathscr{P}$ and $\mathscr{Q}$ are statements in our language, the statement **if $\mathscr{P}$, then $\mathscr{Q}$** formed from them is called an *implication*. The statement means that $\mathscr{Q}$ is true whenever $\mathscr{P}$ is true. **If $\mathscr{P}$, then $\mathscr{Q}$** is therefore defined to be true in the following cases:

$\mathscr{P}$ true and $\mathscr{Q}$ true

$\mathscr{P}$ false and $\mathscr{Q}$ false

$\mathscr{P}$ false and $\mathscr{Q}$ true

The only case in which **if $\mathscr{P}$, then $\mathscr{Q}$** is false is where

$\mathscr{P}$ is true and $\mathscr{Q}$ is false

Notice in the first three cases above that $\mathscr{Q}$ is true whenever $\mathscr{P}$ is true. This leads to our rule for proving implications. Since the only way for **if $\mathscr{P}$, then**

Chapter 1 Sets and Rules of Inference

Q to be false is for $\mathscr{P}$ to be true and Q false, we can prove **if $\mathscr{P}$, then Q** is true by ruling out this possibility, that is, by assuming $\mathscr{P}$ to be true and showing that Q is true in this case.

Inference Rule

*Proving **if-then** statements: In order to prove a statement of the form if $\mathscr{P}$, then Q, assume that $\mathscr{P}$ is true and show that Q is true. Abbreviation: "pr. $\Rightarrow$".*

A proof scheme establishing **if $\mathscr{P}$, then Q** as Step k would be:

$i-1$. $\cdots$
    i. Assume $\mathscr{P}$.
        $\cdot$
        $\cdot$
    j. Q
    k. **if $\mathscr{P}$, then Q**   (i—j; pr. $\Rightarrow$)

Example 1:

1. *
2. $\cdots$
3. $\cdots$
4. *
5. **if $A \subseteq B$, then $A \subseteq C$**   (1—4; pr. $\Rightarrow$)

Solution:

1. Assume $A \subseteq B$.
2. $\cdots$
3. $\cdots$
4. $A \subseteq C$
5. **if $A \subseteq B$, then $A \subseteq C$**   (1—4; pr. $\Rightarrow$)

Theorem 1.7.1

Let A, B, and C be sets. If $A \subseteq B$, then $A \cap C \subseteq B \cap C$.

In specifying hypotheses and conclusion from the metalanguage of a theorem, the language statements used in the theorem must remain as whole units (so as to admit "interpretation" only by the rules of inference). Thus the entire **if-then** statement in Theorem 1.7.1 must be in the conclusion.

1.7 Implications

Proof:

Assume: A, B, C sets
Show: If $A \subseteq B$, then $A \cap C \subseteq B \cap C$

The rule for proving **if–then** gives us the following steps:

1. Assume $A \subseteq B$.
   .
   .
k. $A \cap C \subseteq B \cap C$
k+1. if $A \subseteq B$, then $A \cap C \subseteq B \cap C$ (1—k; pr. ⇒)

The statement of Step 1, $A \subseteq B$, would lead us by the definition of $\subseteq$ to write as Step 2: **for all $x \in A$: $x \in B$**. It is best not to write this down, however. (Treat $A \subseteq B$ as a hypothesis we ignore for a while.) Instead, we will apply the idea of $\subseteq$ to the symbols introduced by analyzing Step k, the thing we are after:

Scrap Paper

We seek   k. $A \cap C \subseteq B \cap C$ ⎞
That is,   **for all $x \in A \cap C$: $x \in B \cap C$** ⎠ by definition of $\subseteq$

We therefore are led to the following steps:

1. Assume $A \subseteq B$.
2. Let $x \in A \cap C$ be arbitrary.
   .
   .
k−1. $x \in B \cap C$
k. $A \cap C \subseteq B \cap C$ (2—k−1; def. $\subseteq$)
k+1. if $A \subseteq B$, then $A \cap C \subseteq B \cap C$ (1—k; pr. ⇒)

A complete proof is therefore given by:

> *Show:* if $A \subseteq B$, then $A \cap C \subseteq B \cap C$
> 1. Assume $A \subseteq B$.
> 2. Let $x \in A \cap C$ be arbitrary.
> 3. $x \in A$ (2; def. $\cap$)
> 4. $x \in C$ (2; def. $\cap$)
> 5. $x \in B$ (1, 3; def. $\subseteq$)
> 6. $x \in B \cap C$ (4, 5; def. $\cap$)
> 7. $A \cap C \subseteq B \cap C$ (2—6; def. $\subseteq$)
> 8. if $A \subseteq B$, then $A \cap C \subseteq B \cap C$ (1—7; pr. $\Rightarrow$)

Suppose Theorem 1.7.1 had been written in the following way:

Theorem   Let A, B, and C be sets. If $A \subseteq B$, then $A \cap C \subseteq B \cap C$.

It is possible to interpret this metalanguage statement so as to get the conclusion:

     Show: if $A \subseteq B$, then $A \cap C \subseteq B \cap C$

exactly as in our proof. It is clear that the meaning of the metalanguage statement allows for this. Another way of interpreting the statement is to get:

     Assume: A, B, C sets
                 $A \subseteq B$
     Show: $A \cap C \subseteq B \cap C$

The rest of the proof would then look like Steps 2 through 7 of the proof of Theorem 1.7.1. Either of these approaches would then be satisfactory since we allow any setup of hypotheses and conclusion that faithfully reflects the meaning of the metalanguage. In fact, either form of the theorem statement could be translated into:

     for all sets A, B, C: if $A \subseteq B$, then $A \cap B \subseteq B \cap C$

Now that we have rules for proving and using **if-then, and, or,** and **for all** statements, it would be possible to write all our theorems so far as single statements in our language and then prove them using the formal rules. This option (no hypothesis, all conclusion) is always open to *you* when deciding on hypotheses and conclusion. However, the process of breaking theorems into hypotheses and conclusion is closer to the way most mathematicians think about proofs than is the procedure of viewing all theorems as formal language statements. The hypotheses-conclusion model translates well into a paragraph style:

1.7 Implications

first you tell your readers what you are assuming, then you tell them what you will show:

Inference Rule Using *if-then* statements: If $\mathscr{P}$ is a step in a proof and if *if* $\mathscr{P}$, *then* $\mathcal{Q}$ is a step, then we may infer that $\mathcal{Q}$ is a step. Abbreviation: "us. $\Rightarrow$".

Format:

    i. $\mathscr{P}$

    .

    .

    j. if $\mathscr{P}$, then $\mathcal{Q}$

  j+1. $\mathcal{Q}$ (i, j; us. $\Rightarrow$)

Example 2:

  1. if $x < 2$, then $x \in A$
  2. $x < 2$
  3. *   (1, 2; us. $\Rightarrow$)

Solution:

  1. if $x < 2$, then $x \in A$
  2. $x < 2$
  3. $x \in A$ (1, 2; us. $\Rightarrow$)

Example 3:

  1. $a \in S$
  2. *
  3. $a \in T$ (1, 2; us. $\Rightarrow$)

Solution:

  1. $a \in S$
  2. if $a \in S$, then $a \in T$
  3. $a \in T$ (1, 2; us. $\Rightarrow$)

Example 4:

  1. $a \in S$
  2. *
  3. $a \in T$ (1, 2; def.[2] $\subseteq$)

Chapter 1 Sets and Rules of Inference

Solution:

1. $a \in S$
2. $S \subseteq T$
3. $a \in T$ (1, 2; def.^{2} $\subseteq$)

Our next rule of inference states that we can insert a step

$$\mathscr{P} \text{ or } \sim\mathscr{P}$$

in any proof.

Inference Rule

$\mathscr{P}$ *or* $\sim\mathscr{P}$ *rule: For any* $\mathscr{P}$, $\mathscr{P}$ *or* $\sim\mathscr{P}$ *is true.*

This rule might be used as follows:

1. $x \in A$ or $x \notin A$ ($\mathscr{P}$ or $\sim\mathscr{P}$)

Case 1   2. $x \in A$
           3. $\cdots$
           4.   desired result follows

Case 2   5. $x \notin A$
           6. $\cdots$
           7.   desired result follows
           8.   desired result follows in any case   (1—7; us. **or**)

Example 5:
For a real number x, the absolute value of x, written $|x|$, is defined by

$$|x| = \begin{cases} x \text{ if } x \geq 0 \\ -x \text{ if } x < 0 \end{cases}$$

Theorem: *For all* $x \in \mathbb{R}$: $|x|^2 = x^2$.

Proof:

1. Let x be arbitrary.
2. $x \geq 0$ or $\sim(x \geq 0)$ ($\mathscr{P}$ or $\sim\mathscr{P}$)

Case 1   3. $x \geq 0$
           4. $|x| = x$ (3; def. $|\ |$)
           5. $|x|^2 = x^2$ (4; prop. $\mathbb{R}$)

Case 2   6. $\sim(x \geq 0)$
           7. $x < 0$ (6; prop. $\mathbb{R}$)

1.7 Implications

8.	$\|x\| = -x$	(7; def. $\| \ \|$)
9.	$\|x\|^2 = (-x)^2$	(8; prop. **R**)
10.	$(-x)^2 = x^2$	(prop. **R**)
11.	$\|x\|^2 = x^2$	(9, 10; sub.)
12.	$\|x\|^2 = x^2$	(1–11; us. **or**)
13.	for all $x \in \mathbb{R}$: $\|x\|^2 = x^2$	(1–12; pr. ∀)

Our next theorem applies not to sets or numbers, the mathematical objects we have considered so far, but to statements in our language and proofs themselves. Such theorems can be considered as *logical theorems* and are differentiated from theorems about our mathematical objects by the suffix "L" after their numbers.

Theorem 1.7.2L
(a) If *if ~𝒫, then 𝒬* has been established as a step in a proof, then *𝒫 or 𝒬* can be established.
(b) If *𝒫 or 𝒬* has been established as a step in a proof, then *if ~𝒫, then 𝒬* can be established.

Proof of (a):

Assume: *if ~𝒫, then 𝒬*
Show: *𝒫 or 𝒬*

	1.	𝒫 or ~𝒫	(𝒫 or ~𝒫)
Case 1	2.	Assume 𝒫.	
	3.	𝒫 or 𝒬	(2; pr. **or**)
Case 2	4.	Assume ~𝒫.	
	5.	𝒬	(hyp., 4; us. ⇒)
	6.	𝒫 or 𝒬	(5; pr. **or**)
	7.	𝒫 or 𝒬	(1—6; us. **or**)

Proof of (b):

Assume: *𝒫 or 𝒬*
Show: *if ~𝒫, then 𝒬*

1.	Assume ~𝒫.	
2.	𝒫 or 𝒬	(hyp.)

> Case 1 3. Assume $\mathscr{P}$. # Step 1
> Case 2 4. Assume $\mathcal{Q}$.
> 5. $\mathcal{Q}$ (2—4; us. **or**)
> 6. if $\sim\mathscr{P}$, then $\mathcal{Q}$ (1—5; pr. $\Rightarrow$)

Note that in Steps 3 and 4, $\mathcal{Q}$ holds in all cases not leading to a contradiction. Therefore we can pop up one level and assert that $\mathcal{Q}$ holds by the rule for using **or**. Theorem 1.7.2La provides a good way of proving statements of the form $\mathscr{P}$ **or** $\mathcal{Q}$. Namely, assume $\sim\mathscr{P}$ and show that $\mathcal{Q}$ must follow. We then have that **if** $\sim\mathscr{P}$**, then** $\mathcal{Q}$ holds so that, by Theorem 1.7.2L, $\mathscr{P}$ **or** $\mathcal{Q}$ holds.

Definition *Statements $\mathscr{P}$ and $\mathcal{Q}$ in our language are called **equivalent** if it is possible to show that $\mathscr{P}$ follows from $\mathcal{Q}$ and that $\mathcal{Q}$ follows from $\mathscr{P}$ using our rules of inference and proof conventions.*

Theorem 1.7.2L can therefore be reworded:

Theorem 1.7.2L *The statements $\mathscr{P}$ **or** $\mathcal{Q}$ and **if** $\sim\mathscr{P}$**, then** $\mathcal{Q}$ are equivalent.*

In order to formalize the procedure we used to prove Theorem 1.7.2L and legitimize its future use, we have the following rule:

Inference Rule *Proving equivalence: In order to show that $\mathscr{P}$ is equivalent to $\mathcal{Q}$, first assume $\mathscr{P}$ and show $\mathcal{Q}$, and then assume $\mathcal{Q}$ and show $\mathscr{P}$. Abbreviation: "pr. $\Leftrightarrow$".*

Format:

1. Assume $\mathscr{P}$.
   .
   .
   i. $\mathcal{Q}$
   i+1. Assume $\mathcal{Q}$.
   .
   .
   j. $\mathscr{P}$
   j+1. $\mathscr{P}$ is equivalent to $\mathcal{Q}$. (1—j; pr. $\Leftrightarrow$)

1.7 Implications

Theorem 1.7.3L — The statement $\mathscr{P}$ or $\mathscr{Q}$ is equivalent to the statement $\mathscr{Q}$ or $\mathscr{P}$.

Proof:
1. Assume $\mathscr{P}$ or $\mathscr{Q}$.
- Case 1
  2. Assume $\mathscr{P}$.
  3. $\mathscr{Q}$ or $\mathscr{P}$ (2; pr. or)
- Case 2
  4. Assume $\mathscr{Q}$.
  5. $\mathscr{Q}$ or $\mathscr{P}$ (4; pr. or)
6. $\mathscr{Q}$ or $\mathscr{P}$ (1—5; us. or)
7. Assume $\mathscr{Q}$ or $\mathscr{P}$.
8. $\mathscr{P}$ or $\mathscr{Q}$ (1—7; symmetry)
9. $\mathscr{P}$ or $\mathscr{Q}$ is equivalent to $\mathscr{Q}$ or $\mathscr{P}$.   (1—9; pr. ⇔)

Theorem 1.7.4L — The statement $\mathscr{P}$ and $\mathscr{Q}$ is equivalent to the statement $\mathscr{Q}$ and $\mathscr{P}$.

Proof: Exercise 6.

Examples of the use of Theorems 1.7.3L and 1.7.4L are given in future sections.

Inference Rule — *Using equivalence: Any statement may be substituted for an equivalent statement. Abbreviation: "us. ⇔".*

The rule for using equivalence is used implicitly in the following examples.

Example 6:
1. $x \in A$ or $x \in B$
2. *   (1; Thm. 1.7.2L)

Solution:
1. $x \in A$ or $x \in B$
2. if $x \notin A$, then $x \in B$ (1; Thm. 1.7.2L)

Example 7:
1. if $x \in A$, then $x \in B$
2. *   (1; Thm. 1.7.2L)

Solution:
1. if $x \in A$, then $x \in B$
2. $x \notin A$ or $x \in B$ (1; Thm. 1.7.2L)

(What is $\mathcal{P}$ of Theorem 1.7.2L here?)

Example 8:
1. for all x such that $|x| \geq 1$: $x \leq -1$ or $x \geq 1$
2. *   (1; Thm. 1.7.2L)

Solution:
1. for all x such that $|x| \geq 1$: $x \leq -1$ or $x \geq 1$
2. for all x such that $|x| \geq 1$: if $\sim(x \leq -1)$, then $x \geq 1$ (1; Thm. 1.7.2L)

Additional Ideas

Define $\mathcal{P} \Rightarrow \mathcal{Q}$ to mean "if $\mathcal{P}$ is a proof step, then $\mathcal{Q}$ can be written as a proof step". Define $\mathcal{P} \leftrightarrow \mathcal{Q}$ provided that both $\mathcal{P} \Rightarrow \mathcal{Q}$ and $\mathcal{Q} \Rightarrow \mathcal{P}$. Then $\leftrightarrow$ denotes what we have called equivalence.

Theorem 1.7.5L *The statement if $\mathcal{P}$, then $\mathcal{Q}$ is true if and only if $\mathcal{P} \Rightarrow \mathcal{Q}$.*

> **Proof:**
>
> Note that **if $\mathcal{P}$, then $\mathcal{Q}$** is a language statement, whereas $\mathcal{P} \Rightarrow \mathcal{Q}$ is a metalanguage statement about proof steps. Assuming $\mathcal{P} \Rightarrow \mathcal{Q}$ allows us to prove **if $\mathcal{P}$, then $\mathcal{Q}$** by the rule for proving **if-then**. Conversely, assuming **if $\mathcal{P}$, then $\mathcal{Q}$** allows us (using **if-then**) to show $\mathcal{P} \Rightarrow \mathcal{Q}$.

Theorem 1.7.5L shows why we used the symbol $\Rightarrow$ in our abbreviations for proving and using **if-then** statements. Metalanguage statements of the form $\mathcal{P} \Rightarrow \mathcal{Q}$, as well as **if-then** language statements, are called "implications". Observe that $\Rightarrow$ is a transitive relation on statements. By this we mean that from $\mathcal{P} \Rightarrow \mathcal{Q}$ and $\mathcal{Q} \Rightarrow \mathcal{R}$, we can infer $\mathcal{P} \Rightarrow \mathcal{R}$. This leads to chains of $\Rightarrow$ statements. For example, a proof of $\mathcal{P} \Rightarrow \mathcal{Q}$ could be given by a chain $\mathcal{P}_0 \Rightarrow \mathcal{P}_1 \Rightarrow \mathcal{P}_2 \Rightarrow \cdots \Rightarrow \mathcal{P}_n$, where $\mathcal{P}_0$ is $\mathcal{P}$ and $\mathcal{P}_n$ is $\mathcal{Q}$. This makes for a compact way of writing proofs that several statements are equivalent. For example, the next theorem could be proved by showing **(a)** $\Rightarrow$ **(b)** $\Rightarrow$ **(c)** $\Rightarrow$ **(a)**—or with any other chain of implications that cycles through all statements **(a)**, **(b)**, and **(c)**.

1.7 Implications

Theorem 1.7.6L *The following are equivalent:*

(a) ($\mathcal{P}$ or $\mathcal{Q}$) or $\mathcal{R}$
(b) $\mathcal{P}$ or ($\mathcal{Q}$ or $\mathcal{R}$)
(c) $\mathcal{P}$ or $\mathcal{Q}$ or $\mathcal{R}$

Proof: Exercise 8.

Theorem 1.7.7L *The following are equivalent:*

(a) ($\mathcal{P}$ and $\mathcal{Q}$) and $\mathcal{R}$
(b) $\mathcal{P}$ and ($\mathcal{Q}$ and $\mathcal{R}$)
(c) $\mathcal{P}$ and $\mathcal{Q}$ and $\mathcal{R}$

Proof: Exercise 9.

Exercises 1.7

Practice Exercises

1. (a) 1. *
        2. ...
        3. *
        4. if $q \in R$, then $q \in S$ (1—3; pr. $\Rightarrow$)

   (b) 1. *
        2. if $q \in R$, then $q \in S$
        3. $q \in S$ (1, 2; us. $\Rightarrow$)

   (c) 1. $A \subseteq B$
        2. *
        3. $B \subseteq C$ (1, 2; us. $\Rightarrow$)

   (d) 1. *
        2. *
        3. if $x \in B$, then $A \subseteq B$ (1, 2; pr. $\Rightarrow$)

   (e) 1. *
        2. *   (1; pr. *)
        3. if $x \in B$, then $A \subseteq B$ (2; Thm. 1.7.2L)

   (f) 1. *
        2. $x < 5$ or $A \subseteq B$ (1; Thm. 1.7.2L)

(g) 1. $\mathscr{P}$ and $\mathcal{Q}$
     2. if $\mathscr{P}$, then $\mathscr{R}$
     3. *   (*)
     4. $\mathscr{R}$ (*)

2. Suppose **if $x \in A$, then $x < 10$** is a step you need to show in some proof. Give all the "inevitable" proof steps that lead up to this statement.

Straightforward Problems

3. Let A, B, and C be sets and $B \subseteq C$. Prove **if $A \subseteq B$, then $A \subseteq C$**.
4. Prove that for sets A and B, if $A \cap B = A$ then $A \subseteq B$.
5. Use the $\mathscr{P}$ or $\sim\mathscr{P}$ rule to prove **for all $x \in \mathbf{R}$: $|x| \geq x$**.
6. Prove Theorem 1.7.4L.

Supplementary Problems

7. Theorem 1.3.1 has the following language interpretation:

   for all sets A, B, C: if $A \subseteq B$ and $B \subseteq C$, then $A \subseteq C$

   Prove this statement without using def.² Compare your proof with the proof of Theorem 1.3.1 (which also does without def.²).

8. Prove Theorem 1.7.6L. Use a proof format such as the following:

   Proof:

   We first show $(a) \Rightarrow (b)$:

   Assume: ($\mathscr{P}$ or $\mathcal{Q}$) or $\mathscr{R}$
   Show: $\mathscr{P}$ or ($\mathcal{Q}$ or $\mathscr{R}$)

   •

   •

   Hence $(a) \Rightarrow (b)$.

   We now show $(b) \Rightarrow (c)$:

   •

   •

   Hence $(b) \Rightarrow (c)$.

   We now show $(c) \Rightarrow (a)$:

   •

   •

1.8   Proof by Contradiction   65

Therefore (c) ⇒ (a).

By the implications above, (a), (b), and (c) are equivalent.

The rules for using and proving **or** statements in their general form (pages 36 and 37) are needed to relate part (c) to the other parts.

9. Prove Theorem 1.7.7L.

1.8   PROOF BY CONTRADICTION

In order to prove a statement $\mathscr{P}$, it is frequently useful to assume the negation of $\mathscr{P}$ and then argue to a step that contradicts some known fact. This fact may be a theorem or, more often, a previous step or current hypothesis in the proof. A proof scheme for this would be:

    i. $\mathcal{Q}$

    j. Assume $\sim\mathscr{P}$ (to get a contradiction).

       ·
       ·

    k. $\sim\mathcal{Q}$ contradicting Step 1.

$k+1$. $\mathscr{P}$ (j—k; contradiction)

The symbol # is used to denote contradiction in both steps and justifications. We will do proofs by contradiction using the format above since this is close to the paragraph style most writers use, but we are not introducing any new rules of inference here since the logic comes from inference rules we already have. The argument above could have been stated:

       i. $\mathcal{Q}$
       j. $\mathscr{P}$ **or** $\sim\mathscr{P}$ ($\mathscr{P}$ **or** $\sim\mathscr{P}$)
Case 1 $j+1$. $\sim\mathscr{P}$
               ·
               ·

       k. $\sim\mathcal{Q}$ # Step i.
Case 2 $k+1$. $\mathscr{P}$
$k+2$. $\mathscr{P}$ (us. **or**)

Thus $\mathscr{P}$ is proved in all cases not leading to a contradiction, so that, by our rule, $\mathscr{P}$ can be inferred.

We emphasize that the validity of doing proofs by contradiction depends on the $\mathscr{P}$ **or** $\sim\mathscr{P}$ rule.

Proofs by contradiction can be employed to use the negations of **and, or,** and **if-then** statements. For example, suppose we knew $x \notin A \cap B$. This states that x is not a member of both A and B. We feel that logically this must mean that either x is not a member of A or x is not a member of B (or, of course, that x is a member of neither). Thus we feel that from $x \notin A \cap B$ we ought to be able to infer $x \notin A$ **or** $x \notin B$. So our problem is to:

> *Assume:* $x \notin A \cap B$
> *Show:* $x \notin A$ or $x \notin B$

From Theorem 1.7.2L, the condition $x \notin A$ **or** $x \notin B$ is equivalent to **if** $x \in A$, **then** $x \notin B$. So we have:

> 1. Assume $x \in A$.
>   .
>   .
>   .
> k. $x \notin B$
> k+1. if $x \in A$, then $x \notin B$ (1—k; pr. ⇒)
> k+2. $x \notin A$ or $x \notin B$ (k+1; Thm. 1.7.2L)

To get Step k, assume the contrary and attempt to contradict the hypothesis:

> 1. Assume $x \in A$.
> 2. To get #, assume $x \in B$.
> 3. $x \in A \cap B$ # hyp.   (1, 2; def. ∩)
> 4. $x \notin B$ (2, 3; #)
> 5. if $x \in A$, then $x \notin B$ (1—4; pr. ⇒)
> 6. $x \notin A$ or $x \notin B$ (5; Thm. 1.7.2L)

Here, as in general, Theorem 1.7.2L provides the most useful way of proving **or** statements.

A proof by contradiction can also be employed to use the negation of an element's being in a set. For example, suppose $A = \{x \in \mathbb{R} \mid x < 1\}$. From $x \in \mathbb{R}$ and $x \notin A$, we would like to infer from our rules that $x \geq 1$. This can be done as follows:

1.8 Proof by Contradiction

> 1. $x \notin A$ (hyp.)
> 2. Assume $x < 1$ to get #.
> 3. $x \in A$ # Step 1.   (2; def. A)
> 4. $\sim(x < 1)$ (2, 3; #)
> 5. $x \geq 1$ (4; prop. R)

To shortcut this trivial process, we will extend our rule for using definitions so that we can infer that, if an element is not in a set, then it satisfies the negation of the defining property.

Inference Rule   *Set definition rule (negation version): If $S = \{x \mid \mathscr{P}(x)\}$, then from $x \notin S$ we may infer $\sim \mathscr{P}(x)$. Conversely, from $\sim \mathscr{P}(x)$ we may infer $x \notin S$.*

Using this negation version, the proof steps above can be written:

> 1. $x \notin A$ (hyp.)
> 2. $x \geq 1$ (1; def. A)

The validity of the negation version depends on the $\mathscr{P}$ or $\sim \mathscr{P}$ rule.

Example 1:
Define $C = \{x \mid x > 5\}$. The definition of C tells us in both cases below why Step 2 follows from Step 1. Keep in mind that it is the negation version of the rule that is being used.

> 1. $a \notin C$
> 2. $a \leq 5$ (1; def. C)
>
> 1. $b \leq 5$
> 2. $b \notin C$ (1; def. C)

Definition   *For sets A and B, the **complement** of B in A (also called the **difference**) is the set $A - B$ (read "A minus B") defined by $A - B = \{x \mid x \in A \text{ and } x \notin B\}$.*

Theorem 1.8.1   *DeMorgan's Laws: For sets A, B, and C:*

(a) $A - (B \cup C) = (A - B) \cap (A - C)$

(b) $A - (B \cap C) = (A - B) \cup (A - C)$

The proof of part (b) is Exercise 3. We now prove part (a):

Proof:
  Assume: A, B, C sets
  Show: $A - (B \cup C) = (A - B) \cap (A - C)$

Scrap Paper
  Show: $A - (B \cup C) = (A - B) \cap (A - C)$
  That is, show: 1. $A - (B \cup C) \subseteq (A - B) \cap (A - C)$ $\Big)$ def. =
            and 2. $(A - B) \cap (A - C) \subseteq A - (B \cup C)$

When our analysis shows that the proof will consist of two or more parts (for example, if the conclusion is an **and** statement), we will henceforth introduce each part by a simple sentence. This will provide the broad outline for what will later become a paragraph proof. (Once again we are reminded of top-down design.) Thus, what we have on scrap paper gives us the following start of a proof:

Proof:
  Show: $A - (B \cup C) = (A - B) \cap (A - C)$
We first show $A - (B \cup C) \subseteq (A - B) \cap (A - C)$:
  1. Let $x \in A - (B \cup C)$ be arbitrary.
    ·
    ·
  k. $x \in (A - B) \cap (A - C)$
Therefore $A - (B \cup C) \subseteq (A - B) \cap (A - C)$ by definition of containment.
  Next we show $(A - B) \cap (A - C) \subseteq A - (B \cup C)$:
  1. Let $x \in (A - B) \cap (A - C)$ be arbitrary.
    ·
    ·
  k. $x \in A - (B \cup C)$

1.8 Proof by Contradiction

> Therefore $(A - B) \cap (A - C) \subseteq A - (B \cup C)$ by definition of containment. Hence, by definition of equality,
>
> $$A - (B \cup C) = (A - B) \cap (A - C)$$

For a while the extent of our use of paragraph format will be limited to showing the major pieces of the proof. It is already obvious, however, that the major change in going from a proof with numbered steps to paragraph format is doing no more than replacing the numbers with words like "thus", "therefore", "so", "hence", and "whence".

Now consider the first part of the proof outline above. Line 1 is:

> 1. Let $x \in A - (B \cup C)$ be arbitrary.

The information we have about x depends on the definition of set union and set complement. By now you should know which definition will be the reason for going to Step 2. The two grammatically big pieces in $A - (B \cup C)$ are A and $(B \cup C)$. Thus line 1 states that x is in the difference of two sets, A and $B \cup C$. At this top level, $B \cup C$ is considered a single set (the function of the parentheses). Therefore we use the definition of complement (abbreviated "def. $-$") to get:

> 2. $x \in A$ and $x \notin B \cup C$ (def. $-$)

Or, using the extended definition rule:

> 2. $x \in A$ (def. $-$)
> 3. $x \notin B \cup C$ (def. $-$)

One is tempted to say immediately from Step 3 that $x \notin B$ and $x \notin C$. The logic behind this is as follows. Step 3 is by our definition rule (negation version) $\sim(x \in B$ or $x \in C)$ and the logical negation $\sim(\mathcal{P}$ or $\mathcal{Q})$ is the same as $\sim\mathcal{P}$ **and** $\sim\mathcal{Q}$. In fact, we will show (Theorem 1.8.4L) that $\sim(\mathcal{P}$ or $\mathcal{Q})$ and $\sim\mathcal{P}$ **and** $\sim\mathcal{Q}$ are equivalent.

One way to proceed from Step 3 therefore would be first to derive, from our rules of inference, the logical equivalence of $\sim(\mathcal{P}$ or $\mathcal{Q})$ and $\sim\mathcal{P}$ **and** $\sim\mathcal{Q}$

and then use this equivalence in the proof steps. Instead, we will use this theorem to illustrate the use of contradiction in proofs. Consider Step 3:

> 3. $x \notin B \cup C$

and the two facts $x \notin B$ and $x \notin C$, which seem to be so logical. Why do they seem logical? Is it not because we are really doing a small proof by contradiction in our heads? We say to ourselves, "if $x \in B$, then $x \in B \cup C$, but $x \notin B \cup C$, therefore $x \notin B$". Our proof proceeds by merely writing this down:

> 1. Let $x \in A - (B \cup C)$ be arbitrary.
> 2. $x \in A$ (1; def. −)
> 3. $x \notin B \cup C$ (1; def. −)
>   4. To get #, assume $x \in B$.
>   5. $x \in B \cup C$ # Step 3.   (4; def. ∪)
> 6. $x \notin B$ (4, 5; #)
> 7. $x \notin C$ (3—6; symmetry, Thm. 1.5.2b)

We have now got all information out of Step 1. This information is given in Steps 2, 6, and 7. It is now our task to put this together to produce Step k:

> k. $x \in (A - B) \cap (A - C)$

This is fairly clear now. Steps 8 through 10 are:

> 8. $x \in A - B$ (2, 6; def. −)
> 9. $x \in A - C$ (2, 7; def. −)
> 10. $x \in (A - B) \cap (A - C)$ (8, 9; def. ∩)

There are two quite different things that could be called "reasons" for writing down steps. The first would be *legitimacy* reasons—why the steps are true based on previous steps, rules of inference, definitions, and proven theorems. Such legitimacy reasons are put in parentheses after the step. The second could be called *tactical* reasons—why *these* steps, and not others, were written down. We have given the tactical reasons for our proof steps in the text as each proof was developed. The complete proof of Theorem 1.8.1a is given below. Be sure

1.8 Proof by Contradiction

you know the tactical reasons for Steps 1 to 12 in the second part. Checking the legitimacy reasons will add almost nothing to your understanding. Reading from the bottom up will clarify the tactical reasons for the steps in a proof.

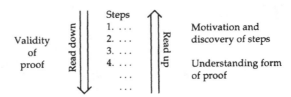

Proof:

    Show: $A - (B \cup C) = (A - B) \cap (A - C)$

We first show $A - (B \cup C) \subseteq (A - B) \cap (A - C)$:

1. Let $x \in A - (B \cup C)$ be arbitrary.
2. $x \in A$ (1; def. −)
3. $x \notin (B \cup C)$ (1; def. −)
   4. To get #, assume $x \in B$.
   5. $x \in B \cup C$ # Step 3.   (4; def. ∪)
6. $x \notin B$ (4, 5; #)
7. $x \notin C$ (3—6; symmetry, Thm. 1.5.2b)
8. $x \in A - B$ (2, 6; def. −)
9. $x \in A - C$ (2, 7; def. −)
10. $x \in (A - B) \cap (A - C)$ (8, 9; def. ∩)

Therefore $A - (B \cup C) \subseteq (A - B) \cap (A - C)$ by definition of set containment.

    Next we show $(A - B) \cap (A - C) \subseteq A - (B \cup C)$:

1. Let $x \in (A - B) \cap (A - C)$ be arbitrary.
2. $x \in A - B$ (1; def. ∩)
3. $x \in A - C$ (1; def. ∩)
4. $x \in A$ (2; def. −)
5. $x \notin B$ (2; def. −)
6. $x \notin C$ (3; def. −)

> 7. To get #, assume $x \in B \cup C$.
> 8. $x \in B$ or $x \in C$ (7; def. $\cup$)
> Case 1   9. $x \in B$ # Step 5.
> Case 2   10. $x \in C$ # Step 6.
> 11. $x \notin B \cup C$ (7—10; #)
> 12. $x \in A - (B \cup C)$ (4, 11; def. $-$)
>
> Therefore $(A - B) \cap (A - C) \subseteq A - (B \cup C)$ by definition of set containment. Hence by definition of set equality,
>
> $$A - (B \cup C) = (A - B) \cap (A - C)$$ ∎

Notice that when we break up a proof into two or more pieces, step numbers and element symbols (such as x above) are used independently (redefined) in the distinct pieces. Either contradiction or the rule for using **or** statements would be justification for Step 11 above. Because we get a contradiction in each case, the rule for using **or** statements allows us to infer the negation of the assumed $x \in B \cup C$. This negation is written on the indentation level of the step immediately preceding the most recent assumption (Step 6 here).

Recall that all our sets are formed with elements from a universal set $\cup$. For any set B, the complement of B in $\cup$, $\cup - B$, is the set of every element in the universe not in B. This is called the *complement of B* and is denoted by $\overline{B}$.

Definition   For any set B, $\overline{B} = \{x \in \cup \mid x \notin B\}$.

By taking A in the statement of DeMorgan's Laws to be the universal set and using the definition above, we have the following corollary:

Corollary 1.8.2   For sets B and C:

(a) $\overline{B \cup C} = \overline{B} \cap \overline{C}$

(b) $\overline{B \cap C} = \overline{B} \cup \overline{C}$

Theorem 1.8.3   For any sets A and B, $A - B = A \cap \overline{B}$.

Proof: Exercise 4.

The following three theorems are useful for doing proofs by contradiction. The proofs of 1.8.4L and 1.8.5L look like the proofs of DeMorgan's Laws.

1.8 Proof by Contradiction

Theorem 1.8.4L The statement $\sim(\mathscr{P} \text{ or } \mathcal{Q})$ is equivalent to $\sim\mathscr{P}$ and $\sim\mathcal{Q}$.

> **Proof:**
> We first assume $\sim(\mathscr{P} \text{ or } \mathcal{Q})$ and show $\sim\mathscr{P}$ and $\sim\mathcal{Q}$:
>
> 1. $\sim(\mathscr{P} \text{ or } \mathcal{Q})$ (hyp.)
> 2. To get #, assume $\mathscr{P}$.
> 3. $\mathscr{P}$ or $\mathcal{Q}$ # Step 1. (2; pr. or)
> 4. $\sim\mathscr{P}$ (1, 3; #)
> 5. $\sim(\mathcal{Q} \text{ or } \mathscr{P})$ (1; Thm. 1.7.3L)
> 6. $\sim\mathcal{Q}$ (1—4, 5; symmetry)
> 7. $\sim\mathscr{P}$ and $\sim\mathcal{Q}$ (5, 6; pr. &)
>
> Next we assume $\sim\mathscr{P}$ and $\sim\mathcal{Q}$ and show $\sim(\mathscr{P} \text{ or } \mathcal{Q})$:
>
> 8. $\sim\mathscr{P}$ and $\sim\mathcal{Q}$ (hyp.)
> 9. $\sim\mathscr{P}$ (8; us. &)
> 10. $\sim\mathcal{Q}$ (8; us. &)
> 11. To get #, assume $\mathscr{P}$ or $\mathcal{Q}$.
> Case 1 12. $\mathscr{P}$ # Step 2.
> Case 2 13. $\mathcal{Q}$ # Step 3.
> 14. $\sim(\mathscr{P} \text{ or } \mathcal{Q})$ (11—13; #)
> 15. $\sim(\mathscr{P} \text{ or } \mathcal{Q})$ is equivalent to $\sim\mathscr{P}$ and $\sim\mathcal{Q}$. (1—14; pr. $\leftrightarrow$) ∎

Note that in Steps 11, 12, and 13 all cases have led to a contradiction. We may therefore infer the negation of the most recently assumed statement (Step 11).

Theorem 1.8.5L The statement $\sim(\mathscr{P} \text{ and } \mathcal{Q})$ is equivalent to $\sim\mathscr{P}$ or $\sim\mathcal{Q}$.

Proof: Exercise 6.

Theorem 1.8.6L The statement $\sim(\text{if } \mathscr{P}, \text{ then } \mathcal{Q})$ is equivalent to $\mathscr{P}$ and $\sim\mathcal{Q}$.

Proof: Exercise 8.

The following theorem is the basis for yet another extension of the definition rule:

Chapter 1 Sets and Rules of Inference

Theorem 1.8.7L If $\mathcal{P}$ is equivalent to Q, then $\sim\mathcal{P}$ is equivalent to $\sim Q$.

> **Proof:**
>
> Assume $\mathcal{P}$ is equivalent to Q. In order to show $\sim\mathcal{P}$ is equivalent to $\sim Q$, we assume $\sim\mathcal{P}$ (in addition to the hypothesis), and then show $\sim Q$:
>
> 1. Assume $\sim\mathcal{P}$.
>   2. To get #, assume Q.
>   3. $\mathcal{P}$ # Step 1.   (2, hyp.; us. $\leftrightarrow$)
> 4. $\sim Q$ (2—3; #)
>
> By symmetry, if we assume $\sim Q$ we can show $\sim\mathcal{P}$, so that $\sim Q$ and $\sim\mathcal{P}$ are equivalent. ∎

According to our definition rule, if a relationship holds, then its defining condition may be inferred. Also, from the defining condition we may infer that the relationship holds. Thus the relationship and its defining condition are *equivalent*. As a corollary to Theorem 1.8.9L, we therefore have the following:

Inference Rule *Definition rule (negation version): Suppose some relationship has been defined. If the negation of the relationship holds, then the negation of the defining condition holds. If the negation of the defining condition holds, then the negation of the relationship may be inferred.*

Example 2:

1. $S \not\subseteq T$
2. $\sim$(for all $x \in S$: $x \in T$)   (1; def. $\subseteq$)

1. $x \notin A \cap B$
2. $\sim(x \in A$ and $x \in B)$ (1; def. $\cap$)

In Example 2, it is the negation version of the definition rule that is being used.

Additional Proof Ideas

Theorem 1.7.2L, which states that $\mathcal{P}$ **or** Q is equivalent to **if** $\sim\mathcal{P}$, **then** Q, is useful for proving **or** statements. This theorem can be combined with Theorem 1.8.4L to obtain a useful way of proving statements of the form $\mathcal{P}$ **or** Q **or** $\mathcal{R}$.

Theorem 1.8.8L *The statement $\mathscr{P}$ or $\mathscr{Q}$ or $\mathscr{R}$ is equivalent to if $\sim\mathscr{P}$ and $\sim\mathscr{Q}$, then $\mathscr{R}$.*

Proof:

First, $\mathscr{P}$ or $\mathscr{Q}$ or $\mathscr{R}$ follows from if $\sim\mathscr{P}$ and $\sim\mathscr{Q}$, then $\mathscr{R}$:

   Assume: if $\sim\mathscr{P}$ and $\sim\mathscr{Q}$, then $\mathscr{R}$
   Show: $\mathscr{P}$ or $\mathscr{Q}$ or $\mathscr{R}$

1. if $\sim\mathscr{P}$ and $\sim\mathscr{Q}$, then $\mathscr{R}$ (hyp.)
2. if $\sim(\mathscr{P}$ or $\mathscr{Q})$, then $\mathscr{R}$ (1; Thm. 1.8.4L)
3. $(\mathscr{P}$ or $\mathscr{Q})$ or $\mathscr{R}$ (2; Thm. 1.7.2L)
4. $\mathscr{P}$ or $\mathscr{Q}$ or $\mathscr{R}$ (3; Thm. 1.7.5L)

The second part of the equivalence follows from these steps in reverse order:

   Assume: $\mathscr{P}$ or $\mathscr{Q}$ or $\mathscr{R}$
   Show: if $\sim\mathscr{P}$ and $\sim\mathscr{Q}$, then $\mathscr{R}$

1. $\mathscr{P}$ or $\mathscr{Q}$ or $\mathscr{R}$ (hyp.)
2. $(\mathscr{P}$ or $\mathscr{Q})$ or $\mathscr{R}$ (1; Thm. 1.7.5L)
3. if $\sim(\mathscr{P}$ or $\mathscr{Q})$, then $\mathscr{R}$ (2; Thm. 1.7.2L)
4. if $\sim\mathscr{P}$ and $\sim\mathscr{Q}$, then $\mathscr{R}$ (3; Thm. 1.8.4L) ∎

Notice the implicit use of equivalence in the proof of Theorem 1.8.8L. DeMorgan's Laws and other set theoretical statements can be proved using similar proofs. In the following example, we need the obvious lemma "$\mathscr{P}$ is equivalent to $\mathscr{P}$ and $\mathscr{P}$". ("Obvious" means it is clear what proof steps are needed to show the assertion.)

Example 3:

Show $A - (B \cup C) \subseteq (A - B) \cap (A - C)$.

Proof:

1. Let $x \in A - (B \cup C)$.
2. $x \in A$ and $\sim(x \in B \cup C)$ (1; def. $-$)
3. $x \in A$ and $\sim(x \in B$ or $x \in C)$ (2; def. $\cup$)
4. $x \in A$ and $(x \notin B$ and $x \notin C)$ (3; Thm. 1.8.4L)

> 5. $(x \in A$ and $x \in A)$ and $\qquad$ (4; lemma)
> $(x \notin B$ and $x \notin C)$
> 6. $(x \in A$ and $x \notin B)$ and $\qquad$ (5; Thm. 1.7.6L repeated)
> $(x \in A$ and $x \notin C)$
> 7. $(x \in A - B)$ and $(x \in A - C)$ $\qquad$ (6; def. $-$)
> 8. $x \in (A - B) \cap (A - C)$ $\qquad$ (7; def. $\cap$)
> 9. $A - (B \cup C) \subseteq (A - B) \cap (A - C)$ $\qquad$ (1–8; def. $\subseteq$) ∎

This proof has the advantage that, since each step from 2 through 8 depends only on the preceding step, the order of the steps may be reversible (as it indeed is) to give the reverse set inclusion. More will be said of this in the next section. In the proof of Example 3, the logical complexity is displayed in the steps, whereas in the given proof of Theorem 1.8.1, the logical complexity is inherent in the form of the steps—dictated by the inference rules. The steps themselves are simple assertions about mathematical objects. The given proof of Theorem 1.8.1 tends to the more advanced paragraph style toward which we aim—where logic is implicit.

Exercises 1.8

Practice Exercise

1. (a) 1. $x \notin A$
       2. $x \in B$
       3. $x \in *$ $\qquad$ (1, 2; *)
   (b) 1. $C = \{x \mid x \leq 2\}$
       2. $y \notin C$
       3. * $\qquad$ (1, 2; *)
   (c) 1. $D = \{x \mid \mathcal{P}(x)\}$
       2. $y \notin D$
       3. * $\qquad$ (1, 2; *)
       4. $z \in D$
       5. * $\qquad$ (1, 4; *)
   (d) 1. $A \subseteq B$
       2. * $\qquad$ (1; *)

1.8 Proof by Contradiction

(e) 1. $A \not\subseteq B$
    2. *   (1; *)
(f) 1. $x \notin A \cap B$
    2. *   (1; *)

Straightforward Problems

2. Apply Theorem 1.7.2L to prove the following:
   Let $A = \{x \in \mathbb{R} | x < 2\}$ and $B = \{x \in \mathbb{R} | x > 0\}$.
   Show that $\mathbb{R} \subseteq A \cup B$.

3. Prove Theorem 1.8.1b.

4. Prove Theorem 1.8.3.

5. Prove that for sets A and B, $(A - B) \cup B = A \cup B$.

6. Prove Theorem 1.8.5L. Use Theorem 1.7.2L to prove **or** statements.

7. Establish one of the relations $=$, $\subseteq$, or $\supseteq$ between $A \cup (B - C)$ and $(A \cup B) - C$. Prove your assertion. ($x \supseteq y$ has the same meaning as $y \subseteq x$.)

8. Prove Theorem 1.8.6L. Use Theorem 1.7.2L and the rule for using equivalence.

Supplementary Problems

9. Consider the tactical reasons for Steps 1 to 12 in the second part of the proof of Theorem 1.8.1a. Which steps were inevitable? Which required choosing a productive line of thought?

10. Prove Theorem 1.8.8L along the following lines:

    Assume: if $\sim \mathscr{P}$ and $\sim \mathscr{Q}$, then $\mathscr{R}$
    Show: $\mathscr{P}$ or $\mathscr{Q}$ or $\mathscr{R}$

    | | 1. | $\mathscr{P}$ or $\sim \mathscr{P}$ |
    Case 1 | 2. | $\mathscr{P}$ |
    | | 3. | $\mathscr{P}$ or $\mathscr{Q}$ or $\mathscr{R}$ |
    Case 2 | 4. | $\sim \mathscr{P}$ |
    | | 5. | $\mathscr{Q}$ or $\sim \mathscr{Q}$ |
    Case 2a | 6. | $\mathscr{Q}$ |
    | | 7. | $\cdots$ |

Compare the complexity of your proof with the given proof of Theorem 1.8.8L. Note that Theorem 1.7.2L is almost always easier to use than the rule for proving **or**.

1.9 IFF

Definition *The statement $\mathcal{P}$ **if and only if** $\mathcal{Q}$ is defined to mean the same as (**if** $\mathcal{P}$, **then** $\mathcal{Q}$) and (**if** $\mathcal{Q}$, **then** $\mathcal{P}$). The words "if and only if" are frequently contracted to "iff".*

Proving $\mathcal{P}$ **iff** $\mathcal{Q}$ is therefore done in two pieces, first showing **if** $\mathcal{P}$, **then** $\mathcal{Q}$ and then showing **if** $\mathcal{Q}$, **then** $\mathcal{P}$. By our recent convention, each piece will be introduced by a simple sentence. In order to prove the piece **if** $\mathcal{P}$, **then** $\mathcal{Q}$, we assume $\mathcal{P}$ and establish $\mathcal{Q}$. This will also be stated in paragraph form rather than in formal proof steps.

A partial paragraph format for proving $\mathcal{P}$ **iff** $\mathcal{Q}$ might be:

> **Proof:**
>
> Show: $\mathcal{P}$ **iff** $\mathcal{Q}$
>
> We first show **if** $\mathcal{P}$, **then** $\mathcal{Q}$. Thus, assume $\mathcal{P}$.
>
> (get $\mathcal{Q}$ here)
>
> Therefore $\mathcal{Q}$ holds. Hence we have **if** $\mathcal{P}$, **then** $\mathcal{Q}$.
> We now show **if** $\mathcal{Q}$, **then** $\mathcal{P}$. Thus suppose $\mathcal{Q}$.
>
> (get $\mathcal{P}$ here)
>
> Therefore $\mathcal{P}$ is true, proving **if** $\mathcal{Q}$, **then** $\mathcal{P}$.
> By the first and second parts of the proof, we have $\mathcal{P}$ **iff** $\mathcal{Q}$ (by the definition of **iff**). ■

Theorem 1.9.1 For sets A, B, and C:

(a) $A \subseteq B$ **iff** $A \cap B = A$

(b) $A \subseteq B$ **iff** $A \cup B = B$

We will prove part (a) and leave (b) as Exercise 1. In the future we will omit stating facts like "A, B, and C are sets" in our hypothesis. We will be able to

1.9 Iff

work with A, B, and C in the proof since they are identified in the statement of the theorem.

> **Proof:**
>
> *Show:* $A \subseteq B$ iff $A \cap B = A$
>
> We first show if $A \subseteq B$, then $A \cap B = A$. Thus suppose $A \subseteq B$. In order to show $A \cap B = A$, we first prove $A \cap B \subseteq A$:
>
> 1. Let $x \in A \cap B$ be arbitrary.
> 2. $x \in A$ (1; def. $\cap$)
> 3. $A \cap B \subseteq A$ (1, 2; def. $\subseteq$)
>
> Next we show $A \subseteq A \cap B$:
>
> 1. Let $x \in A$ be arbitrary.
> 2. $x \in B$ (1, hyp.; def. $\subseteq$)
> 3. $x \in A \cap B$ (1, 2; def. $\cap$)
> 4. $A \subseteq A \cap B$ (1–3; def. $\subseteq$)
>
> Therefore $A \cap B = A$ by definition of set equality. We have now shown if $A \subseteq B$, then $A \cap B = A$.
>
> Second, the implication if $A \cap B = A$, then $A \subseteq B$ was shown in Exercise 1.7.4. The first and second parts of this proof show that $A \cap B = A$ iff $A \subseteq B$. ∎

A proof is not just a sequence of logical steps. It is a communication between writer and reader. The format we have used for doing **iff** proofs will keep the reader informed about the aims and assumptions for the various steps. This is just good writing style.

If a given statement is of the form **if** $\mathcal{P}$, **then** $\mathcal{Q}$, the new statement derived from the first by interchanging $\mathcal{P}$ and $\mathcal{Q}$ is called the *converse* of the first. Thus the converse of **if** $\mathcal{P}$, **then** $\mathcal{Q}$ is **if** $\mathcal{Q}$, **then** $\mathcal{P}$. The converse of the statement "If Karen has a B.S. in mathematics, then she is more than two years old" is "If Karen is more than two years old, then she has a B.S. in mathematics". It is easy to see from this example that the converse of a statement need not be true just because the statement itself is true. The statement $\mathcal{P}$ **iff** $\mathcal{Q}$ is true if and only if both **if** $\mathcal{P}$, **then** $\mathcal{Q}$ and its converse, **if** $\mathcal{Q}$, **then** $\mathcal{P}$, are true.

Theorem 1.9.2L *The statement $\mathcal{P}$ iff $\mathcal{Q}$ is true if and only if $\mathcal{P}$ is equivalent to $\mathcal{Q}$.*

Proof: Exercise 3.

Theorem 1.9.3 For sets A and B, $A = B$ if and only if *for all x: $x \in A$ iff $x \in B$.*

Proof:

Show: 1. If $A = B$, then **for all x: $x \in A$ iff $x \in B$.**
2. If **for all x: $x \in A$ iff $x \in B$**, then $A = B$.

Proof of 1:

Assume: $A = B$
Show: for all x: $x \in A$ iff $x \in B$

1. Let x be arbitrary.
2. Assume $x \in A$.
3. $A \subseteq B$ (hyp.; def.² =)
4. $x \in B$ (2, 3; def.² $\subseteq$)
5. if $x \in A$, then $x \in B$ (2, 4; pr. $\Rightarrow$)
6. Assume $x \in B$.
7. $B \subseteq A$ (hyp.; def.² =)
8. $x \in A$ (6, 7; def.² $\subseteq$)
9. if $x \in B$, then $x \in A$ (6—8; pr. $\Rightarrow$)
10. $x \in A$ iff $x \in B$ (5, 9; def.² iff)
11. for all x: $x \in A$ iff $x \in B$ (1—10; pr. $\forall$)

Proof of 2:

Assume: for all x: $x \in A$ iff $x \in B$
Show: $A = B$

1. Let $t \in A$ be arbitrary.
2. $t \in A$ iff $t \in B$ (hyp.; us. $\forall$)
3. if $t \in A$, then $t \in B$ (2; def.² iff)
4. $t \in B$ (1, 3; us. $\Rightarrow$)
5. $A \subseteq B$ (1—4; def.² $\subseteq$)
6. $B \subseteq A$ (1—5; symmetry)
7. $A = B$ (5, 6; def. =)

∎

The proof of Theorem 1.9.3 illustrates the translation of an "if and only if" statement into two "if . . . then" statements. Each of these is proved by assuming the hypothesis and showing the conclusion. This is an extension of

1.9 Iff

our convention of obtaining hypotheses and conclusion from metalanguage statements and will now be considered standard procedure.

Additional Proof Ideas

Theorem 1.9.3 gives us a way of proving or using set equality other than by the definition of equality, and is the basis for a condensed way of writing a proof that two sets are equal. First observe that **iff** is a transitive relation on statements. By this we mean that from $\mathcal{P}$ **iff** $\mathcal{Q}$ and $\mathcal{Q}$ **iff** $\mathcal{R}$, we can infer $\mathcal{P}$ **iff** $\mathcal{R}$. This leads to chains of **iff** statements. For example, a proof of $\mathcal{P}$ **iff** $\mathcal{Q}$ could be given by a chain $\mathcal{P}_0$ **iff** $\mathcal{P}_1$ **iff** $\mathcal{P}_2$ **iff** $\cdots$ **iff** $\mathcal{P}_n$, where $\mathcal{P}_0$ is $\mathcal{P}$ and $\mathcal{P}_n$ is $\mathcal{Q}$. This makes for a compact way of writing the proof.

Example 1:
Example 3 in the preceding section was a proof that $A - (B \cup C) \subseteq (A - B) \cap (A - C)$. Reasons for all steps were given, but equivalence was used implicitly. We can show set equality here by an **iff** chain using the same steps and Theorem 1.9.3.

> Let x be arbitrary.
> $x \in A - (B \cup C)$
> iff $x \in A$ and $\sim(x \in B \cup C)$
> iff $x \in A$ and $\sim(x \in B$ and $x \in C)$
> iff $x \in A$ and $(x \notin B$ and $x \notin C)$
> iff $(x \in A$ and $x \in A)$ and $(x \notin B$ and $x \notin C)$
> iff $(x \in A$ and $x \notin B)$ and $(x \in A$ and $x \notin C)$
> iff $(x \in A - B)$ and $(x \in A - C)$
> iff $x \in (A - B) \cap (A - C)$
>
> Hence **for all** x: $x \in A - (B \cup C)$ iff $x \in (A - B) \cap (A - C)$.
> By Theorem 1.9.3, $A - (B \cup C) = (A - B) \cap (A - C)$.

In order to verify an argument like the one above, it is necessary to verify each stage $\mathcal{P}_i$ **iff** $\mathcal{P}_{i+1}$ both ways, that is, **if** $\mathcal{P}_i$, **then** $\mathcal{P}_{i+1}$ and **if** $\mathcal{P}_{i+1}$, **then** $\mathcal{P}_i$. This involves proving a total of $2n$ implications. The danger for beginning students is that some stage will be inadvertently checked in only one way. It is safer to do all the implications one way at a time. That is, prove **if** $\mathcal{P}_0$, **then** $\mathcal{P}_1$, **if** $\mathcal{P}_1$, **then** $\mathcal{P}_2$, ..., **if** $\mathcal{P}_{n-1}$, **then** $\mathcal{P}_n$. Note that this has the effect of proving set containment one way ($A \subseteq B$) and involves n implications. Proving the converses of these implications in the reverse order involves another n implications and shows $B \subseteq A$. Thus, unless something is overlooked, the amount of *thought* that goes into an **iff** chain proof of set equality is at least as great as that required for showing containment both ways. One can avoid repeated writing of obvious steps with **iff** chain proofs, however, so that these can be a useful proof abbreviation.

Example 2:
Use an **iff** chain proof to prove: For sets A, B, and C, $A \cup (B \cup C) = (A \cup B) \cup C$ (Theorem 1.6.1b). This is made easy by using the equivalence of $\mathcal{P}$ or ($\mathcal{Q}$ or $\mathcal{R}$) and ($\mathcal{P}$ or $\mathcal{Q}$) or $\mathcal{R}$ (Theorem 1.7.6L).

> Let x be arbitrary.
> $x \in A \cup (B \cup C)$
> iff $x \in A$ or $x \in B \cup C$ (def. $\cup$)
> iff $x \in A$ or ($x \in B$ or $x \in C$)   (def. $\cup$)
> iff ($x \in A$ or $x \in B$) or $x \in C$ (Thm. 1.7.6L)
> iff $x \in A \cup B$ or $x \in C$ (def. $\cup$)
> iff $x \in (A \cup B) \cup C$ (def. $\cup$)
> $A \cup (B \cup C) = (A \cup B) \cup C$ (Thm. 1.9.3) ∎

Summary of the Basic Set Theory Properties

We summarize here some useful theorems about sets. Theorems already given in the text are noted parenthetically. The other statements are simple and can be proved as exercises. Also, by using theorems on equivalence, **iff** chain proofs can be provided (as exercises) for those theorems already proved.

For arbitrary sets A, B, and C we have:

Union Laws

$A \cup B = B \cup A$ (Thm. 1.5.2b)
$A \cup (B \cup C) = (A \cup B) \cup C$ (Thm. 1.6.1b)
$A \cup A = A$
$A \cup \emptyset = A$
$A \subseteq A \cup B$ (Exercise 1.4.3)

1.9 Iff

Intersection Laws

$A \cap B = B \cap A$ (Thm. 1.5.2a)
$A \cap (B \cap C) = (A \cap B) \cap C$ (Thm. 1.6.1a)
$A \cap A = A$
$A \cap \varnothing = \varnothing$
$A \cap B \subseteq A$

Distributive Laws

$A \cap (B \cup C) = (A \cap B) \cup (A \cap C)$ (Thm. 1.6.2a)
$A \cup (B \cap C) = (A \cup B) \cap (A \cup C)$ (Thm. 1.6.2b)

DeMorgan's Laws

$A - (B \cup C) = (A - B) \cap (A - C)$ (Thm. 1.8.1a)
$A - (B \cap C) = (A - B) \cup (A - C)$ (Thm. 1.8.1b)

Complement Laws

$\overline{\overline{A}} = A$ (Note: $\overline{\overline{A}}$ means $\overline{(\overline{A})}$)
$\overline{A \cup B} = \overline{A} \cap \overline{B}$ (Cor. 1.8.2a)
$\overline{A \cap B} = \overline{A} \cup \overline{B}$ (Cor. 1.8.2b)
$\overline{U} = \varnothing$
$\overline{\varnothing} = U$
$A - B = A \cap \overline{B}$
$A \cup \overline{A} = U$
$A \cap \overline{A} = \varnothing$

Equality Laws

$A = A$ (reflexivity of set equality)
if $A = B$, then $B = A$ (symmetry of set equality)
if $A = B$ and $B = C$, then $A = C$ (transitivity of set equality)

Subset Laws

$A \subseteq A$
if $A \subseteq B$ and $B \subseteq A$, then $A = B$ (def. =)
if $A \subseteq B$ and $B \subseteq C$, then $A \subseteq C$ (Thm. 1.3.1)
$\varnothing \subseteq A$

Exercises 1.9

Straightforward Problems

1. Prove Theorem 1.9.1b. Use our partial paragraph style.
2. If A and B are sets, prove $A - (A - B) = B$ iff $B \subseteq A$.
3. Prove Theorem 1.9.2L.

Supplementary Problems

4. Prove each of the laws of sets in the summary. Write **iff** proof chains for those statements already proved in the text.

1.10 *THERE EXISTS* STATEMENTS

Let $A_1, A_2, \ldots, A_n$ be n sets. Define the **union** of these sets, $\bigcup_{i=1}^{n} A_i$, to be the set of all x such that x is in at least one of the n sets; formally, we say $x \in A_i$ **for some** i. Define the **intersection** of these sets, $\bigcap_{i=1}^{n} A_i$, to be the set of all x such that x is in every one of the n sets. The i in $\bigcup_{i=1}^{n} A_i$ and $\bigcap_{i=1}^{n} A_i$ is frequently called a dummy variable since $\bigcup_{i=1}^{n} A_i$ and $\bigcap_{i=1}^{n} A_i$ do not depend on i. Thus i serves only as an index and has no meaning outside each of these expressions. In this sense it is a local variable. The ideas of union and intersection are given in the following definition:

Definition   Let $A_1, A_2, \ldots, A_n$ be n sets.

$$\bigcup_{i=1}^{n} A_i = \{x \mid x \in A_j \text{ for some } j \text{ such that } 1 \leq j \leq n\}$$

$$\bigcap_{i=1}^{n} A_i = \{x \mid x \in A_j \text{ for all } j \text{ such that } 1 \leq j \leq n\}$$

First consider the statement $x \in A_j$ **for all** j such that $1 \leq j \leq n$. It is part of our language and means the same as:

    (2) **for all** j such that $1 \leq j \leq n$: $x \in A_j$

or   (3) **for all** $1 \leq j \leq n$: $x \in A_j$

The last form can be used if it is clear that j and not n is the local variable (that is, if n has already been defined). In this section we will understand that

1.10 There Exists Statements

$1 \leq j \leq n$ means that j is an integer from 1 to n. In order to prove Statement (2) or (3), we would make the step:

> Let j be arbitrary such that $1 \leq j \leq n$.

or Let $1 \leq j \leq n$ be arbitrary.

We would then argue to show $x \in A_j$. From now on we will always use the word "let" in one of two senses. The first is in defining a symbol that has been picked arbitrarily such that a certain condition holds. For example, instead of writing

> Let $x \in A$ be arbitrary.

we will merely write

> Let $x \in A$.

This will mean "let x be an arbitrarily chosen element of A". Similarly, "let $1 \leq j \leq n$" will mean "let j be arbitrarily chosen such that $1 \leq j \leq n$". The word "let" will therefore be useful in proving **for all** statements—the only statements we have so far that involve a local variable. By our convention, the statement **if $x \in A$, then $x \in B$** does not involve a local variable. To prove such a statement, we would make the step "assume $x \in A$". This does not define the symbol "x" but assumes the truth of a statement involving symbols that have been already defined.

Our second use of the word "let" will be in defining a new symbol in terms of previously defined symbols—similar to an assignment statement in computer programming. An example of this use of "let" would be "let $x = 3y + 2$", where y has been previously defined.

The other type of statement in our language that involves a local variable is exemplified by the condition $z \in A_j$ **for some j such that $1 \leq j \leq n$**, used to define the union of the sets A_j. This statement will mean the same as

	(2)	$x \in A_j$ for some $1 \leq j \leq n$
or	(3)	for some $1 \leq j \leq n$: $x \in A_j$
or	(4)	there exists j such that $1 \leq j \leq n$ and such that $x \in A_j$
or	(5)	there exists j such that $(1 \leq j \leq n$ and $x \in A_j)$
or	(6)	there exists $1 \leq j \leq n$ such that $x \in A_j$

We will refer to statements like (2) through (6) as **for some** statements. Any **for some** statement may also be called a **there exists** statement. The meaning of Statements (2) through (6) is that $x \in A_j$ is true for at least one of the j.

In this section, we will consider only **there exists** statements of the form

> $\mathcal{P}(j)$ for some $1 \leq j \leq n$

or for some $1 \leq j \leq n$: $\mathcal{P}(j)$

and **for all** statements of the form

$$\mathcal{P}(j) \text{ for all } 1 \leq j \leq n$$

or   for all $1 \leq j \leq n$: $\mathcal{P}(j)$

We will generalize things in the next section.

Inference Rule

Using **there exists** statements: *To use the statement $\mathcal{P}(j)$ for some $1 \leq j \leq n$ in a proof, immediately follow it with the step*

$$\text{Pick } 1 \leq j_0 \leq n \text{ such that } \mathcal{P}(j_0).$$

This defines the symbol j_0. The truth of both $1 \leq j_0 \leq n$ and $\mathcal{P}(j_0)$ may be used in the remainder of the proof (in appropriate steps).

"There exists" is abbreviated ∃. The rule above is abbreviated "us. ∃".

Example 1:

1. $x \in A_i$ for some $1 \leq i \leq 10$
2. Pick $1 \leq i_0 \leq 10$ such that $x \in A_{i_0}$. (1; us. ∃)
3. $1 \leq i_0 \leq 10$ (Step 2)
4. $x \in A_{i_0}$ (Step 2)

In Step 2 we choose one number i_0 with the property that $x \in A_{i_0}$ and give it a name so that we can refer to it later. Thus in Step 2 the symbol "i_0" is defined for future steps. That is, whereas i in Step 1 is a local variable indicating a range of values, i_0 becomes a global variable and is fixed for the remaining steps. It is not necessary to write Steps 3 and 4. This information is already considered in usable form in Step 2.

In Example 1 we used the symbol "i_0" to denote a fixed value of the variable i. It is not necessary, of course, to use a subscripted variable. In Example 2 the symbol "j" represents the fixed values of the variable i in Step 1.

Example 2:
For each $i = 1, 2, \ldots, 10$, define $A_i = \{t \in \mathbb{R} \mid 0 < t < \frac{1}{i}\}$.

1. $x \in A_i$ for some $1 \leq i \leq 10$
2. Pick $1 \leq j \leq 10$ such that $x \in A_j$. (1; us. ∃)
3. $j < 100$ (2; prop. **N**)
4. $0 < x < \dfrac{1}{j}$ (2; def.[1] A_j)

Theorem 1.10.1

Let $B, A_1, A_2, \ldots, A_n$ be sets.

(a) If $B \subseteq A_i$ for all $1 \leq i \leq n$, then $B \subseteq \bigcap\limits_{i=1}^{n} A_i$.

(b) If $A_i \subseteq B$ for all $1 \leq i \leq n$, then $\bigcup\limits_{i=1}^{n} A_i \subseteq B$.

1.10 *There Exists* Statements

Proof of (a):

Assume: $B \subseteq A_i$ for all $1 \leq i \leq n$
Show: $B \subseteq \bigcap_{i=1}^{n} A_i$

1. Let $x \in B$.
2. Let $1 \leq j \leq n$ be arbitrary.
3. $B \subseteq A_j$ (2, hyp.; us. $\forall$)
4. $x \in A_j$ (1, 3; def.^{2} $\subseteq$)
5. $x \in A_i$ for all $1 \leq i \leq n$ (2—4; pr. $\forall$)
6. $x \in \bigcap_{i=1}^{n} A_i$ (5; def. $\cap$)
7. $B \subseteq \bigcap_{i=1}^{n} A_i$ (1—6; def. $\subseteq$)

∎

Proof of (b):

Assume: $A_i \subseteq B$ for all $1 \leq i \leq n$
Show: $\bigcup_{i=1}^{n} A_i \subseteq B$

1. Let $x \in \bigcup_{i=1}^{n} A_i$.
  .
  .
  .
  k. $x \in B$
  k+1. $\bigcup_{i=1}^{n} A_i \subseteq B$ (1—k; def. $\subseteq$)

The proof so far is routine. We know nothing about B, and so Step $k-1$ is not evident at this time. To get Step 2 from Step 1, we must use the definition of union:

    2. $x \in A_i$ for some $1 \leq i \leq n$ (1; def. $\cup$)

According to our rule, the next step must be:

    3. Pick $1 \leq j \leq n$ such that $x \in A_j$.   (2; us. $\exists$)

Now that we have exhausted our sequence of steps dictated by the conclusion, it is time to consider the hypothesis. Since $1 \leq j \leq n$ we have, by the rule for using **for all** statements,

> 4. $A_j \subseteq B$ (3, hyp.; us. ∀)
> 5. $x \in B$ (3, 4; def. ⊆)

which was our goal—so that Step k is Step 5. Note that the "$1 \leq j \leq n$" part of Step 3 is used to get Step 4 and the "$x \in A_j$" part of Step 3 is used to get Step 5. A complete proof of Theorem 1.10.1b is given by:

> **Proof:**
>
> Assume: $A_i \subseteq B$ for all $1 \leq i \leq n$
> Show: $\bigcup_{i=1}^{n} A_i \subseteq B$
>
> 1. Let $x \in \bigcup_{i=1}^{n} A_i$.
> 2. $x \in A_i$ for some $1 \leq i \leq n$ (1; def. ∪)
> 3. Pick $1 \leq j \leq n$ such that $x \in A_j$.   (2; us. ∃)
> 4. $A_j \subseteq B$ (3, hyp.; us. ∀)
> 5. $x \in B$ (3, 4; def. ⊆)
> 6. $\bigcup_{i=1}^{n} A_i \subseteq B$ (1—5; def. ⊆)

Inference Rule

*Proving **there exists** statements: If $1 \leq i \leq n$ and $\mathcal{P}(i)$ are steps in a proof, then **for some** $1 \leq j \leq n$: $\mathcal{P}(j)$ can be written as a proof step. Abbreviation: "pr. ∃".*

This rule also means that if $1 \leq i \leq n$ and $\mathcal{P}(i)$ are steps in a proof, then **for some** $1 \leq i \leq n$: $\mathcal{P}(i)$ can be written as a proof step—as well as **for some** $1 \leq j \leq n$: $\mathcal{P}(j)$ using any local variable j. Note that this rule does just the opposite of the rule for using a **for some** statement, which gets the statements $1 \leq j_0 \leq n$ and $\mathcal{P}(j_0)$ from the statement **for some** $1 \leq j \leq n$: $\mathcal{P}(j)$.

Example 3:

> 1. *
> 2. $x \in A_i$ for some $1 \leq i \leq 10$ (1; pr. ∃)

1.10 *There Exists* Statements

Solution:
1. $x \in A_3$
2. $x \in A_i$ for some $1 \le i \le 10$ (1; pr. $\exists$)

If the step $1 \le j \le n$ leading to **for some** $1 \le j \le n$: $\mathscr{P}(j)$ is obvious, it is not written down. Thus $1 \le 3 \le 10$ was not written down as a step in Example 3.

Example 4:
1. $1 \le j \le n$
2. $x \in A_j$
3. *   (1, 2; pr. $\exists$)

Solution:
1. $1 \le j \le n$
2. $x \in A_j$
3. $x \in A_i$ for some $1 \le i \le n$ (1, 2; pr. $\exists$)

The rule for proving a **for some** (that is, **there exists**) statement dictates the previous steps or step needed to infer the statement. For example, to prove $x \in A_i$ for some $1 \le i \le n$, we need to know $1 \le j \le n$ and $x \in A_j$ for some particular j. These two statements involve a symbol, j, that has not yet appeared in the proof. You will have to produce j. In general, the way to do this is to define j in your proof in terms of previously defined symbols and then show that $1 \le j \le n$ and $x \in A_j$ hold for your j.

Example 5:
Let n be a fixed even natural number. For every i from 1 to n, define the set

$$A_i = \{x \in \mathbf{R} \mid 0 \le x \le 1 - |\tfrac{i}{n} - \tfrac{1}{2}|\}$$

For example, if $n = 6$, then the sets A_1 through A_6 are:

$$A_1 = \{x \in \mathbf{R} \mid 0 \le x \le \tfrac{2}{3}\}$$
$$A_2 = \{x \in \mathbf{R} \mid 0 \le x \le \tfrac{5}{6}\}$$
$$A_3 = \{x \in \mathbf{R} \mid 0 \le x \le 1\}$$
$$A_4 = \{x \in \mathbf{R} \mid 0 \le x \le \tfrac{5}{6}\}$$
$$A_5 = \{x \in \mathbf{R} \mid 0 \le x \le \tfrac{2}{3}\}$$
$$A_6 = \{x \in \mathbf{R} \mid 0 \le x \le \tfrac{1}{2}\}$$

Chapter 1 Sets and Rules of Inference

Theorem: *Let n be a fixed even natural number. For every i from 1 to n, define the set*

$$A_i = \{x \in \mathbf{R} \mid 0 \leq x \leq 1 - |\tfrac{i}{n} - \tfrac{1}{2}|\}$$

Then for some $1 \leq i \leq n$: $1 \in A_i$.

Scrap Paper

Notice that, for any n, the largest set A_i corresponds to the smallest value of $\left|\dfrac{i}{n} - \dfrac{1}{2}\right|$. This happens when $i = \dfrac{n}{2}$.

Proof:

1. Let $j = \tfrac{n}{2}$.
2. $1 \leq j \leq n$ (1, hyp.; prop. **N**)
3. $A_j = \{x \in \mathbf{R} \mid 0 \leq x \leq 1\}$ (1, hyp., def. A_i; sub., prop. **N**)
4. $1 \in A_j$ (3; def. A_j)
5. **for some $1 \leq i \leq n$: $1 \in A_i$**   (2, 4; pr. ∃)

■

Step 1 in this proof illustrates our second use of the word "let". The "j" is newly defined in terms of symbols already given. Note that we have established Steps 2 and 4 for the j we defined in Step 1. We now give a less formal statement of the rule for proving **for some**, a statement that better indicates what needs to be done.

Inference Rule

Proving **there exists** *statements: To prove the statement $\mathscr{P}(j)$ for some $1 \leq j \leq n$, define j in your proof (in terms of previously defined symbols) and show that $\mathscr{P}(j)$ and $1 \leq j \leq n$ hold for your j.*

Theorem 1.10.2

Let $A_1, A_2, \ldots, A_n$ be sets.

(a) *for all $1 \leq j \leq n$:* $\bigcap_{i=1}^{n} A_i \subseteq A_j$

(b) *for all $1 \leq j \leq n$:* $A_j \subseteq \bigcup_{i=1}^{n} A_i$

Proof of (a): Exercise 3.

1.10 *There Exists* Statements

Proof of (b):

Show: for all $1 \leq j \leq n$: $A_j \subseteq \bigcup_{i=1}^{n} A_i$

    1. Let j be such that $1 \leq j \leq n$.
        ·
        ·
    k. $A_j \subseteq \bigcup_{i=1}^{n} A_i$
$k+1$. for all $1 \leq j \leq n$: $A_j \subseteq \bigcup_{i=1}^{n} A_i$ (1—k; pr. ∀)

The steps above are dictated by the **for all** rule. The next steps are dictated by Step k:

    1. Let j be such that $1 \leq j \leq n$.
    2. Let $x \in A_j$.
        ·
        ·
    $k-1$. $x \in \bigcup_{i=1}^{n} A_i$
    k. $A_j \subseteq \bigcup_{i=1}^{n} A_i$ (1—$k-1$; def. ⊆)
$k+1$. for all $1 \leq j \leq n$: $A_j \subseteq \bigcup_{i=1}^{n} A_i$ (1—k; pr. ∀)

From Step $k-1$ we get Step $k-2$:

    $k-2$. $x \in A_i$ for some $1 \leq i \leq n$
    $k-1$. $x \in \bigcup_{i=1}^{n} A_i$ ($k-2$; def. ∪)

Our job is now to prove the **for some** statement of Step $k-2$. To do this, we must define i (any way we want) so as to get $1 \leq i \leq n$ and $x \in A_i$. It is clear from the things we know about j from Steps 1 and 2 that defining $i = j$ will do the job. Of course, there is no need to write:

    3. Let $i = j$.

for Step 3. The symbol "j" we already have will do fine. The rule for proving **for some** statements states that $k-2$ follows from Steps 1 and 2. Since i is a local variable in Step $k-2$, this step as written means exactly the same thing as $x \in A_j$ **for some** $1 \leq j \leq n$. Our final proof is:

Proof of (b):

Show: for all $1 \leq j \leq n$: $A_j \subseteq \bigcup_{i=1}^{n} A_i$

1. Let $1 \leq j \leq n$.
2. Let $x \in A_j$.
3. $x \in A_i$ for some $1 \leq i \leq n$ (1, 2; pr. $\exists$)
4. $x \in \bigcup_{i=1}^{n} A_i$ (3; def. $\cup$)
5. $A_j \subseteq \bigcup_{i=1}^{n} A_i$ (2—4; def. $\subseteq$)
6. for all $1 \leq j \leq n$: $A_j \subseteq \bigcup_{i=1}^{n} A_i$ (1—5; pr. $\forall$)

In Step 1 (note the abbreviated version), j is picked arbitrarily (to prove the **for all** statement in Step 6) and is fixed thereafter. It is this fixed j that is used to prove the **for some** statement in Step 3.

Additional Proof Ideas

If $\mathcal{Q}(i)$ is equivalent to $\mathcal{P}(i)$ for some free variable i, then $\mathcal{P}(i)$ may be substituted for $\mathcal{Q}(i)$ in the statement **for all** i: $\mathcal{Q}(i)$, obtaining **for all** i: $\mathcal{P}(i)$. Similarly, substitution in **for all** i: $\mathcal{P}(i)$ gets **for all** i: $\mathcal{Q}(i)$. Thus, if $\mathcal{Q}(i)$ is equivalent to $\mathcal{P}(i)$, then **for all** i: $\mathcal{Q}(i)$ is equivalent to **for all** i: $\mathcal{P}(i)$. This idea can be extended to making other inferences within a **for all** statement.

Theorem 1.10.3L   If from $\mathcal{P}(i)$ and $\mathcal{Q}(i)$ we may infer $\mathcal{R}(i)$, then

from $\mathcal{P}(i)$ for all $1 \leq i \leq n$

and $\mathcal{Q}(i)$ for all $1 \leq i \leq n$

we may infer $\mathcal{R}(i)$ for all $1 \leq i \leq n$

Proof: Exercise 4.

In Exercise 5 you are asked to give examples of statements $\mathcal{P}(i)$, $\mathcal{Q}(i)$, and $\mathcal{R}(i)$ such that Theorem 1.10.3L becomes false when "**for all**" is replaced by "**for some**". Theorem 1.10.3L allows us to write an alternate proof of 1.10.1a:

1.10 *There Exists* Statements

Theorem 1.10.1 Let $B, A_1, A_2, \ldots, A_n$ be sets.

(a) If $B \subseteq A_i$ for all $1 \leq i \leq n$, then $B \subseteq \bigcap_{i=1}^{n} A_i$.

> **Proof:**
>
> Assume: $B \subseteq A_i$ for all $1 \leq i \leq n$
> Show: $B \subseteq \bigcap_{i=1}^{n} A_i$
>
> 1. Let $x \in B$.
> 2. $B \subseteq A_i$ for all $1 \leq i \leq n$ (hyp.)
> 3. $x \in A_i$ for all $1 \leq i \leq n$ (1, 2; def.$^2 \subseteq$, Thm. 1.10.3L)
> 4. $x \in \bigcap_{i=1}^{n} A_i$ (3; def. $\cap$)
> 5. $B \subseteq \bigcap_{i=1}^{n} A_i$ (1—4; def. $\subseteq$)

In going from Step 2 to Step 3, we are doing the following: $x \in B$ is true for all $1 \leq i \leq n$ since $x \in B$ is true and does not involve i. Thus from $x \in B$ and $B \subseteq A_i$ we may infer $x \in A_i$ (def. $\subseteq$). Thus, from $x \in B$ **for all** $1 \leq i \leq n$ and $B \subseteq A_i$ **for all** $1 \leq i \leq n$, we may infer $x \in A_i$ for all $1 \leq i \leq n$ (Thm. 1.10.3L). Compare this proof of Theorem 1.10.1a with the earlier proof in the text.

Recall that $\mathscr{P} \Rightarrow \mathscr{Q}$ means that if $\mathscr{P}$ is a proof step, then $\mathscr{Q}$ can be written as a proof step.

Theorem 1.10.4L Let $\mathscr{P}(i)$ and $\mathscr{Q}(i)$ be language statements involving the free variable i. If $\mathscr{P}(i) \Rightarrow \mathscr{Q}(i)$, then $\mathscr{P}(i)$ for all $1 \leq i \leq n \Rightarrow \mathscr{Q}(i)$ for all $1 \leq i \leq n$.

Proof: Exercise 6.

Using the ideas in Theorems 1.10.3L and 1.10.4L, we may construct proofs of the following form:

$$\mathscr{P}(i) \text{ for all } 1 \leq i \leq n$$
$$\mathscr{Q}(i) \text{ for all } 1 \leq i \leq n$$
$$\vdots$$
$$\mathscr{R}(i) \text{ for all } 1 \leq i \leq n$$

In a proof, mathematicians more often pick an arbitrary i and then make the assertions $\mathscr{P}(i), \mathscr{Q}(i), \ldots, \mathscr{R}(i)$.

Exercises 1.10

Practice Exercise

1. (a) 1. **there exists j such that $1 \leq j \leq 9$ such that $x \in A_j$**
      2. *   (1; us. ∃)
   (b) 1. *
      2. Pick $x \in A$ such that $1 \leq x \leq n$.   (1; us. ∃)
   (c) 1. Let $1 \leq j \leq n$ be arbitrary.
      .
      .
      k. $x \in A_j$
      k+1. *   (*; def.² ∩)
   (d) 1. *
      2. **there exists $1 \leq i \leq 9$ such that $x \in A_j$**   (1; pr. ∃)

Straightforward Problems

2. Prove the following extended distribution laws. For all sets $B, A_1, \ldots, A_n$:
   (a) $B \cap (\bigcup_{i=1}^{n} A_i) = \bigcup_{i=1}^{n} (B \cap A_i)$
   (b) $B \cup (\bigcap_{i=1}^{n} A_i) = \bigcap_{i=1}^{n} (B \cup A_i)$
3. Prove Theorem 1.10.2a.

Supplementary Problems

4. Prove Theorem 1.10.3L.
5. Give examples of statements $\mathcal{P}(i), \mathcal{Q}(i)$, and $\mathcal{R}(i)$ such that Theorem 1.10.3L becomes false when "**for all**" is replaced by "**for some**".
6. Prove Theorem 1.10.4L.

1.11 NEGATIONS

The truth of the statement **for all $1 \leq i \leq n$**: $\mathcal{P}(i)$ is equivalent to $\mathcal{P}(1), \mathcal{P}(2), \mathcal{P}(3), \ldots,$ up to $\mathcal{P}(n)$ *all* being true statements. In order that the **for all** statement fail to be true, it is only necessary that *one* of the statements $\mathcal{P}(1), \mathcal{P}(2),$ and so forth be false. This leads to the following formal rule of inference:

1.11 Negations

Inference Rule

*Negation: The negation of **for all** $1 \leq i \leq n$: $\mathcal{P}(i)$ is **for some** $1 \leq i \leq n$: $\sim\mathcal{P}(i)$.*

The truth of the statement **for some $1 \leq i \leq n$: $\mathcal{P}(i)$** is equivalent to there existing at least *one* integer i from 1 to n such that $\mathcal{P}(i)$ is true. In order that this one integer not exist, it is necessary that $\mathcal{P}(i)$ fail to be true for *all* integers from 1 through n. This leads to the following rule of inference:

Inference Rule

*Negation: The negation of **for some** $1 \leq i \leq n$: $\mathcal{P}(i)$ is **for all** $1 \leq i \leq n$: $\sim\mathcal{P}(i)$.*

Example 1:

1. $\sim(x \in A_i$ for all $1 \leq i \leq n)$
2. * (1; negation rule)

Solution:

1. $\sim(x \in A_i$ for all $1 \leq i \leq n)$
2. **for some $1 \leq i \leq n$: $x \notin A_i$** (1; negation rule)

Example 2:

1. $\sim(x \in A_i$ for some $1 \leq i \leq n)$
2. * (1; neg.)

Solution:

1. $\sim(x \in A_i$ for some $1 \leq i \leq n)$
2. $x \notin A_i$ for all $1 \leq i \leq n$ (1; neg.)

Theorem 1.11.1

Let $B, A_1, A_2, \ldots, A_n$ be sets.

(a) $B - \bigcup_{i=1}^{n} A_i = \bigcap_{i=1}^{n} (B - A_i)$ (DeMorgan's Laws)

(b) $B - \bigcap_{i=1}^{n} A_i = \bigcup_{i=1}^{n} (B - A_i)$

Proof of (a): Exercise 2.

Proof of (b):

Show: $B - \bigcap_{i=1}^{n} A_i = \bigcup_{i=1}^{n} (B - A_i)$

We first show $B - \bigcap_{i=1}^{n} A_i \subseteq \bigcup_{i=1}^{n} (B - A_i)$:

1. Let $x \in B - \bigcap_{i=1}^{n} A_i$.
2. $x \in B$ (1; def. $-$)
3. $x \notin \bigcap_{i=1}^{n} A_i$ (1; def. $-$)

    .
    .

k. $x \in \bigcup_{i=1}^{n} (B - A_i)$

k+1. $B - \bigcap_{i=1}^{n} A_i \subseteq \bigcup_{i=1}^{n} (B - A_i)$ (k; def. $\subseteq$)

By definition of union, we need

k−1. $x \in B - A_j$ for some $1 \leq j \leq n$

Our rule states that, in order to establish Step k−1, we need to define j somehow in the proof and then show $x \in B - A_j$ for that j. We can proceed from Step 3 using definitions and the negation rule:

3. $x \notin \bigcap_{i=1}^{n} A_i$ (1; def. $-$)
4. $\sim(x \in A_i$ for all $1 \leq i \leq n)$ (3; def. $\cap$)
5. $x \notin A_i$ for some $1 \leq i \leq n$ (4; neg.)
6. Pick $1 \leq j \leq n$ such that $x \notin A_j$.   (5; us. $\exists$)

This j is just what we need to get Step k−1. Note that Step k−1 (now Step 7) follows from Steps 2 and 6. We continue with the reverse set inclusion:

Next we show $\bigcup_{i=1}^{n} (B - A_i) \subseteq B - \bigcap_{i=1}^{n} A_i$.

1. Let $x \in \bigcup_{i=1}^{n} (B - A_i)$.
2. $x \in B - A_i$ for some $1 \leq i \leq n$ (1; def. $\cup$)

1.11 Negations

    3. Pick $1 \leq j \leq n$ such that $x \in B - A_j$.   (2; us. $\exists$)

    .

    .

    $k.$ $x \in B - \bigcap_{i=1}^{n} A_i$

$k+1.$ $\bigcup_{i=1}^{n} (B - A_i) \subseteq B - \bigcap_{i=1}^{n} A_i$ (1—k; def. $\subseteq$)

Scrap Paper

    to show $x \in B - \bigcap_{i=1}^{n} A_i$

    show $x \in B$ ⎫

    and $x \notin \bigcap_{i=1}^{n} A_i$ ⎭ def. $-$

    to show $x \notin \bigcap_{i=1}^{n} A_i$

    show $\sim(x \in A_i$ for all $1 \leq i \leq n)$ def. $\cap$

    that is, $x \notin A_i$ for some $1 \leq i \leq n$ negation rule

Our scrap-paper analysis has lead us backward to what we already know from Step 3. Reversing the steps in this analysis gives a finished proof:

Proof of (b):

    Show: $B - \bigcap_{i=1}^{n} A_i = \bigcup_{i=1}^{n} (B - A_i)$

We first show $B - \bigcap_{i=1}^{n} A_i \subseteq \bigcup_{i=1}^{n} (B - A_i)$:

    1. Let $x \in B - \bigcap_{i=1}^{n} A_i$.

    2. $x \in B$ (1; def. $-$)

    3. $x \notin \bigcap_{i=1}^{n} A_i$ (1; def. $-$)

    4. $\sim(x \in A_i$ for all $1 \leq i \leq n)$ (3; def. $\cap$)

Chapter 1 Sets and Rules of Inference

5. $x \notin A_i$ for some $1 \leq i \leq n$ (4; neg.)
6. Pick $1 \leq j \leq n$ such that $x \notin A_j$.   (5; us. $\exists$)
7. $x \in B - A_j$ (2, 6; def. $-$)
8. $x \in B - A_j$ for some $1 \leq j \leq n$ (6, 7; pr. $\exists$)
9. $x \in \bigcup_{i=1}^{n} (B - A_i)$ (8; def. $\cup$)
10. $B - \bigcap_{i=1}^{n} A_i \subseteq \bigcup_{i=1}^{n} (B - A_i)$ (1—9; def. $\subseteq$)

Next we show $\bigcup_{i=1}^{n} (B - A_i) \subseteq B - \bigcap_{i=1}^{n} A_i$:

1. Let $x \in \bigcup_{i=1}^{n} (B - A_i)$.
2. $x \in B - A_i$ for some $1 \leq i \leq n$ (1; def. $\cup$)
3. Pick $1 \leq j \leq n$ such that $x \in B - A_j$.   (2; us. $\exists$)
4. $x \in B$ (3; def. $-$)
5. $x \notin A_j$ (3; def. $-$)
6. $x \notin A_j$ for some $1 \leq j \leq n$.   (3, 5; pr. $\exists$)
7. $\sim(x \in A_i$ for all $1 \leq i \leq n)$ (6; neg.)
8. $\sim(x \in \bigcap_{i=1}^{n} A_i)$ (7; def. $\cap$)
9. $x \notin \bigcap_{i=1}^{n} A_i$ (8; def. $\notin$)
10. $x \in B - \bigcap_{i=1}^{n} A_i$ (4, 9; def. $-$)
11. $\bigcup_{i=1}^{n} (B - A_i) \subseteq B - \bigcap_{i=1}^{n} A_i$ (1—10; def. $\subseteq$)

By the first and second parts of the proof we have $B - \bigcap_{i=1}^{n} A_i = \bigcup_{i=1}^{n} (B - A_i)$ by definition of set equality. ∎

Exercises 1.11

Practice Exercise

1. (a) 1. *
       2. $x \in A_j$ for some $1 \leq j \leq 3$ (1; pr. $\exists$)

1.11 Negations

(b) 1. $x \in A_j$ for some $1 \leq j \leq 3$
    2. *   (1; us. $\exists$)

(c) 1. $\sim(x \in A_j$ for some $1 \leq j \leq 3)$
    2. *   (1; neg.)

(d) 1. $\sim(x \in A_j$ for all $1 \leq j \leq 3)$
    2. *   (1; neg.)

(e) 1. $\sim(x \notin A_j$ for all $1 \leq i \leq n)$
    2. *   (1; neg.)

(f) 1. $\sim(x \notin A_i$ for some $1 \leq i \leq n)$
    2. *   (1; neg.)

Straightforward Problems

2. Prove Theorem 1.11.1a.
3. Prove the following extended complement laws. For sets $A_1, \ldots, A_n$:

(a) $\overline{\bigcup_{i=1}^{n} A_i} = \bigcap_{i=1}^{n} \overline{A_i}$

(b) $\overline{\bigcap_{i=1}^{n} A_i} = \bigcup_{i=1}^{n} \overline{A_i}$

Challenging Problem

4. Given sets $A_1, A_2, A_3,$ and $A_4,$ define:

$$B_1 = A_1$$
$$B_2 = A_2 - A_1$$
$$B_3 = A_3 - (A_1 \cup A_2)$$
$$B_4 = A_4 - \bigcup_{i=1}^{3} A_i$$

Prove (a) $\bigcup_{i=1}^{4} B_i = \bigcup_{i=1}^{4} A_i$

      (b) for all $1 \leq i, j \leq 4, i \neq j: B_i \cap B_j = \emptyset$

This problem is generalized in Chapter 3, after a closer study of **N**. The more powerful ideas introduced will allow us to do much simpler proofs.

1.12 INDEX SETS

Given a fixed positive integer n, define the set $\mathscr{I} = \{i \in \mathbf{N} \mid 1 \leq i \leq n\}$. The set $\mathscr{I}$ is also written $\{1, 2, \ldots, n\}$. The statement

$$\mathscr{P}(i) \text{ for all } 1 \leq i \leq n$$

can also be written

$$\mathscr{P}(i) \text{ for all } i \in \mathscr{I}$$

In the statement $\mathscr{P}(i)$ **for all** $1 \leq i \leq n$, i is assumed (as in the preceding section) to be in **N**. Similarly,

$$\mathscr{P}(i) \text{ for some } i \in \mathscr{I}$$

means the same as

$$\mathscr{P}(i) \text{ for some } 1 \leq i \leq n$$

The set $\mathscr{I}$ is called an *index set*. Unions and intersections can also be written in terms of index sets:

$$\bigcup_{i=1}^{n} A_i = \bigcup_{i \in \mathscr{I}} A_i$$

$$\bigcap_{i=1}^{n} A_i = \bigcap_{i \in \mathscr{I}} A_i$$

Index sets can be quite general—not merely the set of integers from 1 to n. For example, the sets of real numbers, positive real numbers, and positive integers are frequently used as index sets.

Definition   Let $\mathscr{I}$ be any nonempty set whatever. For every $i \in \mathscr{I}$, suppose there is defined a set A_i. Define

$$\bigcup_{i \in \mathscr{I}} A_i = \{x \mid x \in A_i \text{ for some } i \in \mathscr{I}\}$$

$$\bigcap_{i \in \mathscr{I}} A_i = \{x \mid x \in A_i \text{ for all } i \in \mathscr{I}\}$$

$\{A_i \mid i \in \mathscr{I}\}$ is called an indexed family of sets.

If $\mathscr{I} = \mathbf{N}$, then $\bigcup_{i \in \mathscr{I}} A_i$ is written $\bigcup_{i=1}^{\infty} A_i$ and similarly for intersections.

The rules for using and proving **for some** statements can be generalized by using index sets, making them parallel to the **for all** rules we already have.

1.12 Index Sets

Inference Rule

*Using **for some** statements:* To use the statement $\mathscr{P}(j)$ *for some* $j \in \mathscr{I}$ in a proof, write immediately following this statement:

Pick $j_0 \in \mathscr{I}$ such that $\mathscr{P}(j_0)$.

This defines the symbol "j_0". The truth of $\mathscr{P}(j_0)$ and $j_0 \in \mathscr{I}$ may be used in the remainder of the proof.

Inference Rule

*Proving **for some** statements:* To prove the statement $\mathscr{P}(j)$ *for some* $j \in \mathscr{I}$, define j in your proof (in terms of previously defined symbols) and show that $\mathscr{P}(j)$ and $j \in \mathscr{I}$ hold for your j.

Inference Rules

Negation: The negation of $\mathscr{P}(i)$ *for all* $i \in \mathscr{I}$ is $(\sim \mathscr{P}(i))$ *for some* $i \in \mathscr{I}$.
The negation of $\mathscr{P}(i)$ *for some* $i \in \mathscr{I}$ is $(\sim \mathscr{P}(i))$ *for all* $i \in \mathscr{I}$.

In the next example, the wild-card Steps 1 and 2 are obtained by reading up from Step 3. That is, from Step 3 and its reason there is only one possibility for Step 2—which in turn determines Step 1 (but not uniquely).

Example 1:

1. *
2. * (1; pr. ∃)
3. $x \in \bigcup_{i=1}^{\infty} A_i$ (2; def.^{1} ∪)

Solution:

1. $x \in A_{17}$
2. $x \in A_i$ **for some** $i \in \mathbb{N}$ (1; pr. ∃)
3. $x \in \bigcup_{i=1}^{\infty} A_i$ (2; def.^{1} ∪)

Example 2:

1. Let $y = *$.
2. * (prop. **R**)
3. $|y - 2\frac{1}{4}| < \frac{1}{2}$ **for some** $y \in \{1, 2, 3\}$ (1, 2; pr. ∃)

Solution:

1. Let $y = 2$.
2. $|y - 2\frac{1}{4}| < \frac{1}{2}$ (prop. **R**)

3. $|y - 2\frac{1}{4}| < \frac{1}{2}$ **for some $y \in \{1, 2, 3\}$** (1, 2; pr. ∃)

Note: There are no other possibilities for Step 1.

Example 3:
For real numbers a and b, define $[a, b] = \{x \in \mathbb{R} \mid a \leq x \leq b\}$ and $(a, b) = \{x \in \mathbb{R} \mid a < x < b\}$. The sets $[a, b]$ and (a, b) are called, respectively, the closed and open intervals from a to b (for reasons given in Chapter 4). Suppose that for each $i \in \mathbb{N}$, a set A_i is defined below. Find $\bigcup_{i \in \mathbb{N}} A_i$ or $\bigcap_{i \in \mathbb{N}} A_i$ as indicated. Prove your assertions.

(a) If $A_i = [0, \frac{1}{i}]$, find $\bigcup_{i \in \mathbb{N}} A_i$.

(b) If $A_i = [0, \frac{1}{i}]$, find $\bigcap_{i \in \mathbb{N}} A_i$.

(c) If $A_i = (0, \frac{1}{i})$, find $\bigcup_{i \in \mathbb{N}} A_i$.

(d) If $A_i = (0, \frac{1}{i})$, find $\bigcap_{i \in \mathbb{N}} A_i$.

(e) If $A_i = (0, i)$, find $\bigcup_{i \in \mathbb{N}} A_i$.

(f) If $A_i = (0, i)$, find $\bigcap_{i \in \mathbb{N}} A_i$.

We will do parts (a) and (d) and leave the other parts as exercises.

Discussion of (a)

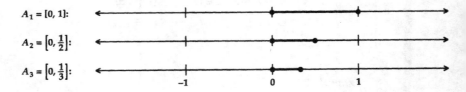

By sketching the intervals $[0, \frac{1}{i}]$ for a few $i = 1, 2, 3, \ldots,$ we see that

$$(1) \quad \bigcup_{i \in \mathbb{N}} A_i = [0, 1]$$

To prove this, we show containment in both directions:

1.12 Index Sets

Proof of (1):

First show $\bigcup_{i \in \mathbb{N}} A_i \subseteq [0, 1]$:

1. Let $x \in \bigcup_{i \in \mathbb{N}} A_i$.
2. $x \in A_i$ for some $i \in \mathbb{N}$ (1; def. $\cup$)
3. Pick $j \in \mathbb{N}$ such that $x \in A_j$.   (2; us. $\exists$)
4. $0 \leq x \leq \dfrac{1}{j}$ (3; def. A_j)
5. $\dfrac{1}{j} \leq 1$ (3; prop. $\mathbb{R}$)
6. $0 \leq x \leq 1$ (4, 5; prop. $\mathbb{R}$)
7. $x \in [0, 1]$ (6; def. $[0, 1]$)

Thus $\bigcup_{i \in \mathbb{N}} A_i \subseteq [0, 1]$. To show the reverse inclusion:

1. Let $x \in [0, 1]$.
2. $0 \leq x \leq 1$ (1; def. $[0, 1]$)
   .
   .
   .
k. $x \in A_i$ for some $i \in \mathbb{N}$ (; pr. $\exists$)
k+1. $x \in \bigcup_{i \in \mathbb{N}} A_i$ (k; def. $\cup$)

Thus $[0, 1] \subseteq \bigcup_{i \in \mathbb{N}} A_i$. Set equality follows from both inclusions. ■

In order to show Step k, we use the rule for proving a **for some** statement; that is, define i, show $i \in \mathbb{N}$ and show $x \in A_i$ for the i you have defined. This is easy: $i = 1$ will do fine. Of course, there is no need to use the symbol "i" for "1" because "1" itself will do. We need therefore to show $1 \in \mathbb{N}$ and $x \in A_1$. While it is necessary that the i you have defined be in the index set, this fact is never proved or even asserted except in cases where it is not clear. Since $1 \in \mathbb{N}$ is obvious, we don't even bother to write it down as a proof step. (In similar cases you need not bother either.)

The second condition, $x \in A_1$, needed to show Step k follows right from Step 1 by substitution since A_1 is by definition $[0, 1]$. Thus our tentative Step 2, $0 \leq x \leq 1$, is not needed, even though it would be the obvious step in most situations. We have:

1. Let $x \in [0, 1]$.
2. $x \in A_1$ (1; def. A_1)
3. $x \in A_i$ for some $i \in \mathbb{N}$ (2; pr. $\exists$)
4. $x \in \bigcup_{i \in \mathbb{N}} A_i$ (3; def. $\cup$)

.
.

(as before)

In the next proof we will need a property of $\mathbb{R}$ (given in Appendix 1) that is not as well known as the computational properties we have used in examples. Because of this, we cite it again here.

Property of R

Archimedean property: For any $r \in \mathbb{R}$, there exists $n \in \mathbb{N}$ such that $r < n$.

Discussion of (d)

The open intervals $(0, \frac{1}{j})$ are here sketched for $j = 1, 2, 3$:

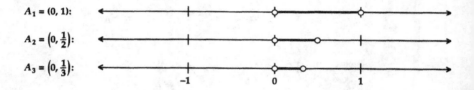

$A_1 = (0, 1)$:

$A_2 = \left(0, \frac{1}{2}\right)$:

$A_3 = \left(0, \frac{1}{3}\right)$:

If y is any real number greater than zero, then there is an integer j such that $\frac{1}{j}$ is smaller than y. (Pick j to be any integer greater than $\frac{1}{y}$ by the Archimedean property; since $j > \frac{1}{y}$, $y > \frac{1}{j}$.) Therefore y will not be in $(0, \frac{1}{j})$ for this j. We thus see that

$$(2) \quad \bigcap_{i \in \mathbb{N}} A_i = \emptyset \quad \text{for} \quad A_i = (0, \tfrac{1}{i})$$

Proving (2) requires that we show a set is empty. In order to show $X = \emptyset$ for some set X, we can show that X satisfies the property defining the specific set $\emptyset$; that is, show X has no elements.[†] The format we will always use

---

[†] If we attempt to show $X = \emptyset$ by getting containment each way, we have the following: $\emptyset \subseteq X$ is true, but formally attempting to prove $X \subseteq \emptyset$ by our usual procedure is inappropriate for reasons given on page 13. If we select an arbitrary element from X, we need to show that we get a contradiction. Doing this in the first place shows $X = \emptyset$ by the definition of $\emptyset$.

1.12 Index Sets

to show a set is equal to the empty set is to assume there is some element in the set and then obtain a contradiction.

Proof of (2):

1. Assume $x \in \bigcap_{i \in \mathbb{N}} A_i$ for some element x.
2. $x \in A_i$ for all $i \in \mathbb{N}$ (1; def. $\cap$)
3. $x \in A_1$ (2; us. $\forall$)
4. $0 < x$ (3; def. A_1)
5. **there exists** $n \in \mathbb{N}$ such that $n > \frac{1}{x}$ (Arch. prop.)
6. Pick $j \in \mathbb{N}$ such that $j > \frac{1}{x}$.   (5; us. $\exists$)
7. $x > \frac{1}{j}$ (6; prop. $\mathbb{R}$)
8. $x \notin A_j$ (7; def. A_j)
9. $x \in A_j \# 7$ (2, 6; us. $\forall$)
10. $\bigcap_{i \in \mathbb{N}} A_i = \emptyset$ (1—9; def. $\emptyset$)

∎

Theorem 1.12.1   Let A_i, $i \in \mathscr{I}$, be an indexed family of sets and let B be any set. Then

(a) $B - \bigcup_{i \in \mathscr{I}} A_i = \bigcap_{i \in \mathscr{I}} (B - A_i)$ *(DeMorgan's Laws)*

(b) $B - \bigcap_{i \in \mathscr{I}} A_i = \bigcup_{i \in \mathscr{I}} (B - A_i)$

Proof: Exercise 2.

There exists statements are extended naturally to involve more than one variable, exactly as **for all** statements are. For example, the negation of

$$\mathscr{P}(a, b, c) \text{ for all } a, b \in X \text{ and } c \in Y$$

is

$$(\sim \mathscr{P}(a, b, c)) \text{ for some } a, b \in X \text{ and some } c \in Y$$

To prove $\mathscr{P}(x, y)$ **for some** $x, y \in \mathscr{I}$, define both x and y, show that they are in $\mathscr{I}$, and then show that $\mathscr{P}(x, y)$ holds. We will consider our rules of inference as applying to several variables in such straightforward ways.

It is necessary to formally specify the negations of all **for all** and **there exists** statements we encounter. In this section we have considered **for all** statements of the form

$$\text{for all } i \text{ such that } i \in \mathscr{I}: \mathcal{Q}(i)$$

which have the negations

$$\text{there exists } i \in \mathcal{I}: \sim Q(i)$$

The most general **for all** statements we have considered, however, are of the form

$$\text{for all } i \text{ such that } \mathcal{P}(i): Q(i)$$

In order to extend our negation rules to such statements, we consider the following statements to be equivalent:

$$\text{for all } i \text{ such that } \mathcal{P}(i): Q(i)$$

$$\text{for all } i: \text{if } \mathcal{P}(i), \text{ then } Q(i)$$

$$\text{for all } i \in \mathfrak{U}: \text{if } \mathcal{P}(i), \text{ then } Q(i)$$

The negation of any one of these is any one of the following:

$$\text{there exists } i \in \mathfrak{U}: \sim(\text{if } \mathcal{P}(i), \text{ then } Q(i))$$

$$\text{there exists } i \in \mathfrak{U}: (\mathcal{P}(i) \text{ and } \sim Q(i))$$

$$\text{there exists } i \text{ such that } (\mathcal{P}(i) \text{ and } \sim Q(i))$$

The negation rules we already have apply to the most general type of **there exists** statements.

Additional Proof Ideas

From $x \in \bigcup_{i=1}^{n} A_i$ we can get $x \in A_1$ or $x \in A_2$ or $\cdots$ or $x \in A_n$ by our rule for proving **or**:

1. $x \in \bigcup_{i=1}^{n} A_i$
2. $x \in A_j$ for some $1 \leq j \leq n$ (1; def. ∪)
3. Pick $1 \leq k \leq n$ such that $x \in A_k$.   (2; us. ∃)
4. $x \in A_1$ or $x \in A_2$ or $\cdots$ or $x \in A_n$ (3; pr. or)

However, to write a formal proof that $x \in \bigcup_{i=1}^{n} A_i$ follows from $x \in A_1$ or $x \in A_2$ or $\cdots$ or $x \in A_n$ requires a "proof" by cases—where we couldn't write all the cases down. This is not allowed. Similarly, from $x \in A_1$ and $x \in A_2$ and $\cdots$ and $x \in A_n$ we can show $x \in \bigcap_{i=1}^{n} A_i$ by employing the rule for using **and**. However, to get $x \in A_1$ and $x \in A_2$ and $\cdots$ and $x \in A_n$ from $x \in \bigcap_{i=1}^{n} A_i$ requires the n separate steps $x \in A_1, x \in A_2, \ldots, x \in A_n$—which again could not all be written down.

1.12 Index Sets

Exercise 1.11.4 can be generalized by substituting a general $n \in \mathbb{N}$ for the 4. We will need the material in Chapter 3, however, in order to avoid problems like those above.

One might wish that from $x \in \bigcup_{i=1}^{\infty} A_i$ we could get the infinitely long statement $x \in A_1$ or $x \in A_2$ or $x \in A_3$ or $\cdots$. Such infinitely long statements are not allowed. Other ways (like our definition of $\bigcup_{i=1}^{\infty} A_i$) must be used to convey the idea of what we mean by this.

Exercises 1.12

Practice Exercise

1. (a) 1. *
   2. $x \in \bigcup_{i \in \mathscr{I}} A_i$ (1; def. $\cup$)

   (b) 1. $x \notin \bigcup_{i \in \mathscr{I}} A_i$
   2. *   (1; def. $\cup$)
   3. *   (2; neg.)

   (c) 1. $\sim(x \in A$ for all $A \subseteq B)$
   2. *   (1; neg.)

   (d) 1. *
   2. $\sim(x \notin A_i$ for some $i \in \mathscr{I})$ (1; neg.)

Straightforward Problems

2. Prove Theorem 1.12.1

3. Prove the following general complement laws. For a family of sets A_i, $i \in \mathscr{I}$:

   (a) $\overline{\bigcup_{i \in \mathscr{I}} A_i} = \bigcap_{i \in \mathscr{I}} \overline{A_i}$

   (b) $\overline{\bigcap_{i \in \mathscr{I}} A_i} = \bigcup_{i \in \mathscr{I}} \overline{A_i}$

4. Prove the following general distribution laws. For sets B, A_i, $i \in \mathscr{I}$:

   (a) $B \cap (\bigcup_{i \in \mathscr{I}} A_i) = \bigcup_{i \in \mathscr{I}} (B \cap A_i)$

   (b) $B \cup (\bigcap_{i \in \mathscr{I}} A_i) = \bigcap_{i \in \mathscr{I}} (B \cup A_i)$

5. Given sets $B, A_i, i \in \mathscr{I}$, prove that if $B \subseteq A_i$ for all $i \in \mathscr{I}$, then $B \subseteq \bigcap_{i \in \mathscr{I}} A_i$.

Chapter 1 Sets and Rules of Inference

6. Given sets $B, A_i, i \in \mathscr{I}$, prove that if $A_i \subseteq B$ for all $i \in \mathscr{I}$, then $\bigcup_{i \in \mathscr{I}} A_i \subseteq B$.

Harder Problem

7. Write out the proofs involved in Example 3 in this section.

Supplementary Problem

8. Recall the notation of Section 1.10, additional proof ideas. Show that **(there exists x such that for all y: $\mathscr{P}(x, y)$) $\Rightarrow$ (for all y: there exists x such that $\mathscr{P}(x, y)$)**.

Chapter 2

Functions

2.1 FUNCTIONS AND SETS

In the previous chapter we considered the idea of a set. Although this term was not formally defined, a particular set could be defined by our giving a rule (itself an undefined term) for deciding which elements are in the particular set and which are not. Another fundamental idea in mathematics is the idea of a function. Informally, a *function* is a rule of correspondence between two sets: a function f from a nonempty set A to a set B is a rule that associates to each element x of A a uniquely determined element, denoted $f(x)$, of B. $f(x)$ is called the *image* of x under f. You can think of f as "sending" or "mapping" x in A to $f(x)$ in B. Thus to know a particular function f from A to B, you must know a rule for getting $f(x)$ in B given any x in A. The set A is called the *domain* of f, and B is called the *codomain* of f. The fact that f is a function from A to B is written $f: A \to B$.

Our format for defining a specific function will be to give (1) the function name together with the domain and codomain, (2) the rule that specifies what the function does to each element in the domain, and (3) a "for all elements in the domain" clause. A formal, set-theoretic definition of function is given in Section 2.8.

Example 1:
Define
(a) $f: \mathbb{Z} \to \mathbb{N}$ by $f(x) = x^2 + 1$ for all $x \in \mathbb{Z}$
(b) $g: \mathbb{R} \to \mathbb{R}$ by $g(x) = 0$ for all $x \in \mathbb{R}$
(c) $h: \mathbb{N} \to \mathbb{N}$ by $h(x) = x + 1$ for all $x \in \mathbb{N}$

Chapter 2 Functions

(d) $j: \mathbb{R} \to \mathbb{Z}$ by $j(x) = \begin{cases} 1 \text{ if } x \text{ is irrational} \\ 0 \text{ if } x \text{ is rational} \end{cases}$

(e) $k: \mathbb{R} \to \mathbb{R}$ by $k(x) = \begin{cases} x \text{ for } x \geq 0 \\ -x \text{ for } x < 0 \end{cases}$

Note that the rules in (a), (b), and (c) for specifying the function are given by formulas but that the rules in (d) and (e) also specify functions. The **for all** quantification is optional since it is implied when the domain is specified. To use information in the definition of a function, use the (perhaps implicit) **for all** statement.

The rule that defines a function is sometimes given by listing the images of the elements in the domain.

Example 2:
Let $A = \{a, b, c, d\}$ and $B = \{1, 2, 3\}$. Define $f: A \to B$ by $f(a) = 1$, $f(b) = 3$, $f(c) = 3$, $f(d) = 1$. (Sometimes arrows are used to give the same information: $a \to 1, b \to 3, c \to 3, d \to 1$.)

Definition

Let $f: A \to B$. Suppose $X \subseteq A$, $Y \subseteq B$, and $y \in B$.
(a) $f(X) = \{b \in B \mid b = f(x) \text{ for some } x \in X\}$
(b) $f^{-1}(Y) = \{a \in A \mid f(a) \in Y\}$
(c) $f^{-1}(y) = \{a \in A \mid f(a) = y\}$

$f(A)$ is called the **range** of f, $f(X)$ is called the **image** of X under f, $f^{-1}(Y)$ is called the **preimage** of Y under f, and $f^{-1}(y)$ is called the **preimage** of y under f.

We note that $f(\emptyset) = \emptyset$ and $f^{-1}(\emptyset) = \emptyset$.

Example 3:
Let $f: \mathbb{R} \to \mathbb{R}$ be given by the rule $f(x) = 3x + 1$. Let $X = [1, 2]$ and $Y = [9, 10]$. Then

$$f(1) = 4$$
$$f(2) = 7$$
$$f(X) = [4, 7]$$

Proof:

We first show $f(X) \subseteq [4, 7]$:

1. Let $y \in f(X)$.
2. $y = f(x)$ for some $x \in X$ (1; def. $f(X)$)

2.1 Functions and Sets

> 3. Pick $z \in X$ such that $y = f(z)$. (2; us. $\exists$)
> 4. $1 \leq z \leq 2$ (3; def. X)
> 5. $4 \leq 3z + 1 \leq 7$ (4; prop. $\mathbb{R}$)
> 6. $4 \leq f(z) \leq 7$ (5; def. f)
> 7. $4 \leq y \leq 7$ (3, 6; sub.)
> 8. $y \in [4, 7]$ (7; def. $[4, 7]$)
> 9. $f(X) \subseteq [4, 7]$ (1—9; def. $\subseteq$)
>
> We show next that $[4, 7] \subseteq f(X)$:
>
> 1. Let $y \in [4, 7]$.
> 2. $4 \leq y \leq 7$ (1; def. $[4, 7]$)
> 3. $3 \leq y - 1 \leq 6$ (2; prop. $\mathbb{R}$)
> 4. $1 \leq \dfrac{y-1}{3} \leq 2$ (3; prop. $\mathbb{R}$)
> 5. Let $x = \dfrac{y-1}{3}$.
> 6. $x \in X$ (4, 5; def. X)
> 7. $f(x) = y$ (5, hyp.; def. x, f)
> 8. $y = f(t)$ for some $t \in X$ (6, 7; pr. $\exists$)
> 9. $y \in f(X)$ (8; def. $f(X)$)
> 10. $[4, 7] \subseteq f(X)$ (1—9; def. $\subseteq$)
>
> By both containments above, $[4, 7] = f(X)$. ∎

$f^{-1}(9) = \{\frac{8}{3}\}$ (since if $f(x) = 9$, then $3x + 1 = 9$ and so $x = \frac{8}{3}$)

$f^{-1}(10) = \{3\}$ (since if $f(x) = 10$, then $3x + 1 = 10$ and so $x = 3$)

$f^{-1}(Y) = [\frac{8}{3}, 3]$ (Exercise 3)

The range of f is $\mathbb{R}$ since for any $y \in \mathbb{R}$, there is some $x \in \mathbb{R}$ such that $3x + 1 = y$, namely, $x = \frac{y-1}{3}$.

Example 4:
Let $f: \mathbb{R} \to \mathbb{R}$ be given by $f(x) = x^2 + 1$.

$f(1) = 2$

$f(2) = 5$

$f([1, 2]) = [2, 5]$ (Exercise 4)

$f^{-1}(2) = \{-1, 1\}$ (since if $f(x) = 2$, then $x = \pm 1$)
$f^{-1}(5) = \{-2, 2\}$ (since if $f(x) = 5$, then $x = \pm 2$)
$f^{-1}([2, 5]) = [1, 2] \cup [-2, -1]$ (Exercise 5)
$f^{-1}(-1) = \emptyset$ (since $-1 = f(x) = x^2 + 1$ has no solution for x)

The range of f is $\{x \in \mathbf{R} \mid x \geq 1\}$.

Theorem 2.1.1 Let $f: A \to B$; $X, Y \subseteq A$; and $W, Z \subseteq B$.
 (a) If $x \in X$, then $f(x) \in f(X)$.
 (b) If $z \in Z$, then $f^{-1}(z) \subseteq f^{-1}(Z)$.
 (c) If $X \subseteq Y$, then $f(X) \subseteq f(Y)$.
 (d) If $W \subseteq Z$, then $f^{-1}(W) \subseteq f^{-1}(Z)$.
 (e) If $c, d \in B$, $c \neq d$, then $f^{-1}(c) \cap f^{-1}(d) = \emptyset$.

Proof of (b) and (c): Exercise 6.

Proof of (a):
 Assume: $f: A \to B$
  $X \subseteq A, x \in X$
 Show: $f(x) \in f(X)$

 1. $x \in X$ (hyp.)
 2. Let $y = f(x)$ (defining the new symbol "y")
 3. $y = f(x)$ for some $x \in X$ (1, 2; pr. $\exists$)
 4. $y \in f(X)$ (3; def. $f(X)$)
 5. $f(x) \in f(X)$ (2, 4; sub.)

∎

Proof of (d):
 Assume: $f: A \to B$
  $W, Z \subseteq B, W \subseteq Z$
 Show: $f^{-1}(W) \subseteq f^{-1}(Z)$

 1. Let $a \in f^{-1}(W)$.
 2. $f(a) \in W$ (1; def. $f^{-1}(W)$)
 3. $f(a) \in Z$ (2, hyp.; def. $\subseteq$)
 4. $a \in f^{-1}(Z)$ (3; def. $f^{-1}(Z)$)
 5. $f^{-1}(W) \subseteq f^{-1}(Z)$ (1—4; def. $\subseteq$)

∎

2.1 Functions and Sets

Proof of (e):

Assume: $f: A \to B;\ c, d \in B$
$c \neq d$

Show: $f^{-1}(c) \cap f^{-1}(d) = \emptyset$

1. Assume $a \in f^{-1}(c) \cap f^{-1}(d)$.
2. $a \in f^{-1}(c)$ (1; def. $\cap$)
3. $f(a) = c$ (2; def. $f^{-1}(c)$)
4. $a \in f^{-1}(d)$ (1; def. $\cap$)
5. $f(a) = d$ (4; def. $f^{-1}(d)$)
6. $c = d$ # hyp.   (3, 5; sub.)
7. $f^{-1}(c) \cap f^{-1}(d) = \emptyset$ (1—6; def. $\emptyset$)

In part (e) above, we used the idea that the image of the element a in the domain of f is uniquely determined; that is, because $f(a) = c$ and $f(a) = d$, we must have $c = d$.

If $f: A \to B$, then f^{-1}, as it has been defined, is not a function from B to A since, given $b \in B$, $f^{-1}(b)$ is a set and not a single (uniquely determined) element of B. The symbol "f^{-1}" is applied to subsets Y of B in order to denote their preimages $f^{-1}(Y)$ in A. Even for a single $y \in B$, however, $f^{-1}(y)$ may be empty or contain more than one element—as in Example 4. Later we will be able, under certain conditions, to define a function from B to A that people also call f^{-1}. This will explain the notation $f^{-1}(Y)$ for the subset of A.

Theorem 2.1.2   Let $f: A \to B$, $C \subseteq B$, and $D \subseteq B$.

(a) $f^{-1}(C \cap D) = f^{-1}(C) \cap f^{-1}(D)$
(b) $f^{-1}(C \cup D) = f^{-1}(C) \cup f^{-1}(D)$

Proof of (b): Exercise 7.

Proof of (a):

Assume: $C \subseteq B$
$D \subseteq B$

Show: $f^{-1}(C \cap D) = f^{-1}(C) \cap f^{-1}(D)$

To begin our efforts toward establishing the conclusion, we again consider the top-level form of the statement $f^{-1}(C \cap D) = f^{-1}(C) \cap f^{-1}(D)$. This expression asserts the equality of two sets and, by definition, is shown by getting set containment both ways:

First we show $f^{-1}(C \cap D) \subseteq f^{-1}(C) \cap f^{-1}(D)$:

1. Let $x \in f^{-1}(C \cap D)$.
   .
   .
   .
   k. $x \in f^{-1}(C) \cap f^{-1}(D)$
   k+1. $f^{-1}(C \cap D) \subseteq f^{-1}(C) \cap f^{-1}(D)$ (1—k; def. $\subseteq$)

Step 1 states that x is an arbitrarily chosen but henceforth fixed element about which we know nothing except that it is in the set $f^{-1}(C \cap D)$. And we know nothing about the set $f^{-1}(C \cap D)$ other than its defining property. (The definition here at the top level is seen to be the definition of preimage—abbreviated $f^{-1}(\ )$—not the definition of $\cap$). Thus Step 2 is inevitable:

1. Let $x \in f^{-1}(C \cap D)$.
2. $f(x) \in C \cap D$ (1; def. $f^{-1}(\ )$)

The word "inevitable" will become inappropriate as we proceed to more complicated proofs, since many options will be available. Among all these options, however, there is one that is always there: proceeding by definition and the rules of inference. So, rather than calling such steps "inevitable", we will use the word "straightforward".

Obtaining the steps leading to Step k (by working backward) is similarly "straightforward":[†]

k−2. $x \in f^{-1}(C)$
k−1. $x \in f^{-1}(D)$
k. $x \in f^{-1}(C) \cap f^{-1}(D)$ (k−2, k−1; def. $\cap$)

Continuing from Step 2 by a straightforward application of the definition of $\cap$, we get Steps 3 and 4:

2. $f(x) \in C \cap D$
3. $f(x) \in C$ (2; def. $\cap$)
4. $f(x) \in D$ (2; def. $\cap$)

[†] It is ironic that mathematicians customarily use the word "straightforward" to describe proofs that are discovered by working backward. Perhaps you could think of the word as being applied to the routine method of proof analysis. In a straightforward proof, one need only proceed through the algorithmic type of analysis to find a complete proof. Thus "forward" means forward through the "algorithm".

2.1 Functions and Sets

A look at the definition of $f^{-1}(\ )$ (do it!) shows that Step $k-2$ follows from Step 3 by definition. Similarly, $k-1$ follows from Step 4. Collecting our results so far and continuing, we get:

First we show $f^{-1}(C \cap D) \subseteq f^{-1}(C) \cap f^{-1}(D)$:

1. Let $x \in f^{-1}(C \cap D)$.
2. $f(x) \in C \cap D$ (1; def. $f^{-1}(\ )$)
3. $f(x) \in C$ (2; def. $\cap$)
4. $f(x) \in D$ (2; def. $\cap$)
5. $x \in f^{-1}(C)$ (3; def. $f^{-1}(\ )$)
6. $x \in f^{-1}(D)$ (4; def. $f^{-1}(\ )$)
7. $x \in f^{-1}(C) \cap f^{-1}(D)$ (5, 6; def. $\cap$)
8. $f^{-1}(C \cap D) \subseteq f^{-1}(C) \cap f^{-1}(D)$ (1—7; def. $\subseteq$)

Next we show $f^{-1}(C) \cap f^{-1}(D) \subseteq f^{-1}(C \cap D)$:

1. Let $x \in f^{-1}(C) \cap f^{-1}(D)$.
2. $x \in f^{-1}(C)$ (1; def. $\cap$)
3. $x \in f^{-1}(D)$ (1; def. $\cap$)

The way to proceed from here should be clear to you. Straightforward, right?

4. $f(x) \in C$ (2; def. $f^{-1}(\ )$)
5. $f(x) \in D$ (3; def. $f^{-1}(\ )$)
6. $f(x) \in C \cap D$ (4, 5; def. $\cap$)
7. $x \in f^{-1}(C \cap D)$ (6; def. $f^{-1}(\ )$)
8. $f^{-1}(C) \cap f^{-1}(D) \subseteq f^{-1}(C \cap D)$ (1—7; def. $\subseteq$)

Thus $f^{-1}(C \cap D) = f^{-1}(C) \cap f^{-1}(D)$, since we have shown containment both ways. ∎

In order to illustrate the relationship between our formal proof style and the paragraph form toward which we are headed, we rewrite the proof of Theorem 2.1.1a in paragraph form:

Proof:

Assume $C \subseteq B$ and $D \subseteq B$. We will show $f^{-1}(C \cap D) = f^{-1}(C) \cap f^{-1}(D)$. First we show $f^{-1}(C \cap D) \subseteq f^{-1}(C) \cap f^{-1}(D)$. Let $x \in f^{-1}(C \cap D)$.

> Then $f(x) \in C \cap D$ by definition of $f^{-1}(\ )$, so that by the definition of intersection $f(x) \in C$ and $f(x) \in D$. Hence, by the definition of $f^{-1}(\ )$, we have $x \in f^{-1}(C)$ and $x \in f^{-1}(D)$. Therefore $x \in f^{-1}(C) \cap f^{-1}(D)$. It follows that $f^{-1}(C \cap D) \subseteq f^{-1}(C) \cap f^{-1}(D)$.
>
> Next we show $f^{-1}(C) \cap f^{-1}(D) \subseteq f^{-1}(C \cap D)$. Let $x \in f^{-1}(C) \cap f^{-1}(D)$. Then $x \in f^{-1}(C)$ and $x \in f^{-1}(D)$, so that $f(x) \in C$ and $f(x) \in D$. Thus $f(x) \in C \cap D$ and $x \in f^{-1}(C \cap D)$, which shows $f^{-1}(C) \cap f^{-1}(D) \subseteq f^{-1}(C \cap D)$.
>
> Thus $f^{-1}(C \cap D) = f^{-1}(C) \cap f^{-1}(D)$, since we have shown containment both ways. ∎

At the end of this section, we will give a more compact proof of this theorem.

A glance at the first proof of Theorem 2.1.2a shows that, for the indented steps, we used as reasons only the definitions of preimage and intersection in a straightforward way. The content of Theorem 2.1.2 is that the operations of taking set intersection and taking preimages *commute*. That is, given two subsets C and D of the codomain of f, we can either take their intersection first and then take the preimage of this intersection, or take the preimages of sets C and D first and then form the intersection of these preimages. The resulting set is the same both ways. Briefly: $f^{-1}(\ )$ commutes with $\cap$. Theorem 2.1.2b (Exercise 7) shows that $f^{-1}(\ )$ commutes with $\cup$.

We could wonder whether the operation of taking images commutes with union and intersection. If it did, we would get the following statement, which we call a *conjecture*.

Conjecture 2.1.3 Let $f: A \to B$, $E \subseteq A$, $F \subseteq A$.
(a) $f(E \cap F) = f(E) \cap f(F)$
(b) $f(E \cup F) = f(E) \cup f(F)$

The proof of (b) is Exercise 8. We here attempt a proof of (a). Note that in the proof of Theorem 2.1.2, the hypotheses $C \subseteq B$ and $D \subseteq B$ were neither introduced as proof steps nor used as justifications. It is necessary to have $C \subseteq B$ and $D \subseteq B$ in order that the sets $f^{-1}(C)$, $f^{-1}(D)$, and $f^{-1}(C \cap D)$ be well defined (look at the definition). Thus the facts that f is a function from A to B and that $C \subseteq B$ and $D \subseteq B$ give us a *context* for our proof but don't provide specific justifications for obtaining new steps. These facts *could* be considered the hypotheses under which the definition can be made. Instead we will not bother to mention, as hypotheses, those pieces of information that are clearly useful only to establish a context where definitions apply. You may include all possible information in the hypotheses if the role of this information is not clear from the start. We will, however, list no hypotheses for Conjecture 2.1.3.

2.1 Functions and Sets

Attempted Proof of (a):

Show: $f(E \cap F) = f(E) \cap f(F)$

First we show $f(E \cap F) \subseteq f(E) \cap f(F)$:

1. Let $x \in f(E \cap F)$.

Of course, we immediately go to the definition of image—abbreviated $f(\ )$—to get Step 2. Step 1 takes x as an arbitrary element of the set $f(E \cap F)$, which has been defined by its property (rule for set membership). So Step 2 must be this property applied to x:

2. $x = f(z)$ for some $z \in E \cap F$ (1; def. $f(\ )$)

According to the rule for using **for some** statements, Step 3 follows. Steps 4 and 5 are also straightforward:

3. Pick $a \in E \cap F$ such that $x = f(a)$.   (2; us. $\exists$)
4. $a \in E$ (3; def. $\cap$)
5. $a \in F$ (3; def. $\cap$)

We have neglected to fill in the steps toward which we are heading. This is only because by now they should be clear in the back of our minds. They are:

$k-2$. $x \in f(E)$
$k-1$. $x \in f(F)$
k. $x \in f(E) \cap f(F)$ ($k-2, k-1$; def. $\cap$)
$k+1$. $f(E \cap F) \subseteq f(E) \cap f(F)$ (1—k; def. $\subseteq$)

From Step 3 we have $x = f(a)$, and from Step 4 $a \in E$. These two statements give us $x \in f(E)$ by definition. So we get Step $k-2$ and in a similar manner $k-1$. The proof so far is:

Attempted Proof of (a):

Show: $f(E \cap F) = f(E) \cap f(F)$

First we show $f(E \cap F) \subseteq f(E) \cap f(F)$:

1. Let $x \in f(E \cap F)$.

> 2. $x = f(z)$ for some $z \in E \cap F$ (1; def. $f(\ )$)
> 3. Pick $a \in E \cap F$ such that $x = f(a)$. (2; us. $\exists$)
> 4. $a \in E$ (3; def. $\cap$)
> 5. $a \in F$ (3; def. $\cap$)
> 6. $x \in f(E)$ (3, 4; def. $f(\ )$)
> 7. $x \in f(F)$ (3, 5; def. $f(\ )$)
> 8. $x \in f(E) \cap f(F)$ (6, 7; def. $\cap$)
> 9. $f(E \cap F) \subseteq f(E) \cap f(F)$ (1—8; def. $\subseteq$)

Now to the second half of the proof:

> Next we show $f(E) \cap f(F) \subseteq f(E \cap F)$:
> 1. Let $x \in f(E) \cap f(F)$.
> 2. $x \in f(E)$ (1; def. $\cap$)
> 3. $x \in f(F)$ (1; def. $\cap$)
>   .
>   .
> k. $x \in f(E \cap F)$
> k+1. $f(E) \cap f(F) \subseteq f(E \cap F)$ (1—k; def. $\subseteq$)

To get Step k, we need $x = f(a)$ for some $a \in E \cap F$ as Step $k-1$. From Step 2 we get Steps 4 and 5 in a straightforward manner:

> 4. $x = f(b)$ for some $b \in E$ (2; def. $f(\ )$)
> 5. Pick $b_0 \in E$ such that $x = f(b_0)$. (4; us. $\exists$)

and from Step 3 we get Steps 6 and 7 in a straightforward manner:

> 6. $x = f(c)$ for some $c \in F$ (3; def. $f(\ )$)
> 7. Pick $c_0 \in F$ such that $x = f(c_0)$. (6; us. $\exists$)

However, there is no reason the $b_0 \in E$ in Step 5 and the $c_0 \in F$ in Step 7 need be the same element of A. (If they were the same element, it would be in $E \cap F$—which would give us Step $k-1$.)

We suspect that conjecture 2.1.3a may not be true. To prove it is not true,

2.1 Functions and Sets

we need a *counterexample*, an example for which the hypotheses of the conjecture are seen to hold but the conclusion is seen to be false. The attempt at finding a straightforward proof of Conjecture 2.1.3a gives us a clue as to how to construct a counterexample: we need elements in sets E and F that f sends to x but no one element in both E and F that f sends to x.

Counterexample 5:
Let $f: \{1, 2, 3\} \to \{4, 5\}$ be defined by $f(1) = 4$, $f(2) = 5$, and $f(3) = 4$. (This defines the function f since we know the rule for determining where f sends every element of its domain $\{1, 2, 3\}$.) Let $E = \{1, 2\}$ and $F = \{2, 3\}$. Then $E \cap F = \{2\}$ so that $f(E \cap F) = \{5\}$. However, $f(E) = \{4, 5\}$ and $f(F) = \{4, 5\}$, so that $f(E) \cap f(F) = \{4, 5\}$. Therefore $f(E \cap F) \neq f(E) \cap f(F)$.

We can now rename, as a theorem, that part of the conjecture which is true:

Theorem 2.1.4 Let $f: A \to B$, $E \subseteq A$, and $F \subseteq A$.
(a) $f(E \cap F) \subseteq f(E) \cap f(F)$
(b) $f(E \cup F) = f(E) \cup f(F)$

Generally, the best way to proceed in trying to prove or find a counterexample for some conjecture is to first attempt a straightforward proof. If this fails, the attempt may show how to construct a counterexample. If attempts to construct counterexamples always fail, this may in turn exhibit why the proposition is true and lead to a proof. Thus you should alternately attempt to prove the conjecture and to find a counterexample, gaining insight into the problem as you proceed.

Definition *For any set X, the **power set** of X (denoted $\mathbb{P}(X)$) is the set of all subsets of X.*

Example 6
If $X = \{1, 2, 3\}$, $\mathbb{P}(X) = \{\emptyset, \{1\}, \{2\}, \{3\}, \{1, 2\}, \{1, 3\}, \{2, 3\}, X\}$.

If $f: A \to B$ is a function and $C \subseteq A$, then $f(C) \subseteq B$. Thus f can be thought of as taking subsets of A to subsets of B, that is, as a function from $\mathbb{P}(A)$ to $\mathbb{P}(B)$. More properly, we define the *induced function* $\bar{f}: \mathbb{P}(A) \to \mathbb{P}(B)$ by $\bar{f}(C) = f(C)$ for any $C \in \mathbb{P}(A)$. Similarly, we define the induced function $\tilde{f}: \mathbb{P}(B) \to \mathbb{P}(A)$ by $\tilde{f}(D) = f^{-1}(D)$ for any $D \in \mathbb{P}(B)$. Although f^{-1} need not be a function from B to A, it can be thought of as a function from $\mathbb{P}(B)$ to $\mathbb{P}(A)$—that is, the induced function $\tilde{f}$. A subset of A is a single element of $\mathbb{P}(A)$. $\tilde{f}$ is a function since, for any element D in its domain $\mathbb{P}(B)$, $\tilde{f}(D)$ is a uniquely determined single element of $\mathbb{P}(A)$—and similarly for $\bar{f}$.

Additional Proof Ideas

A compact proof of Theorem 2.1.2a can be written using the idea of an **iff** chain (Section 1.9, Additional Proof Ideas).

Theorem 2.1.2 Let $f: A \to B$, $C \subseteq B$, and $D \subseteq B$.

(a) $f^{-1}(C \cap D) = f^{-1}(C) \cap f^{-1}(D)$

> **Proof:**
>
> Assume $C \subseteq B$ and $D \subseteq B$. We will show $f^{-1}(C \cap D) = f^{-1}(C) \cap f^{-1}(D)$:
> Let $x \in A$ be arbitrary. Then
>
> $\qquad x \in f^{-1}(C \cap D)$
> $\qquad$ iff $f(x) \in C \cap D$ $\qquad$ (def. $f^{-1}(\ )$)
> $\qquad$ iff $f(x) \in C$ and $f(x) \in D$ $\qquad$ (def. $\cap$)
> $\qquad$ iff $x \in f^{-1}(C)$ and $x \in f^{-1}(D)$ $\qquad$ (def. $f^{-1}(\ )$)
> $\qquad$ iff $x \in f^{-1}(C) \cap f^{-1}(D)$. $\qquad$ (def. $\cap$)
> For all $x \in A$: $x \in f^{-1}(C \cap D)$
> $\qquad$ iff $x \in f^{-1}(C) \cap f^{-1}(D)$. $\qquad$ (iff chain)
> $f^{-1}(C \cap D) = f^{-1}(C) \cap f^{-1}(D)$ $\qquad$ (Thm. 1.9.3) ■

Substitution of equivalent statements (us. ↔) has been used implicitly in the steps above.

Exercises 2.1

Practice Exercises

1. Define a function $f: A \to B$ where
   (a) A has two elements and B has four.
   (b) B has two elements and A has four.
2. Let $f: \mathbb{R} \to \mathbb{R}$ be defined by $f(x) = x^2$. Find
   (a) $f([1, 2])$
   (b) $f(3)$
   (c) $f^{-1}(4)$
   (d) $f^{-1}([1, 2])$

2.2 Composition

(e) $f(f^{-1}([1, 2]))$

(f) $f^{-1}(f([1, 2]))$

Straightforward Problems

3. Prove $f^{-1}(Y) = [\frac{8}{3}, 3]$ for f and Y defined in Example 3.
4. Prove $f([1, 2]) = [2, 5]$ for the function f in Example 4.
5. Prove $f^{-1}([2, 5]) = [1, 2] \cup [-2, -1]$ for f in Example 4.
6. Prove Theorem 2.1.1b and c.
7. Give a formal proof of Theorem 2.1.2b. Rewrite your formal proof in paragraph form.
8. Give a formal proof of Theorem 2.1.4b. Rewrite your formal proof in paragraph form.
9. Prove or find a counterexample to the following propositions. In cases for which equality fails to hold, does set containment hold in one direction?

   Let $f: A \to B$, $C, D \subseteq B$, $E, F \subseteq A$.

   (a) $f^{-1}(C - D) = f^{-1}(C) - f^{-1}(D))$

   (b) $f(E - F) = f(E) - f(F)$

   (c) $f^{-1}(f(E)) = E$

   (d) $f(f^{-1}(C)) = C$

   Let $f: A \to B$, $A_i \subseteq A$ for $i \in \mathcal{I}$ and $B_j \subseteq B$ for $j \in \mathcal{J}$.

   (e) $f^{-1}(\bigcup_{j \in \mathcal{J}} B_j) = \bigcup_{j \in \mathcal{J}} f^{-1}(B_j)$

   (f) $f^{-1}(\bigcap_{j \in \mathcal{J}} B_j) = \bigcap_{j \in \mathcal{J}} f^{-1}(B_j)$

   (g) $f(\bigcup_{i \in \mathcal{I}} A_i) = \bigcup_{i \in \mathcal{I}} f(A_i)$

   (h) $f(\bigcap_{i \in \mathcal{I}} A_i) = \bigcap_{i \in \mathcal{I}} f(A_i)$

Supplementary Problem

10. Recall Exercise 1.12.8. Give a counterexample to: **(for all y: there exists x such that $\mathcal{P}(x, y)$) $\Rightarrow$ (there exists x such that for all y: $\mathcal{P}(x, y)$)**.

2.2 COMPOSITION

Definition Let $f: A \to B$, $g: B \to C$. Define $g \circ f: A \to C$ by $g \circ f(a) = g(f(a))$ for all $a \in A$.

$g \circ f$ is a new function, called the *composition* of g with f, that has the effect of first applying f to an element in A and then applying g to the result.

Chapter 2 Functions

Note that for this to make sense, the range of f must be contained in the domain of g.

Example 1:
Let $f: \mathbf{R} \to \mathbf{R}$ be given by $f(x) = x - 1$ and $g: \mathbf{R} \to \mathbf{R}$ be given by $g(y) = y^2 + 2$. Then $g \circ f: \mathbf{R} \to \mathbf{R}$ is given by $g \circ f(x) = g(f(x)) = g(x-1) = (x-1)^2 + 2$. Also, $f \circ g: \mathbf{R} \to \mathbf{R}$ is given by $f \circ g(y) = f(g(y)) = f(y^2 + 2) = (y^2 + 2) - 1 = y^2 + 1$.

In this example we used the variable x describing f to also describe $g \circ f$, since in $g \circ f$ we first apply f. Similarly, y describes both g and $f \circ g$. Although this was done to illustrate the way in which functions are composed, it is important to understand functions as rules. Composition of functions should therefore be viewed as a rule and not merely as the substitution of variables. The variables are only local variables needed to describe the rules. In the following example, doing without the aid of using different variables to describe f and g forces us to think of the functions as rules.

Example 2:
Let $f: \mathbf{R} \to \mathbf{R}$ be given by $f(x) = x^2 - 5$ for all $x \in \mathbf{R}$ and $g: \mathbf{R} \to \mathbf{R}$ be given by $g(x) = 3x + 2$ for all $x \in \mathbf{R}$.

(a) $g \circ f(x) = *$
(b) $f \circ g(x) = *$

Solution:

(a) $g \circ f(x) = 3(x^2 - 5) + 2$ for all $x \in \mathbf{R}$
(b) $f \circ g(x) = (3x + 2)^2 - 5$ for all $x \in \mathbf{R}$

Example 3:
Let $f: \{1, 2, 3, 4\} \to \{a, b, c\}$ be given by $1 \to a, 2 \to b, 3 \to b, 4 \to c$. Let $g: \{a, b, c\} \to \{x, y, z\}$ be defined by $a \to x, b \to y, c \to z$. Then $g \circ f: \{1, 2, 3, 4\} \to \{x, y, z\}$ is defined by $1 \to x, 2 \to y, 3 \to y, 4 \to z$. $f \circ g$ is not defined since the domain of f is not the codomain of g.[†]

Our next theorem asserts the associativity of composition, but first we need the idea of equal functions.

[†] In defining $g \circ f$, we have used the usual definition of composition, where the domain of g is equal to the codomain of f. This definition will keep notation simple in future theorems, with no real loss of generality. Alternate definitions sometimes require only that the range of f be a subset of the domain of g. The codomain of any such f can easily be redefined to be the domain of g.

2.2 Composition

Definition *Two functions $f: A \to B$ and $g: A \to B$ are said to be **equal** (written $f = g$) provided that for all $x \in A$: $f(x) = g(x)$.*

Note that for f and g to be equal they must have the same domain and codomain (the context for definition). The definition just given states that functions are equal if the rules defining them yield the same value when applied to each element of their domain. The idea of equality asserts that the expressions on the left and right of the equal sign are just two names for exactly the same object. The reason we need definitions for equal sets and equal functions is that the ideas of set and function are themselves undefined. Therefore "sameness" needs to be defined in these cases.

Theorem 2.2.1 Let $f: A \to B$, $g: B \to C$, and $h: C \to D$. Then $(h \circ g) \circ f = h \circ (g \circ f)$.

> **Proof:**
>
> Show: $(h \circ g) \circ f = h \circ (g \circ f)$

Note that at the top level this statement asserts that two functions are equal. We therefore translate the conclusion using the definition of equality:

> **Scrap Paper**
>
> Show: $(h \circ g) \circ f = h \circ (g \circ f)$
>
> That is, show: 1. $(h \circ g) \circ f$ and $h \circ (g \circ f)$ have the same domain and codomain.
> 2. $[(h \circ g) \circ f](x) = [(h \circ (g \circ f)](x)$ for all $x \in A$

> **Proof:**
>
> Show: $(h \circ g) \circ f = h \circ (g \circ f)$
>
> First observe that $h \circ g: B \to D$ so that $(h \circ g) \circ f: A \to D$. Also, $g \circ f: A \to C$, so that $h \circ (g \circ f): A \to D$. Therefore $(h \circ g) \circ f$ and $h \circ (g \circ f)$ both have domain A and codomain D by definition of composition.
>
> We next show that $[(h \circ g) \circ f](x) = [h \circ (g \circ f)](x)$ for all $x \in A$:
>
> 1. Let $x \in A$.
>   •
>   •

> k. $(h \circ g) \circ f(x) = h \circ (g \circ f)(x)$
>
> k+1. $(h \circ g) \circ f(x) = h \circ (g \circ f)(x)$ for all $x \in A$ $\quad$ (1—k; pr. $\forall$)
>
> Therefore $(h \circ g) \circ f = h \circ (g \circ f)$ by definition of equal functions.

Step k states that $(h \circ g) \circ f$ and $h \circ (g \circ f)$ do exactly the same thing to x. What these functions do to x is given by their definition. By definition, $[(h \circ g) \circ f](x)$ is $(h \circ g)(f(x))$ and $[h \circ (g \circ f)](x)$ is $h(g \circ f(x))$. Applying the definition again, $(h \circ g)(f(x))$ is $h(g(f(x)))$ and $h(g \circ f(x))$ is $h(g(f(x)))$. The left and right sides of Step k are therefore the same. In order to establish Step k, we therefore start with this same thing as a step in our proof:

> 2. $h(g(f(x))) = h(g(f(x)))$

Such steps (obvious identities) need no justification in parentheses. We now have the following proof:

> **Proof:**
>
> Show: $(h \circ g) \circ f = h \circ (g \circ f)$
>
> First observe that $h \circ g: B \to D$ so that $(h \circ g) \circ f: A \to D$. Also, $g \circ f: A \to C$ so that $h \circ (g \circ f): A \to D$. Therefore $(h \circ g) \circ f$ and $h \circ (g \circ f)$ both have domain A and codomain D by definition of composition.
>
> We next show that $[(h \circ g) \circ f](x) = [h \circ (g \circ f)](x)$ for all $x \in A$:
>
> 1. Let $x \in A$.
> 2. $h(g(f(x))) = h(g(f(x)))$
> 3. $h \circ g(f(x)) = h(g \circ f(x))$ $\quad$ (2; def. $\circ$)
> 4. $(h \circ g) \circ f(x) = h \circ (g \circ f)(x)$ $\quad$ (3; def. $\circ$)
> 5. $(h \circ g) \circ f(x) = h \circ (g \circ f)(x)$ for all x $\quad$ (1—4; pr. $\forall$)
> 6. $(h \circ g) \circ f = h \circ (g \circ f)$ $\quad$ (5; def. = fcns.)
>
> Therefore $(h \circ g) \circ f = h \circ (g \circ f)$ by definition of equal functions. ∎

The fact that the domain and codomain of $h \circ (g \circ f)$ and $(h \circ g) \circ f$ are the same is needed for equality. It is the context in which the definition is made. We "observed" this fact in the first few lines of our proof. In general, we will use the word "observe" in asserting, in a proof, the appropriate context for a theorem or definition. It is customary to omit such observations in proofs if they are obvious.

2.2 Composition

Steps like Step 2 above, which appear in proofs seemingly out of a clear blue sky, are almost always determined by thinking backward from a desired result (scrap-paper analysis). They seem mysterious only to those who imagine steps are discovered in the same order in which they appear in the proof. People who memorize proofs (a wholly worthless activity) may memorize steps in this order. People who *think about* proofs never think in this order. This is why, when reading a mathematics text, it is not informative to merely see why each step follows logically from the preceding steps. Instead, try analyzing the proofs yourself and use the text only if you get stuck. Such a do-it-yourself approach will reveal not only that the theorems are true (false theorems are rarely printed in texts) but *why* they are true.

The form of the preceding proof, with Step 2 appearing out of the blue and with different but simultaneous manipulations of each side of the equations, is unnatural both to do and to read. A more natural proof style involves a *chain* of equalities: $a_1 = a_2 = a_3 = \cdots = a_n$, usually written vertically in proofs:

$$a_1$$
$$= a_2 \quad \text{(reason } a_1 = a_2\text{)}$$
$$= a_3 \quad \text{(reason } a_2 = a_3\text{)}$$
$$\cdot$$
$$\cdot$$
$$= a_n \quad \text{(reason } a_{n-1} = a_n\text{)}$$

The steps in the preceding proof would be:

> Let $x \in A$.
> $(h \circ g) \circ f(x)$
> $= (h \circ g)(f(x))$ (def. $\circ$)
> $= h(g(f(x)))$ (def. $\circ$)
> $= h((g \circ f)(x))$ (def. $\circ$)
> $= h \circ (g \circ f)(x)$ (def. $\circ$)
> $(h \circ g) \circ f = h \circ (g \circ f)$ (def.^{2} = fcns.)

Definition   For any set A, define the function $i_A: A \to A$ by $i_A(a) = a$ for each $a \in A$. i_A is called the **identity function** on A.

Theorem 2.2.2   For any $f: A \to B$

(a) $f \circ i_A = f$

(b) $i_B \circ f = f$

Proof of (a):

Show: $f \circ i_A = f$

Observe first that $f \circ i_A$ and f both have domain A and codomain B.

1. Let $x \in A$.
2. $f \circ i_A(x) = f(i_A(x))$ (def. $\circ$)
3. $i_A(x) = x$ (def. i_A)
4. $f \circ i_A(x) = f(x)$ (2, 3; sub.)
5. $f \circ i_A(x) = f(x)$ for all $x \in A$ (1—4; pr. $\forall$)
6. $f \circ i_A = f$ (5; def. = fcns.)

Proof of (b): Exercise 3.

In the proof of Theorem 2.2.2 we used the substitution rule of inference. The definition of composition asserts that the element to which $f \circ i_A$ maps x is the element $f(i_A(x))$. Thus $f \circ i_A(x)$ and $f(i_A(x))$ are the same thing by definition. In Step 2 the equal sign denotes that we have two different names or representations for the same thing, and the same is true in Step 3. Step 4 was obtained by replacing $i_A(x)$ with x in Step 2, these things being equal by Step 3. It is better to use substitution implicitly. The following steps do this for Theorem 2.2.2a:

1. Let $x \in A$.
2. $f \circ i_A(x) = f(i_A(x))$ (def. $\circ$)
3. $f \circ i_A(x) = f(x)$ (2; def. i_A)
4. $f \circ i_A(x) = f(x)$ for all $x \in A$ (1—3; pr. $\forall$)

In these steps the definition of i_A was used as a reason for changing Step 2 to Step 3. In doing this, substitution need not be stated explicitly.

In the first proof of Theorem 2.2.2, information from the appropriate definitions was put down first. (Note that in this proof Steps 2 and 3 do not depend on previous steps.) Then this information was organized in Step 4. It is generally better to organize your thoughts on scrap paper, analyzing and changing steps by definition, than to put the contents of definitions down as steps in a proof and organize things in later proof steps. A proof step with the following justification would be indicative of a poorly organized proof that was difficult to read (see also Exercise 9):

22.   (Steps 2, 4, 7, 18, 21; sub.)

The most natural proof of Theorem 2.2.2, and the easiest to read, involves a chain of equalities:

> Let $x \in A$.
> $f \circ i_A(x)$
> $= f(i_A(x))$ (def. $\circ$)
> $= f(x)$ (def. i_A)
> $f \circ i_A = f$ (def.^{2} = fcns.)

Example 4:
Let $f: \mathbb{R} \to \mathbb{R}$ and $g: \mathbb{R} \to \mathbb{R}$ be functions. Define the function $f+g: \mathbb{R} \to \mathbb{R}$ by the rule $f+g(x) = f(x) + g(x)$ for all $x \in \mathbb{R}$. Prove or find a counterexample to:

(a) For all functions $h: \mathbb{R} \to \mathbb{R}$: $(f+g) \circ h = (f \circ h) + (g \circ h)$.
(b) For all functions $h: \mathbb{R} \to \mathbb{R}$: $h \circ (f+g) = (h \circ f) + (h \circ g)$.

> **Proof of (a):**
> Let $x \in \mathbb{R}$ be arbitrary.
> $(f+g) \circ h(x)$
> $= (f+g)(h(x))$ (def. $\circ$)
> $= f(h(x)) + g(h(x))$ (def. + of fcns.)
> $= f \circ h(x) + g \circ h(x)$ (def. $\circ$)
> $= [f \circ h + g \circ h](x)$ (def. + of fcns.)
> Therefore $(f+g) \circ h = f \circ h + g \circ h$.   (def. = fcns.)

Counterexample to (b):
Let $h(x) = x^2$, $f(x) = x$, and $g(x) = x$. Then $h \circ (f+g)(x) = h(f+g(x)) = h(f(x) + g(x)) = h(x+x) = h(2x) = (2x)^2 = 4x^2$. But $h \circ f + h \circ g(x) = h \circ f(x) + h \circ g(x) = h(f(x)) + h(g(x)) = h(x) + h(x) = x^2 + x^2 = 2x^2$.

Exercises 2.2

Practice Exercises

1. Let $h: \mathbb{R} \to \mathbb{R}$ be defined by $h(z) = z^3 + z$ and $k: \mathbb{R} \to \mathbb{R}$ by $k(z) = z^2 - 2$. Define:
   (a) $h \circ k$
   (b) $k \circ h$

2. Let $f: \{1, 2, 3, 4, 5\} \to \{a, b, c, d, e\}$ be defined by $1 \to a, 2 \to a, 3 \to b, 4 \to b, 5 \to c$. Let $g: \{a, b, c, d, e\} \to \{1, 2, 3, 4, 5\}$ be defined by $a \to 5, b \to 4, c \to 4, d \to 3, e \to 2$. Define:
   (a) $f \circ g$
   (b) $g \circ f$

Straightforward Problems

3. Prove Theorem 2.2.2b.

4. Prove or find a counterexample to the following "cancellation laws" for function composition:
   (a) Let $f: A \to B, g: A \to B$, and $h: B \to C$. If $h \circ f = h \circ g$, then $f = g$.
   (b) Let $f: A \to B, g: A \to B$, and $h: C \to A$. If $f \circ h = g \circ h$, then $f = g$.

5. Let $f: \mathbf{R} \to \mathbf{R}$ and $g: \mathbf{R} \to \mathbf{R}$ be functions. Define the function $f - g: \mathbf{R} \to \mathbf{R}$ by the rule $f - g(x) = f(x) - g(x)$ for all $x \in \mathbf{R}$. Prove or find a counterexample to:
   (a) For all functions $h: \mathbf{R} \to \mathbf{R}$: $(f - g) \circ h = (f \circ h) - (g \circ h)$.
   (b) For all functions $h: \mathbf{R} \to \mathbf{R}$: $h \circ (f - g) = (h \circ f) - (h \circ g)$.

6. Let $f: \mathbf{R} \to \mathbf{R}$ and $g: \mathbf{R} \to \mathbf{R}$ be functions. Define the function $f \cdot g: \mathbf{R} \to \mathbf{R}$ by the rule $f \cdot g(x) = f(x) \cdot g(x)$ for all $x \in \mathbf{R}$, where the dot on the right side of the equation denotes multiplication of real numbers. Prove or find a counterexample to:
   (a) For all functions $h: \mathbf{R} \to \mathbf{R}$: $(f \cdot g) \circ h = (f \circ h) \cdot (g \circ h)$.
   (b) For all functions $h: \mathbf{R} \to \mathbf{R}$: $h \circ (f \cdot g) = (h \circ f) \cdot (h \circ g)$.

7. Let $f: \mathbf{R} \to \mathbf{R}$ be a function and $c \in \mathbf{R}$. Define the function $cf: \mathbf{R} \to \mathbf{R}$ by the rule $cf(x) = c \cdot f(x)$, where the $\cdot$ denotes multiplication of real numbers. Prove or find a counterexample to:
   (a) For all $c \in \mathbf{R}$ and all functions $f, h: \mathbf{R} \to \mathbf{R}$: $(cf) \circ h = c(f \circ h)$.
   (b) For all $c \in \mathbf{R}$ and all functions $f, h: \mathbf{R} \to \mathbf{R}$: $h \circ (cf) = c(h \circ f)$.

8. Let $c \in \mathbf{R}$ be fixed. Define the function $c: \mathbf{R} \to \mathbf{R}$ by $c(x) = c$ for all $x \in \mathbf{R}$. (c is a constant function.) Prove or find a counterexample to:
   (a) For all functions $h: \mathbf{R} \to \mathbf{R}$: $c \circ h = c$.
   (b) For all functions $h: \mathbf{R} \to \mathbf{R}$: $h \circ c = h(c)$.

Supplementary Problem

9. Comment on the following universal proof scheme. Suppose we are given a theorem $\mathscr{P}$. To prove $\mathscr{P}$, write down all the definitions (as steps) of the terms in $\mathscr{P}$ plus the definitions of the terms in those definitions, and so on until only undefined terms (such as set and function) remain. Call these definitions Steps 1 through k. For Step $k+1$, write down $\mathscr{P}$ and give "substitution" as a reason. (Regardless of what your opinion may be as to the validity of this, you should avoid making your proofs look like this.)

2.3 ONE-TO-ONE FUNCTIONS

In Counterexample 2.1.5 we considered a function $f: \{1, 2, 3\} \to \{4, 5\}$ where $f(1) = 4$ and $f(3) = 4$. Such a function is called *many-to-one* since there is an element, 4, in the range of f with at least two different elements mapping to it. Functions with only one element in their domain mapping to each element in the range are called *one-to-one*. We seek a wording for a definition. This wording should be in terms of our standard phrases: **for all; if . . . , then . . . ; and; or;** and so on. Think for a minute of what you could give for a condition on a function $f: A \to B$ that would ensure that f was one-to-one. Here is how we will do it:

Definition   *A function $f: A \to B$ is called **one-to-one** provided that for all $a_1, a_2 \in A$: if $f(a_1) = f(a_2)$, then $a_1 = a_2$.*

The idea in the definition is that we pick two different names, a_1 and a_2, for objects in A. The condition $f(a_1) = f(a_2)$ states that f sends the object named by a_1 to the same place it sends the object named by a_2. Under these conditions, if f is to be a one-to-one function, it must be the case that a_1 and a_2 are two different names for the same object. Hence $a_1 = a_2$. "One-to-one" is abbreviated "1-1".

One-to-one functions have the property that, for each element in their range, there is a *unique* element in their domain mapping to it. The approach above is generally used to prove uniqueness: pick two different names for an object or objects with a property, then show both names are names for the same object. There is therefore only one object with the property.

Many mathematical statements are of the form **there exists a unique x such that $\mathcal{P}(x)$**, where $\mathcal{P}(x)$ is some property that x might have. Proofs of such statements involve two distinct parts—the existence part and the uniqueness part. The existence part is done according to our rule for proving **there exists** statements. The uniqueness part is done according to the scheme mentioned above, which we now formalize.

Inference Rule   *Uniqueness: To show that x with property $\mathcal{P}(x)$ is **unique**, assume x_1 has property $\mathcal{P}$ and x_2 has property $\mathcal{P}$, then show $x_1 = x_2$. To use the fact that x with property $\mathcal{P}(x)$ is unique, we may infer $x = y$ from $\mathcal{P}(x)$ and $\mathcal{P}(y)$.*

The uniqueness rules are abbreviated "pr. !" and "us. !", and the words "there exists a unique" are abbreviated "∃!". We will combine the rules for existence and uniqueness in proving the next theorem. The wording used in a format for proving uniqueness depends on whether or not existence has been previously established.

Format for proving uniqueness:

> i. Let x_1 and x_2 have property $\mathcal{P}$.
> .
> .
> j. $x_1 = x_2$
> j+1. There exists a unique x such that $\mathcal{P}$. (if existence has already been shown)
>
> or j+1. There is at most one x such that $\mathcal{P}$. (if existence has not already been shown)

Theorem 2.3.1 Let $f: A \to B$. Then f is one-to-one iff for each b in the range of f there exists a unique $a \in A$ such that $f(a) = b$.

Proof:

First assume f is one-to-one and show **for all $b \in f(A)$: there exists a unique $a \in A$ such that $f(a) = b$:**

1. Let $b \in f(A)$.

Existence:

2. **there exists $a \in A$ such that $b = f(a)$** (1; def. $f(A)$)

Uniqueness:

3. Let $a_1, a_2 \in A$ have the property that $f(a_1) = b$ and $f(a_2) = b$.
4. $f(a_1) = f(a_2)$ (3; sub.)
5. $a_1 = a_2$ (4, hyp.; def.² 1-1)
6. **there exists a unique $a \in A$ such that $b = f(a)$** (2—6; pr. ∃!)
7. **for all $b \in f(A)$: there exists a unique $a \in A$ such that $b = f(a)$** (1—6; pr. ∀)

Next assume **for all $b \in f(A)$: there exists a unique $a \in A$ such that $f(a) = b$** and show that f is one-to-one:

1. Let $a_1, a_2 \in A$.
2. Assume $f(a_1) = f(a_2)$.
3. Let $b = f(a_1)$.
4. $b = f(a_2)$ (2, 3; sub.)
5. $a_1 = a_2$ (1, 3, 4; us. !)

2.3 One-to-One Functions

> 6. if $f(a_1) = f(a_2)$, then $a_1 = a_2$ (2—5; pr. ⇒)
> 7. for all $a_1, a_2 \in A$: if $f(a_1) = f(a_2)$, then $a_1 = a_2$ (1—6; pr. ∀)
> 8. f is one-to-one.   (7; def. 1-1) ∎

Suppose we wish to prove that a function $f: A \to B$ is one-to-one. Our inference rules dictate the following:

1. Let $a_1, a_2 \in A$.
2. Assume $f(a_1) = f(a_2)$.
   .
   .
   .
k. $a_1 = a_2$
k+1.   if $f(a_1) = f(a_2)$, then $a_1 = a_2$ (2—k; pr. ⇒)
k+2.   for all $a_1, a_2 \in A$: if $f(a_1) = f(a_2)$, then $a_1 = a_2$ (1—k+1; pr. ∀)
k+3. f is one-to-one.   (k+2; def. 1-1)

By our extended definition rule, we may omit Step $k+2$ since the property $\mathscr{P}$, which establishes that f is one-to-one, is just that given in Step $k+2$. Our extended definition rule does not completely remove the strictly logical assertions from the proof, however, since Step $k+1$ serves to express a *part* of the defining condition $\mathscr{P}$. It seems inappropriate that we would need to state a part of $\mathscr{P}$ but not $\mathscr{P}$ itself.

Our next step in proof abbreviation involves combining Steps 1 and 2 above and eliminating Step $k+1$. That is, we give a single rule for proving statements of the form **for all** $x, y \in A$: **if** $\mathscr{P}(x, y)$, **then** $\mathcal{Q}(x, y)$. Note that in the proof fragment above, we prove Step $k+2$ by first choosing arbitrary a_1 and a_2, then assuming $f(a_1) = f(a_2)$ for these, and finally proving $a_1 = a_2$.

Inference Rule

Proving **for all**: **if–then** *statements: In order to prove a statement of the form* **for all** $x \in A$: **if** $\mathscr{P}(x)$, **then** $\mathcal{Q}(x)$, *choose an arbitrary x in A and assume $\mathscr{P}(x)$ is true for this x. (Either $x \in A$ or $\mathscr{P}(x)$ may then be used in future steps.) Then prove that $\mathcal{Q}(x)$ is true. Analogous rules hold for more than one variable. (We use "pr. ∀ ⇒" to abbreviate this rule.)*

Format:

     i.   Let $x \in A$ and $\mathscr{P}(x)$
        (or "Let $x \in A$ and assume $\mathscr{P}(x)$"
        or "Suppose $x \in A$ and $\mathscr{P}(x)$"

or "Let $\mathcal{P}(x)$ for $x \in A$".[†]

  .
  .
  .

j. $\mathcal{Q}(x)$
j+1. **for all $x \in A$: if $\mathcal{P}(x)$, then $\mathcal{Q}(x)$** $\qquad (i\mathrm{-}j; \mathrm{pr.}\ \forall \Rightarrow)$

The extension of the preceding rule to two variables is used in the following example:

Example 1:
Let $f: \mathbb{R} \to \mathbb{R}$ be defined by $f(x) = 2x + 4$ for all $x \in \mathbb{R}$. Prove that f is one-to-one.

Proof:
1. Let $x, y \in \mathbb{R}$ and $f(x) = f(y)$.
2. $2x + 4 = 2y + 4$ $\qquad$ (1; def. f)
3. $2x = 2y$ $\qquad$ (2; prop. $\mathbb{R}$)
4. $x = y$ $\qquad$ (3; prop. $\mathbb{R}$)
5. **for all $x, y \in \mathbb{R}$: if $f(x) = f(y)$, then $x = y$** $\qquad$ (1—4; pr. $\forall \Rightarrow$)
6. f is one-to-one. $\qquad$ (5; def.[1] 1-1)

Step 5 could be omitted by using the extended definition of one-to-one in Step 6.

Example 2:
The proof of Example 1 in paragraph form might be:

Proof:
Let $x, y \in \mathbb{R}$ and $f(x) = f(y)$. Then $2x + 4 = 2y + 4$ by definition of f. Hence $x = y$ so that f is one-to-one by definition.

[†] You will see many other wordings that mean the same thing. It is not the words that count. Readers who know the conclusion you are after will automatically interpret any reasonable words so that their meaning is consistent with obtaining this conclusion. This is the way it is with informal language; the ideas carry us through what would otherwise be ambiguous wordings. Words and phrases are interpreted in context.

2.3 One-to-One Functions

Theorem 2.3.2 Let $f: A \to B$ and $g: B \to C$ be one-to-one functions. Then $g \circ f$ is one-to-one.

Proof: Exercise 5.

Conjecture 2.3.3 Let $f: A \to B$ and $g: B \to C$ be functions.
(a) If $g \circ f$ is one-to-one, then f is one-to-one.
(b) If $g \circ f$ is one-to-one, then g is one-to-one.

Attempted Proof of (a):
Assume: $g \circ f$ is 1-1
Show: f is 1-1

Scrap Paper
Show: f is 1-1
Show: for all $a_1, a_2 \in A$: if $f(a_1) = f(a_2)$, then $a_1 = a_2$ by def. 1-1

1. Let $a_1, a_2 \in A$ and $f(a_1) = f(a_2)$
   .
   .
k. $a_1 = a_2$
k+1. f is one-to-one.   (1—k; def.^{2} 1-1)

Further analysis at this time yields nothing: if we ask what it means for $a_1 = a_2$, we learn nothing. It means only that a_1 and a_2 name the same thing. There is no way to break this down further by definition. So, as usual, it is time to invoke the hypothesis. We are starting to get away from proofs that follow immediately from definitions. Generally, we need to be a little bit clever in the way we apply the hypothesis to the problem at hand. Here, of course, we need not be too clever. We know $f(a_1) = f(a_2)$ and that this element is in B—the domain of g. So we apply g, which as a function can map this element to only one element in C; that is, $g(f(a_1)) = g(f(a_2))$. To justify this step, we say that g is *well defined*, indicating that, as a good function, it maps an element in its domain to a *single* element in its codomain.

1. Let $a_1, a_2 \in A$ and $f(a_1) = f(a_2)$.
2. $g(f(a_1)) = g(f(a_2))$ (1; g well def.)
3. $(g \circ f)(a_1) = (g \circ f)(a_2)$ (2; def. $\circ$)
4. $a_1 = a_2$ (3; hyp. $g \circ f$ 1-1)
5. f is one-to-one.   (1—4; def.² 1-1)

Attempted Proof of (b):

Assume: $g \circ f$ is 1-1
Show: g is 1-1

Scrap Paper

$A \xrightarrow{f} B \xrightarrow{g} C$ (sketch showing function domains and codomains)

Show: for all $b_1, b_2 \in B$: if $g(b_1) = g(b_2)$, then $b_1 = b_2$

---

1. Let $b_1, b_2 \in B$ and $g(b_1) = g(b_2)$
   .
   .
   .
k. $b_1 = b_2$
k+1. g is one-to-one.   (1—k; def. 1-1)

If there were some $a_1, a_2 \in A$ such that $f(a_1) = b_1$ and $f(a_2) = b_2$, then the b_1 and b_2 would be related to the composition $g \circ f$ and we could perhaps proceed. (See Theorem 2.4.2.) If there are no such a_1 and a_2, then there does not seem to be any way the hypotheses will help us proceed. We therefore will try to construct a counterexample. This will be a specific example where the hypothesis is true but the conclusion is false. Inventing counterexamples is generally at least as difficult as proving theorems. So we try to keep things as simple as possible. Here we need to construct functions g and f such that g is not one-to-one but $g \circ f$ is one-to-one. In doing this, we will make some $b \in B$ have the property that f sends no $a \in A$ to this b.

Counterexample 3:
Define $A = \{1\}$, $B = \{2, 3\}$, $C = \{4\}$.
Define $f: A \to B$ by $f(1) = 2$.

2.3 One-to-One Functions

Define $g: B \to C$ by $g(2) = 4$, $g(3) = 4$.
Check that g is not one-to-one: $g(2) = g(3)$ but $2 \neq 3$.
Check that $g \circ f$ is one-to-one:

>   1. Let $a_1, a_2 \in A$ and $g \circ f(a_1) = g \circ f(a_2)$.
>   2. $a_1 = 1$ (def. A)
>   3. $a_2 = 1$ (def. A)
>   4. $a_1 = a_2$ (sub.)
>   5. $g \circ f$ is one-to-one.   (1—4; def.² 1-1)

We can now rewrite as a theorem the part of the conjecture we were able to prove:

Theorem 2.3.4 *Let $f: A \to B$ and $g: B \to C$ be functions. If $g \circ f$ is one-to-one, then f is one-to-one.*

Suppose $f: A \to B$ is a function and $C \subseteq A$. Then f is a rule that states where each element in C maps to, since each element in C is in A. Thus f can be considered a function from C to B—called the *restriction* of f to C and denoted by "$f|_C$".

Definition *Let $f: A \to B$ and $C \subseteq A$. The function $f|_C: C \to B$ defined by $f|_C(x) = f(x)$ for all $x \in C$ is called the **restriction of f to C**.*

Theorem 2.3.5 *If $f: A \to B$ is one-to-one and $C \subseteq A$, then $f|_C: C \to B$ is one-to-one.*

Proof: Exercise 3.

In order that a function $f: A \to B$ not be one-to-one, it must satisfy the negation of the defining condition for one-to-one.

    Condition: **for all $a_1, a_2 \in A$: if $f(a_1) = f(a_2)$, then $a_1 = a_2$**

    Negation: **for some $a_1, a_2 \in A$: $\sim$(if $f(a_1) = f(a_2)$, then $a_1 = a_2$)**

By Theorem 1.8.6L this can be written:

    Negation: **for some $a_1, a_2 \in A$: $f(a_1) = f(a_2)$ and $a_1 \neq a_2$**

In order to prove that some $f: A \to B$ is not one-to-one, then, we need to establish the existence statement above—that is, define a_1 and a_2 and show that they have the required property.

Example 4:
Show that $f: \mathbb{R} \to \mathbb{R}$ defined by $f(x) = x^2$ is not one-to-one.

> **Proof:**
>
> $f(-2) = 4 = f(2)$ and $-2 \neq 2$. ∎

The following theorem provides an alternate formulation for a function's being one-to-one:

Theorem 2.3.6 *A function $f: A \to B$ is one-to-one iff for all $a_1, a_2 \in A$: if $a_1 \neq a_2$, then $f(a_1) \neq f(a_2)$.*

Given a statement **if $\mathscr{P}$, then $\mathcal{Q}$**, we can form another statement **if $\sim\mathcal{Q}$, then $\sim\mathscr{P}$** by negating $\mathscr{P}$ and $\mathcal{Q}$ and reversing their roles. The second statement is called the *contrapositive* of the first. For example, the statement **if $a_1 \neq a_2$, then $f(a_1) \neq f(a_2)$** appearing in Theorem 2.3.6 is the contrapositive of **if $f(a_1) = f(a_2)$, then $a_1 = a_2$** that appears in the definition of one-to-one.

Theorem 2.3.7L *A statement and its contrapositive are equivalent.*

> **Proof:**
>
> First:
>
> > *Assume:* if $\mathscr{P}$, then $\mathcal{Q}$
> > *Show:* if $\sim\mathcal{Q}$, then $\sim\mathscr{P}$
> >
> > 1. Assume $\sim\mathcal{Q}$.
> > 2. To get #, assume $\mathscr{P}$.
> > 3. if $\mathscr{P}$, then $\mathcal{Q}$ (hyp.)
> > 4. $\mathcal{Q}$ # 1   (2, 3; us. ⇒)
> > 5. $\sim\mathscr{P}$ (2—4; #)
> > 6. if $\sim\mathcal{Q}$, then $\sim\mathscr{P}$ (1—5; pr. ⇒)
>
> Thus, if a statement is true, then its contrapositive is true. Since the statement is the contrapositive of its own contrapositive, we have established both parts of our assertion. ∎

Theorem 2.3.6 follows from the definition of one-to-one by substituting **if $a_1 \neq a_2$, then $f(a_1) \neq f(a_2)$** for its contrapositive, **if $f(a_1) = f(a_2)$, then $a_1 = a_2$**.

Additional Proof Ideas

Recall that in our discussions all sets consist of elements from some universal set U, which may be R, N, or any other set that stays fixed for the discussion. All sets under consideration, then, will be subsets of U.

The definition of $A \subseteq B$ is given by the statement:

(1) for all $x \in A$: $x \in B$

Since A and B are both subsets of U, it seems clear that $A \subseteq B$ could be defined by

(2) for all $x \in U$: if $x \in A$, then $x \in B$

It's not difficult to show that (1) is equivalent to (2) (Exercise 12). Using and proving statements in the form of (2) is more complicated than doing the same for statements in the form of (1)—which is why we didn't use (2) to begin our development of proofs. Abbreviations of (2) are commonly used in informal mathematics, however. First, since every element x under consideration must come from U, saying so is not always necessary. Thus (2) can be abbreviated:

(3) for all x: if $x \in A$, then $x \in B$

Secondly, the quantification "**for all** x:" is omitted, giving:

(4) if $x \in A$, then $x \in B$

In (4), x is called a *free variable*, being neither quantified nor previously defined. However, (4) is not considered to be an open sentence (one that could be either true or false depending on what is substituted for x). It is considered to be an abbreviation of (2) or (3).

Many mathematicians, if asked the question "How is $A \subseteq B$ defined?", would reply that it means "If $x \in A$, then $x \in B$"—using an undefined symbol "x". Since x has not been defined previously, what is meant is "If x is an arbitrarily chosen element of A, then $x \in B$." This use of the "if–then" construction departs from our language. We are not allowed to use undefined symbols in proof statements. Thus the only allowable statements involving a new variable x would either define it, as in "let $x \in A$ be arbitrary" or "let $x = 2 + \cdots$", or use it as a local variable in one of the two types of quantified statements **for all x such that** $\mathcal{P}(x)$: $\cdots$ or **for some x such that** $\mathcal{P}(x)$: $\cdots$".

One frequently sees the definition of a function's being one-to-one given informally by, "$f: A \to B$ is one-to-one provided **if** $f(a_1) = f(a_2)$, **then** $a_1 = a_2$." Since a_1 and a_2 have not appeared before, we would tend to think of this as an abbreviation of **for all** a_1, a_2: **if** $f(a_1) = f(a_2)$, **then** $a_1 = a_2$. However, $f(a_1)$ and $f(a_2)$ need be defined not for arbitrary elements of U, but only for elements of A. Thus this definition makes the additional assumption that a_1 and a_2 are restricted to a domain in which the notation ($f(a_1)$ and $f(a_2)$) makes sense, that is, restricted to A. This construction is common in informal mathematics:

If $\mathcal{P}(x, y)$ is an assertion in a proof involving previously undefined symbols "x" and "y", then x and y are taken to be arbitrarily chosen elements subject only to the constraint of having $\mathcal{P}(x, y)$ make sense.

We will postpone the use of **if–then** statements involving free variables until Section 4.2. In Chapters 1 through 3, where our primary focus is on learning to do proofs, we will use the equivalent **for all: if–then** form and thus stick to defining all our symbols. In the paragraph proofs of the remaining chapters, it will be natural to make implicit use of the rule above.

Exercises 2.3

Practice Exercise

1. Suppose $f: A \to B$.
    1. *
    2. · · ·
    3. · · ·
    4. *
    5. *   (1—4; pr. $\forall \Rightarrow$)
    6. f is one-to-one.   (5; def.[1] 1-1)

Straightforward Problems

2. Let $f: [0, 1] \to [a, b]$ for $a, b \in \mathbb{R}$, $a < b$ be defined by $f(x) = a + (b - a)x$. Show that f is one-to-one.
3. Prove Theorem 2.3.5.
4. Let $f: \mathbb{R} \to \mathbb{R}$ be defined by $f(x) = \frac{x}{2}$ for all $x \in \mathbb{R}$. Prove that f is one-to-one.
5. Prove Theorem 2.3.2.
6. Let $f: \mathbb{N} \to \mathbb{N}$ be defined by $f(x) = x + 1$. Prove that $f \circ f$ is one-to-one.
7. Recall Problem 2.2.4a. Prove the following cancellation property of composition: Let $f: A \to B$ and $g: A \to B$. Let $h: B \to C$ be one-to-one. Then if $h \circ f = h \circ g$, then $f = g$.

Harder Problems

8. Show[†] that **arctan**: $\mathbb{R} \to (-\frac{\pi}{2}, \frac{\pi}{2})$ is one-to-one. [Hint: apply the tangent function to both sides of a line in your proof.] [Hint on using hints: never

---

[†] Proofs depend on definitions. Since the arctan function and other functions below have not been defined, you can't give a formal proof. Use the formal proof style, however, and use well-known properties of the functions instead of definitions.

2.4 *Onto* Functions　　　　　　　　　　　　　　　　　　　　　　　　　　　　139

write a hint down to start a problem and then try to squeeze something out of the hint. Stick to the routine method of (1) analyzing the conclusion and (2) using the hypotheses when you get stuck. If you get stuck even *after* using the hypotheses, then is the time to use the hint.]

9. Define $f: [0, \infty) \to \mathbb{R}$ by $f(x) = \sqrt{x}$. Show that f is one-to-one.
10. Show whether or not each of the following functions is one-to-one:
    (a) $f: \mathbb{R} \to \mathbb{R}$ defined by $f(x) = \sqrt{x^2}$
    (b) $\sin: \mathbb{R} \to \mathbb{R}$
    (c) $f: \mathbb{R} \to \mathbb{R}$ defined by $f(x) = e^x$
11. Recall that for any set X, the power set of X (denoted $\mathbb{P}(X)$) is the set of all subsets of X. For any function $f: A \to B$, we defined (Section 2.1) the induced function $\bar{f}: \mathbb{P}(A) \to \mathbb{P}(B)$ by $\bar{f}(C) = f(C)$ for any $C \in \mathbb{P}(A)$. Similarly, we defined the induced function $\tilde{f}: \mathbb{P}(B) \to \mathbb{P}(A)$ by $\tilde{f}(D) = f^{-1}(D)$ for any $D \in \mathbb{P}(B)$. Prove or find a counterexample for:
    (a) If f is one-to-one, then $\bar{f}$ is one-to-one.
    (b) If f is one-to-one, then $\tilde{f}$ is one-to-one.
    (c) If $\bar{f}$ is one-to-one, then f is one-to-one.
    (d) If $\tilde{f}$ is one-to-one, then f is one-to-one.

Supplementary Problem

12. Let A and B be sets and $\mathbb{U}$ the universal set. Prove that **for all** $x \in A: x \in B$ and **for all** $x \in \mathbb{U}$: **if** $x \in A$, **then** $x \in B$ are equivalent statements.

2.4 *ONTO* FUNCTIONS

Conjecture 2.3.3b states: if $f: A \to B$, $g: B \to C$, and $g \circ f$ is one-to-one, then g is one-to-one. Counterexample 3 in the last section shows that this is not true. Recall that this example was manufactured so that there was a $b \in B$ with no $a \in A$ mapping to b by f. In this section, we will see that we can "fix" the conjecture to make it true. That is, we can add another hypothesis that will prevent us from constructing an example like Counterexample 3 of Section 2.3. To do this, we need a definition for functions $f: A \to B$ that have the property that for each $b \in B$ there is some $a \in A$ such that $f(a) = b$.

In this section we abandon the requirement that the defining condition for new definitions be given in our language. This will continue our trend toward informality. Of course, it is absolutely essential that the meaning of the new definitions be clear. This means that proof formats for proving and for using the defining condition should both be evident.

Up to this point, one requirement in dealing with metalanguage has been to determine the hypotheses and conclusion from the statement of a theorem. This determination gave us a format for proving the theorem. The format for using the theorem was determined by finding a language statement equivalent to the metalanguage of the theorem. The formats for proving and using the theorem determine the meaning of the theorem.

Mathematics is written with no distinction between what we have called language and metalanguage. Or, to put it another way, it is written in metalanguage, and it is up to the reader to interpret the meaning. The meaning is to be understood in terms of proof formats for using and proving the statements. Interpretations in terms of formats can be found by translating the metalanguage statements into our formal language and then using our rules of inference for these. Our language statements formalize the meaning in common mathematical metalanguage, and our formal rules of inference copy what mathematicians generally do to prove or use the metalanguage statements. The goal in our approach is to be able to understand metalanguage statements in a very precise way. Thus the formal language and rules are there to build precise mathematical writing and reading habits.

Definition *A function $f: A \to B$ is called **onto** if for each $b \in B$ there exists some $a \in A$ such that $f(a) = b$.*

The metalanguage statement

"for each $b \in B$ there exists some $a \in A$ such that $f(a) = b$"

in the definition above means exactly the same as

for all $b \in B$: $f(a) = b$ for some $a \in A$

or   for all $b \in B$: there exists $a \in A$ such that $f(a) = b$

Recall that $f(a) = b$ for some $a \in A$

and   there exists $a \in A$ such that $f(a) = b$

are two language statements that mean the same thing.

The reason for using the word "each" in the phrase "for each $b \in B$" is that the weight of the word "all" would tend to make some people violate the grammar of the condition defining onto, as if it meant one a worked for all b. Note the difference between

(there exists $a \in A$ such that $f(a) = b$) for all $b \in B$

and   there exists $a \in A$ such that ($f(a) = b$ for all $b \in B$)

The defining condition is given by the first of these statements and not the second. Since the English language does not use parentheses to remove ambiguity,

2.4 Onto Functions

other ways must be used to accomplish this—using the word "each", for instance, to make it clearer that first b is chosen and then some a (that depends on the choice of b) is found.

Suppose we wish to prove a theorem with the conclusion "f is onto". Since "onto" has not been defined in terms of our formal language, the form of the conclusion does not automatically lead to a proof format or suggest proof steps. It is up to us to capture the meaning of "onto" in the proof steps we select. This can be done by following the rules suggested by an equivalent language statement.

Example 1:
The function $f: \mathbb{R} \to \mathbb{R}$ given by $f(x) = 2x + 4$ is onto.

By the definition of "onto", we need to show **for all $c \in \mathbb{R}$: there exists $a \in \mathbb{R}$ such that $f(a) = c$**. The following steps therefore prove that f is onto:

Proof:

1. Let $c \in \mathbb{R}$.   (because $\mathbb{R}$ is codomain of f)
   .
   (Define a in here.)
   .
   k. $f(a) = c$
k+1. f is onto.

Scrap Paper
Since $f(a) = 2a + 4$, we want $c = 2a + 4$. We are given c and want to define a in terms of c. Hence $c - 4 = 2a$, so that $a = \dfrac{c-4}{2}$.

1. Let $c \in \mathbb{R}$.
2. Let $a = \dfrac{c-4}{2}$.
3. $2a + 4 = c$ (2; prop. $\mathbb{R}$)
4. $f(a) = c$ (3; def. f)
5. f is onto.   (1—4; def. onto)

Example 2:
Here is a paragraph form for the proof in Example 1.

Chapter 2 Functions

> **Proof:**
>
> Let $c \in \mathbb{R}$ be arbitrary. Define a to be $\dfrac{c-4}{2}$. Then $f(a) = c$ by the definition of f, so that f is onto. ∎

In order that a function $f: A \to B$ not be onto, it must satisfy the negation of the defining condition for onto:

Condition: **for all $b \in B$: there exists $a \in A$ such that $f(a) = b$**

Negation: **for some $b \in B$: ~(there exists $a \in A$ such that $f(a) = b$)**

That is,   **for some $b \in B$: for all $a \in A$: $f(a) \neq b$**

Thus, to show that $f: A \to B$ is not onto, we must define an element $b \in B$ and then show that there is no $a \in A$ that f sends to b.

Example 3:
Show that $f: \mathbb{R} \to \mathbb{R}$ defined by $f(x) = x^2$ is not onto.

> **Proof:**
>
> $f(a) \neq -1$ for all $a \in \mathbb{R}$. ∎

The assertion that a function is onto amounts to saying no more than that its range is equal to its codomain.

Example 4:
The function $f: \mathbb{R} \to \mathbb{R}$ given by $f(x) = x^2$ in Example 3 is not onto. The only reason it is not is that we have chosen to specify the codomain of f as $\mathbb{R}$. The range of f, $f(\mathbb{R})$, is the set $\{r \in \mathbb{R} \mid r \geq 0\}$, which we call S. The function $h: \mathbb{R} \to S$ given by $h(x) = x^2$ is onto. Although h is onto and f is not, the only reason h is not equal to f by definition is that the two functions have different codomains. One reason for requiring equal functions to have the same codomain is that otherwise we might have two equal functions one of which was onto and the other not.

Example 5:
The function $f: \{1, 2, 3\} \to \{a, b, c\}$ defined by $1 \to a$, $2 \to b$, and $3 \to a$ is not onto since f maps no element of A to the element c in the codomain of f.

2.4 *Onto* Functions

Theorem 2.4.1   *If $f: A \to B$ and $g: B \to C$ are onto, then $g \circ f$ is onto.*

Proof:
  Assume: 1. f onto
             2. g onto
  Show: $g \circ f$ onto

Observe that $g \circ f: A \to C$ by definition of composition.

  1.   Let $c \in C$.
       .
       (Define a in here.)
       .
  k. $g \circ f(a) = c$
  $k+1$. $g \circ f$ is onto.   (1—k; def. onto)

Backing up from Step k, we get:

  1.   Let $c \in C$.
       .
  $k-1$. $g(f(a)) = c$
  k. $g \circ f(a) = c$ ($k-1$; def. $\circ$)
  $k+1$. $g \circ f$ is onto.   (1—k; def. onto)

Since g is onto, it will map something to c; call it b. Then $g(b) = c$. Since f is onto, it will map something to b; call it a. (We have now found a.)

  1. Let $c \in C$.
  2. **there exists $x \in B$ such that $g(x) = c$.**   (1, hyp. 2; def. g onto)
  3. Pick $b \in B$ such that $g(b) = c$.   (2; us. $\exists$)
  4. **there exists $x \in A$ such that $f(x) = b$**   (3, hyp. 1; def. f onto)
  5. Pick $a \in A$ such that $f(a) = b$.   (4; us. $\exists$)
  6. $g(f(a)) = c$ (3, 5; sub.)
  7. $(g \circ f)(a) = c$ (6; def. $\circ$)
  8. $g \circ f$ is onto.   (1—7; def. onto) ∎

In Step 5 we "found" a by using the hypothesis that f is onto. This is the usual pattern for existence proofs.

In the future we will contract steps with **there exists** statements and the "pick . . ." steps that follow from them. For example, in the last proof:

   2. **there exists** $x \in B$ such that $g(x) = c$ (1, hyp. 2; def. g onto)
   3. Pick $b \in B$ such that $g(b) = c$.   (2; us. $\exists$)

will be contracted to

   2. There exists $b \in B$ such that $g(b) = c$.   (1, hyp. 2; def. g onto)

Thus we use this metalanguage statement of existence to define the global variable b for subsequent steps—exactly as if it meant "pick b such that $g(b) = c$". The statement in Step 2 above will therefore have a double meaning: b is considered a newly defined global variable that may also serve as a local variable in the implicit **there exists** statement.

Rewriting the proof of Theorem 2.4.1 using this contraction gives:

Theorem 2.4.1   If $f: A \to B$ and $g: B \to C$ are onto, then $g \circ f$ is onto.

Proof:

   Assume: 1. f onto
                 2. g onto
   Show: $g \circ f$ onto

   1. Let $c \in C$.
   2. There exists $b \in B$ such that $g(b) = c$.   (1, hyp. 2; def. onto)
   3. There exists $a \in A$ such that $f(a) = b$.   (2, hyp. 1; def. onto)
   4. $g(f(a)) = c$ (2, 3; sub.)
   5. $(g \circ f)(a) = c$ (4; def. $\circ$)
   6. $g \circ f$ is onto.   (1—5; def. onto)

A paragraph proof of this theorem amounts to no more than writing these steps down with a few connecting words to smooth the flow. Not all reasons are given in a paragraph proof, but it is a good idea to tell the reader just where you are using the hypotheses.

2.4 Onto Functions

> **Paragraph Proof:**
>
> Assume f and g are onto. We will show $g \circ f$ is onto. Let $c \in C$. Then, since g is onto, there exists $b \in B$ such that $g(b) = c$. Since f is onto, there exists $a \in A$ such that $f(a) = b$. Substituting, $g(f(a)) = c$, so that $(g \circ f)(a) = c$. Thus $g \circ f$ is onto. ∎

Recall from the last section:

Conjecture 2.3.3 Let $f: A \to B$ and $g: B \to C$ be functions.
(a) If $g \circ f$ is one-to-one, then f is one-to-one.
(b) If $g \circ f$ is one-to-one, then g is one-to-one.

Part (a) was proved and renumbered as Theorem 2.3.4; part (b) was found to be false. Our attempted proof of (b), however, can be made to "go through" if we add another hypothesis, namely, that f is onto:

Theorem 2.4.2 Let $f: A \to B$ and $g: B \to C$ be functions. If $g \circ f$ is one-to-one and f is onto, then g is one-to-one.

Proof: Exercise 4.

Conjecture 2.4.3 Let $f: A \to B$ and $g: B \to C$.
(a) If $g \circ f$ is onto, then g is onto.
(b) If $g \circ f$ is onto, then f is onto.

Proof or Disproof: Exercise 6.

Additional Proof Ideas

Recall the definition of onto:

Definition A function $f: A \to B$ is called **onto** if for each $b \in B$ there exists some $a \in A$ such that $f(a) = b$.

In this definition, "if" really means "iff", in order to reflect the reversible way in which the definition rule is used. Since people know definitions work in this way, they use the word "if" with the assumption that, since it serves to set off the defining condition, it is interpreted to mean "iff". We have used the words "provided that" to set off defining conditions. Hence "provided that" is interpreted as "iff" when used this way.

Exercises 2.4

Practice Exercise

1. Let $f: A \to B$. Replace the * with a language statement:
   1. *
   2. f is onto. (1; def. onto)

Straightforward Problems

2. Prove that the function given in Exercise 2.3.2 is onto.
3. Let $X = \{x \in \mathbb{R} \mid x \geq 0\}$. Prove that $f: X \to X$ defined by $f(x) = 3x$ is onto.
4. Prove Theorem 2.4.2.
5. Decide and prove whether or not each of the following functions is onto:
   (a) $f: \mathbb{R} \to \mathbb{R}$ defined by $f(x) = 3x + 2$
   (b) $f: \mathbb{Z} \to \mathbb{Z}$ defined by $f(x) = 3x + 2$
   (c) $f: \mathbb{R} \to \mathbb{R}$ defined by $f(x) = x^2$
   (d) $f: \mathbb{Z} \to \mathbb{Z}$ defined by $f(x) = x^2$
   (e) $f: \mathbb{R} \to \mathbb{R}$ defined by $f(x) = x^3$
   (f) $f: \mathbb{Z} \to \mathbb{Z}$ defined by $f(x) = x^3$

Harder Problems

6. Prove or find a counterexample to Conjecture 2.4.3a and b. If either (a) or (b) fails to hold, add some hypothesis to the conjecture to make it true. (Recall how Conjecture 2.3.3b was "fixed" to give Theorem 2.4.2.)
7. Recall Exercises 2.2.4 and 2.3.7. Fix Exercise 2.2.4b to make it true and prove your assertion.
8. Show that $\textbf{arctan}: \mathbb{R} \to (-\frac{\pi}{2}, \frac{\pi}{2})$ is onto. [Hint: after setting up your proof, use the tangent function.]
9. Show whether or not each of the following functions is onto:
   (a) $f: \mathbb{R} \to \mathbb{R}$ given by $f(x) = e^x$
   (b) $\sin: \mathbb{R} \to \mathbb{R}$
   (c) $f: \{x \mid x > 0\} \to \mathbb{R}$ given by $f(x)$ is the natural log of x.
10. Let $f: A \to B$. The induced functions $\bar{f}: \mathbb{P}(A) \to \mathbb{P}(B)$ and $\tilde{f}: \mathbb{P}(B) \to \mathbb{P}(A)$ were defined in Section 2.1. Prove or find a counterexample to:
    (a) If f is onto, then $\bar{f}$ is onto.
    (b) If f is onto, then $\tilde{f}$ is onto.

Hard Supplementary Problem

11. Let X be a nonempty set. Prove that there is no onto function $f: X \to \mathbb{P}(X)$.

2.5 INVERSES

Definition Let $f: A \to B$. A function $g: B \to A$ is called a **left inverse** of f if $g \circ f = i_A$ and a **right inverse** of f if $f \circ g = i_B$.

> *Example 1:*
> Let $f: \mathbb{N} \to \mathbb{N}$ be given by $f(a) = a + 1$. Let $g: \mathbb{N} \to \mathbb{N}$ be given by:
>
> $$g(b) = \begin{cases} 1 \text{ if } b = 1 \\ b - 1 \text{ if } b \geq 2 \end{cases}$$
>
> Then $g \circ f(a) = g(a+1) = (a+1) - 1 = a$. Thus $g \circ f = i_\mathbb{N}$, so that g is a left inverse of f. But $f \circ g(1) = f(g(1)) = f(1) = 2$, so that g is not a right inverse of f. Note that f is one-to-one.

> *Example 2:*
> Let $A = \{1, 2\}$ and $B = \{a, b, c\}$. Let $f(1) = a$ and $f(2) = b$ define the function $f: A \to B$. Let $g(a) = 1$, $g(b) = 2$, and $g(c) = 1$. Then $g \circ f(1) = g(f(1)) = g(a) = 1$ and $g \circ f(2) = g(f(2)) = g(b) = 2$, so that $g \circ f = i_A$. Hence f has a left inverse g. But $f \circ g(c) = f(g(c)) = f(1) = a$, so that $f \circ g \neq i_B$, and g is therefore not a right inverse of f. Note that f is one-to-one.

In both Example 1 and Example 2, the one-to-one function f has a left inverse. Theorem 2.5.1 gives the general result. We will prove Theorem 2.5.1 with the aid of a *lemma*—a theorem that is useful primarily for proving more important theorems. Some lemmas are mere stepping stones, useful in the proof of a single theorem. Other lemmas have been found to be useful over and over again and are better known than the theorems they aid in proving. Lemmas are frequently invented in the course of finding a proof for a theorem and then used as reasons for steps in the proof. The proof of the lemmas is put off until later.

Theorem 2.5.1 Let $f: A \to B$. Then f has a left inverse iff f is one-to-one.

> **Proof:**
>
> *Show:* 1. If f has a left inverse, then f is one-to-one.
>   2. If f is one-to-one, then f has a left inverse.
>
> We first show that, if f has a left inverse, then f is one-to-one. For this, assume f has a left inverse. That is, there exists a function $g: B \to A$ such that $g \circ f = i_A$.

If we knew that i_A were one-to-one, then, by Theorem 2.3.4, f would be one-to-one. It seems clear that i_A can easily be shown to be one-to-one, and so we will call this fact a lemma, use the lemma in our proof, and prove the lemma later (not using Theorem 2.5.1, of course).

Lemma

Lemma to Theorem 2.5.1: For any set A, $i_A: A \to A$ is one-to-one.

Continuing with the proof of Theorem 2.5.1:

> **Proof:**
>
> *Show:* 1. If f has a left inverse, then f is one-to-one.
> 2. If f is one-to-one, then f has a left inverse.
>
> We first show that, if f has a left inverse, then f is one-to-one. For this, assume f has a left inverse. That is, there exists a function $g: B \to A$ such that $g \circ f = i_A$. Since by the lemma i_A is one-to-one, it follows that f is one-to-one by Theorem 2.3.4.
>
> We next show that, if f is one-to-one, then there exists a function $g: B \to A$ such that $g \circ f = i_A$.

At this stage in the proof, we need to establish an existence statement: there exists g (from B to A) with a property ($g \circ f = i_A$). According to our existence rule, we now need to invent or define g. We are free to define g any way we wish, but after we do so, we will have to show $g \circ f = i_A$. In order to define $g: B \to A$, we need to have a rule for determining $g(b)$ for any $b \in B$. There are two cases:

Case 1: b is in $f(A)$, the range of f

Case 2: b is not in $f(A)$

In Case 1, by definition of $f(A)$ there exists some $a \in A$ such that $f(a) = b$. Since f is one-to-one, there is only one such element a. In this case, let us define g by $g(b) = a$. In Case 2, we can pick any element $a_0 \in A$ (which set is nonempty by our convention for functions) and define g by $g(b) = a_0$. We continue with the proof:

> Let $g: B \to A$ be defined as follows:
>
> 1. Pick any $a_0 \in A$.
> 2. For a given $b \in B$, define $g(b)$ by the rule:
>
> Case 1 $b \in f(A)$: $g(b) =$ the unique $a \in A$ such that $f(a) = b$
>
> Case 2 $b \notin f(A)$: $g(b) = a_0$

2.5 Inverses

> We now show $g \circ f = i_A$ (that is, $g \circ f(a) = i_A(a)$ for all $a \in A$). Let $a \in A$. Then $g \circ f(a) = g(f(a))$ by definition of $\circ$. And $g(f(a)) = a$ by definition of g. Also, $i_A(a) = a$ by definition of i_A.
> Therefore $g \circ f(a) = i_A(a)$. Whence $g \circ f = i_A$ by the definition of equal functions. ∎

This proof was largely a matter of organization—keeping track of hypotheses, conclusions, and what they mean by definition. In our proof style, this is always done informally. Formal proof steps serve only to establish a conclusion. Instead of numbering these steps, we have used words like "then", "also", "therefore", and "whence" and written the steps down as sentences in the proof. We therefore have a natural, informal, paragraph-style proof. The proof will be complete when you show $i_A: A \to A$ is one-to-one (Exercise 2).

In the proof of Theorem 2.5.1, the given function f was used to construct the new function g. It may not be possible to exhibit g as a rule in any way other than that given in the proof of the theorem. For example, if f is described by some fifth-degree polynomial:

$$y = f(x) = ax^5 + bx^4 + cx^3 + dx^2 + ex + k$$

then an explicit expression for g is obtained by solving this equation for x. However, it is known that there is no algorithm for solving the general fifth-degree equation and hence for obtaining x as a function of y. If f is one-to-one, we know that for each y there is a unique x that is mapped to y, but we have, in general, no way of describing this x other than "the unique x that is mapped to y". To a mathematician, this is an acceptable way to go about defining the function g.

In Examples 1 and 2, the focus is on the function f, which has a left inverse but no right inverse. We can rewrite these examples with the focus on g, which has a right inverse but no left inverse.

Example 3:
Let $f: \mathbb{N} \to \mathbb{N}$ be given by $f(a) = a + 1$. Let $g: \mathbb{N} \to \mathbb{N}$ be given by

$$g(b) = \begin{cases} 1 \text{ if } b = 1 \\ b - 1 \text{ if } b \geq 2 \end{cases}$$

Then $g \circ f(a) = g(a + 1) = (a + 1) - 1 = a$. Thus $g \circ f = i_\mathbb{N}$, so that f is a right inverse of g. But $f \circ g(1) = f(g(1)) = f(1) = 2$, so that f is not a left inverse of g. Note that g is onto.

Example 4:
Let $A = \{1, 2\}$ and $B = \{a, b, c\}$. Let $f(1) = a$ and $f(2) = b$ define the function $f: A \to B$. Let $g(a) = 1$, $g(b) = 2$, and $g(c) = 1$. Then $g \circ f(1) = g(f(1)) = g(a) =$

150   Chapter 2 Functions

1 and $g \circ f(2) = g(f(2)) = g(b) = 2$, so that $g \circ f = i_A$ and f is a right inverse of g. But $f \circ g(c) = f(g(c)) = f(1) = a$, so that $f \circ g \neq i_B$, so that f is not a left inverse of g. Note that g is onto.

Each of the onto functions g in Examples 3 and 4 has a right inverse. The next theorem gives a general statement. The theorem is stated with the usual symbol f for functions playing the role of g in the examples.

Theorem 2.5.2   Let $f: A \to B$. Then f has a right inverse iff f is onto.

> **Proof:**
> We first show that if f has a right inverse, then f is onto. Thus assume there is a function $g: B \to A$ such that $f \circ g = i_B$. Let $b \in B$. We need to show that there exists some $a \in A$ such that $f(a) = b$. Define a to be $g(b)$. Then $f(a) = f(g(b)) = f \circ g(b) = i_B(b) = b$. Hence, **for all $b \in B$: $b = f(a)$ for some $a \in A$,** that is, f is onto.
>
> We now show that if f is onto, then there exists a function $g: B \to A$ such that $f \circ g = i_B$. Thus assume f is onto.

> **Scrap Paper**
> Since our aim is to prove a **there exists** statement, we need to define the function g and then show that $f \circ g = i_B$ for the g we have defined. To define g, we need a rule that gives a value $g(b)$ for each $b \in B$. We therefore pick an arbitrary $b \in B$ and describe the element $g(b) \in A$ associated with it. Continuing our second paragraph . . .

> We now show that if f is onto, then there exists a function $g: B \to A$ such that $f \circ g = i_B$. Thus assume f is onto. Let $b \in B$. Since f is onto, we have that **for all $x \in B$: $x = f(z)$ for some $z \in A$.** By the rule for using this **for all** statement (and the fact that $b \in B$), we have $b = f(z)$ **for some** $z \in A$. By the rule for using this **there exists** statement, we pick $a \in A$ such that $b = f(a)$. (We are now free to use this a and the fact that $b = f(a)$ in the rest of the proof.) For our arbitrarily chosen $b \in B$, then, we define $g(b)$ to be the element a given above such that $b = f(a)$.[†] By a chain of equalities:
>
> $$\text{Let } x \in B.$$
> $$f \circ g(x) =$$
>
> ---
> [†] See remarks on the axiom of choice at the end of this section.

2.5 Inverses

$$f(g(x)) = \quad \text{(def. } \circ\text{)}$$
$$x = \quad \text{(def. } g\text{)}$$
$$i_B(x) \quad \text{(def. } i_B\text{)}$$
$$\text{So} \quad f \circ g = i_B \quad \text{(def. = fcns.)}$$

■

Definition   Let $f: A \to B$. A function $g: B \to A$ is called an ***inverse*** of f if g is both a left and a right inverse of f, that is, if $g \circ f = i_A$ and $f \circ g = i_B$.

Recall our uniqueness rule: to show that x with property $\mathscr{P}(x)$ is unique, assume x_1 has property $\mathscr{P}$ and x_2 has property $\mathscr{P}$, then show $x_1 = x_2$. This rule is used in proving the following theorem:

Theorem 2.5.3   If $f: A \to B$ has an inverse, then this inverse is unique.

Proof:
     Assume: g_1, g_2 inverses of f
     Show: $g_1 = g_2$

We must use the definition of inverse to make use of the hypothesis. The properties available are:

$$f \circ g_1 = i_B$$
$$f \circ g_2 = i_B$$
$$g_1 \circ f = i_A$$
$$g_2 \circ f = i_A$$

It is not appropriate to list these as Steps 1 through 4 without having any idea of where they might lead. (Instead, they should be listed on scrap paper.) If we were to consider the meaning of the conclusion (equality of functions), we would be led to introduce the following steps:

     1.   Let $x \in B$.
          .
          .
   $n-1$. $g_1(x) = g_2(x)$
   n. $g_1 = g_2$ (1—$n-1$; def. = fcns.)

We can save ourselves some work here if, instead of going all the way back to definitions, we use Theorems 2.2.1 and 2.2.2:

Proof:

Assume: g_1, g_2 inverses of f
Show: $g_1 = g_2$

1. $(g_1 \circ f) \circ g_2 = g_1 \circ (f \circ g_2)$ (Thm. 2.2.1)
2. $(i_A) \circ g_2 = g_1 \circ (i_B)$ (1, hyp.; def. inverse)
3. $g_2 = g_1$ (2; Thm. 2.2.2) ∎

In this proof, the uniqueness rule was used to obtain the hypothesis and conclusion and Theorems 2.2.1 and 2.2.2 were used to justify steps. Using previous theorems in proofs relieves us of the necessity of repeating—over and over in new contexts—the steps involved in proving these theorems earlier. The definition of inverse involves the ideas of composition of functions and identity functions. It was therefore the theorems involving these ideas that were useful to us.

Notation

If $f: A \to B$ has an inverse, then by Theorem 2.5.3 the inverse is unique. It is denoted by f^{-1}. Thus $f^{-1}: B \to A$.

Suppose $f: A \to B$ and $C \subseteq B$. If f has no inverse, then by the set $f^{-1}(C)$ we mean the preimage of C under f. If f has an inverse, then by $f^{-1}(C)$ we could mean either the preimage of C under f or the image of C under f^{-1}. We can show that both of these sets are the same.

In order to keep our notation straight, let's agree, for the moment, to denote f^{-1} by g. Then the image of C under g is the set $G = \{a \in A \mid a = g(c) \text{ for some } b \in C\}$. The preimage of C under f is the set $H = \{a \in A \mid f(a) \in C\}$. In fact, $G = H$, which we can show as follows.

Let $x \in G$. Then $x = g(c)$ for some $c \in C$. To see if x is in H, apply f: $f(x) = f(g(c)) = f \circ g(c) = i_B(c) = c \in C$. By the definition of H, then, $x \in H$. Therefore $G \subseteq H$. Let $x \in H$. Then $f(x) \in C$ and $x = i_A(x) = g \circ f(x) = g(f(x))$. So x is g of something in C (namely, $f(x)$) and is therefore in G. Therefore $H \subseteq G$. From both inclusions we get $H = G$.

To summarize, if f has no inverse, then the set $f^{-1}(C)$ is unambiguously defined; if f has an inverse, however, then there are two distinct ways in which the set symbolized by "$f^{-1}(C)$" is defined. Since both ways give the same set, we say that the set $f^{-1}(C)$ is *well defined*. This important idea will come up again.

We now give some theorems to be proved as exercises.

Theorem 2.5.4

A function $f: A \to B$ has an inverse iff f is one-to-one and onto.

Proof: Exercise 3.

2.5 Inverses

Theorem 2.5.5 If $f: A \to B$ is one-to-one and onto, then f^{-1} is one-to-one and onto.

Proof: Exercise 4.

Additional Proof Ideas

Corollaries to Proofs

In the proof of Theorem 2.5.3, we used only the fact that g_1 is a left inverse and g_2 a right inverse of f. This means we proved the stronger result in Corollary 2.5.6 (below), which follows from the proof of Theorem 2.5.3 but not from the statement of the theorem. Such a corollary is frequently called a "corollary to the proof".

Corollary 2.5.6 *Corollary to the proof of Theorem 2.5.3:* If $f: A \to B$ has a left inverse g_1 and a right inverse g_2, then $g_1 = g_2$.

The Axiom of Choice

Mathematicians see a problem in formalizing the procedure for specifying the "rule" defining g in Theorem 2.5.2. The problem is that we make, in effect, an infinite number of arbitrary choices. Let's look again at how the function g was defined. For every $b \in B$, $f^{-1}(b)$ is a set. Since f is onto, this set is nonempty for each $b \in B$. But $f^{-1}(b)$ could be very large, perhaps with an infinite number of elements. Our definition of g depends on picking one element in $f^{-1}(b)$ to be $g(b)$. And this is done for all $b \in B$, perhaps an infinite number of bs. We have defined g by picking an arbitrary element out of each one of a (perhaps infinite) collection of nonempty sets. There is an axiom of logic called the Axiom of Choice that allows us to do just this.

Axiom of Choice Let S be a nonempty set and $\{S_i | i \in \mathscr{I}\}$ a family of nonempty subsets of S. There exists a choice function $F: \mathscr{I} \to S$ such that $F(i) \in S_i$ for all $i \in \mathscr{I}$.

The Axiom of Choice gives us the function $g: B \to A$ as a choice function when we take $\{f^{-1}(b) | b \in B\}$ as the given family of nonempty subsets of A. If the proof of Theorem 2.5.2 seems to have been done logically without the Axiom of Choice, it is because we have defined neither "function" nor "rule". We are therefore dealing at this point not with formal mathematics but with our intuitive ideas about "rule" and "function". Attempts at making these ideas precise involve "axioms" for set theory and are done in courses in logic or the foundations of mathematics. The Axiom of Choice is independent of the other usual axioms of set theory. That is, it cannot be proved from these axioms, nor will its use ever lead to a contradiction. Most mathematicians are therefore content to use it freely.

Exercises 2.5

Practice Exercise

1. Let $f: A \to B$
   - (a) 1. g is a left inverse of f.
     - 2. *   (1; def. left inverse)
   - (b) 1. g is a right inverse of f.
     - 2. *   (1; def. right inverse)
   - (c) 1. g is an inverse of f.
     - 2. *   (1; def. inverse)
     - 3. *   (1; def. inverse)

Straightforward Problems

2. (a) Prove the lemma to Theorem 2.5.1.
   (b) Prove that for any set A, $i_A: A \to A$ is onto.
3. Prove Theorem 2.5.4. [Hint: if g is a left and h a right inverse of f, consider $g \circ f \circ h$.]
4. Prove Theorem 2.5.5.
5. The following functions are one-to-one, so that by Theorem 2.5.1 each of them has a left inverse. For each function, determine a left inverse. State whether your left inverse is also a right inverse.
   - (a) arctan: $\mathbb{R} \to \mathbb{R}$
   - (b) $f: \mathbb{R} \to \mathbb{R}$ given by $f(x) = e^x$
   - (c) $f: [0, \infty) \to [0, \infty)$ given by $f(x) = x^2$
   - (d) $f: \{1, 2, 3\} \to \{1, 2, 3, 4\}$ given by $1 \to 1, 2 \to 2, 3 \to 3$
6. The following functions are onto, so that by Theorem 2.5.2 each of them has a right inverse. For each function, determine a right inverse. State whether your right inverse is also a left inverse.
   - (a) arctan: $\mathbb{R} \to (-\frac{\pi}{2}, \frac{\pi}{2})$
   - (b) $f: \mathbb{R} \to (0, \infty)$ given by $f(x) = e^x$
   - (c) $f: [0, \infty) \to [0, \infty)$ given by $f(x) = x^2$
   - (d) $f: \{1, 2, 3\} \to \{1, 2\}$ given by $1 \to 1, 2 \to 2, 3 \to 2$.

Supplementary Problem

7. Prove Theorem 2.5.3 using the definition of equal functions without quoting Theorems 2.2.1 and 2.2.2. Look at the ideas in the proofs of Theorems 2.2.1 and 2.2.2, however, for hints as to which steps to use.

Hard Supplementary Problems

8. Suppose that $f: A \to B$ is one-to-one with left inverse g. Prove or find a counterexample to: f is onto iff f has a unique left inverse. Consider the proof in one direction. If f is one-to-one and onto, then it has a unique inverse. This will also be a left inverse, but this is not what you are to show. You need to show uniqueness *as a left inverse:* Apply the uniqueness rule by letting g_1 and g_2 be left inverses (only) and show $g_1 = g_2$. Fix the assertion if it is false.
9. Suppose $f: A \to B$ is onto with right inverse g. Prove or find a counterexample to: f is one-to-one iff f has a unique right inverse. Work by analogy to Exercise 8 here. Fix the assertion if it is false.

2.6 BIJECTIONS

One-to-one functions are sometimes called *injections*, and onto functions, *surjections.* Thus we have the following:

Definition   A function $f: A \to B$ is called a **bijection** if it is one-to-one and onto.

According to Theorems 2.5.4 and 2.5.5, if f is a bijection, then f^{-1} exists and is also a bijection. By the definition of inverse, $f \circ f^{-1} = i_B$ and $f^{-1} \circ f = i_A$. If $f: A \to B$ and $g: B \to C$ are both bijections, then by Theorems 2.3.2 and 2.4.1 $g \circ f$ is a bijection.

We single out the special case where our bijections have the same domain and codomain.

Definition   If A is any set, then the set of all bijections from A to A is denoted by $\mathbb{B}(A)$.

To say that $f \in \mathbb{B}(A)$, then, is to say that $f: A \to A$ is a bijection.

Theorem 2.6.1   Let A be any set and $\mathbb{B}(A)$ the set of all bijections from A to A. Then:

(a) For all $f, g \in \mathbb{B}(A)$: $f \circ g \in \mathbb{B}(A)$.
(b) For all $f, g, h \in \mathbb{B}(A)$: $f \circ (g \circ h) = (f \circ g) \circ h$.
(c) There exists $i \in \mathbb{B}(A)$ such that for all $f \in \mathbb{B}(A)$: $i \circ f = f$ and $f \circ i = f$.
(d) For each $f \in \mathbb{B}(A)$, there exists $g \in \mathbb{B}(A)$ such that $f \circ g = i$ and $g \circ f = i$.

Proof: Exercise 6.

Example 1:
Let A be the set $\{x_1, x_2, x_3\}$. A bijection f of A is given by a rule for mapping the elements x_i. This can be shown by an arrow. For example, $x_1 \to x_2$,

$x_2 \to x_3$, $x_3 \to x_1$. Thus $f(x_1) = x_2$, and so forth. The same information is conveyed in compact form by listing only the subscripts:

$$f = \begin{pmatrix} 123 \\ 231 \end{pmatrix}$$

An element in the top row is mapped to the element in the bottom row just below it. There are therefore equivalent ways of denoting f:

$$f = \begin{pmatrix} 231 \\ 312 \end{pmatrix} = \begin{pmatrix} 132 \\ 213 \end{pmatrix} = \begin{pmatrix} 312 \\ 123 \end{pmatrix} = \cdots$$

All of these arrays show that $f(1) = 2$, $f(2) = 3$, and $f(3) = 1$ and therefore describe the same function. Thus in our representation of f, we can list the columns in any order.

For a given ordering of the elements in a finite set identified with subscripted variables, any bijection of the set gives a *permutation* of the subscripts. Conversely, a permutation of the subscripts gives a bijection of the set. We therefore view a permutation as a function: to say that 123 is permuted to 231 is to say that $1 \to 2$, $2 \to 3$, and $3 \to 1$. This permutation is the function f above. Suppose

$$f = \begin{pmatrix} 123 \\ 231 \end{pmatrix} \quad \text{and} \quad g = \begin{pmatrix} 123 \\ 213 \end{pmatrix}$$

Then the function $g \circ f$ is given by

$$\begin{pmatrix} 123 \\ 213 \end{pmatrix} \circ \begin{pmatrix} 123 \\ 231 \end{pmatrix}$$

The permutation f on the right is applied first (by our definition of composition) and sends 1 to 2. Then g sends 2 to 1. The composition $g \circ f$ therefore sends 1 to 1. Also, f sends 2 to 3 and g sends 3 to 3, so that $g \circ f$ sends 2 to 3. Similarly, $g \circ f(3) = 2$. Hence

$$g \circ f = \begin{pmatrix} 123 \\ 213 \end{pmatrix} \circ \begin{pmatrix} 123 \\ 231 \end{pmatrix} = \begin{pmatrix} 123 \\ 132 \end{pmatrix}$$

We also see

$$f \circ g = \begin{pmatrix} 123 \\ 231 \end{pmatrix} \circ \begin{pmatrix} 123 \\ 213 \end{pmatrix} = \begin{pmatrix} 123 \\ 321 \end{pmatrix}$$

By rearranging the columns of the left factor in the middle term of this equation, we can put the top row of this factor in the same order as the bottom row of the right factor: hence

$$g \circ f = \begin{pmatrix} 231 \\ 132 \end{pmatrix} \circ \begin{pmatrix} 123 \\ 231 \end{pmatrix} = \begin{pmatrix} 123 \\ 132 \end{pmatrix}$$

2.6 Bijections

and

$$f \circ g = \begin{pmatrix} 213 \\ 321 \end{pmatrix} \circ \begin{pmatrix} 123 \\ 213 \end{pmatrix} = \begin{pmatrix} 123 \\ 321 \end{pmatrix}$$

The composition of the permutations is therefore easily obtained by making the indicated cancellation.

The set of all permutations of the numbers 1, 2, 3 is the same as the set of all bijections of the set $\{x_1, x_2, x_3\}$ except for the names we have called things. The number of bijections of the set $\{1, 2, 3\}$ can be determined as follows. There are three possible images for the element **1**. For each of these possibilities, there remain two possible images for the element **2**. For each of these possibilities, there remains only one possibility for the image of the element **3**. The total number of bijections of $\{1, 2, 3\}$ is therefore $3 \cdot 2 \cdot 1$, or **6**. The elements of $\mathbb{B}(\{1, 2, 3\})$ are the identity permutation

$$i = \begin{pmatrix} 123 \\ 123 \end{pmatrix}$$

and

$$\begin{pmatrix} 123 \\ 132 \end{pmatrix}, \begin{pmatrix} 123 \\ 213 \end{pmatrix}, \begin{pmatrix} 123 \\ 231 \end{pmatrix}, \begin{pmatrix} 123 \\ 312 \end{pmatrix}, \begin{pmatrix} 123 \\ 321 \end{pmatrix}$$

Definition   *For any $n \in \mathbb{N}$, the number **n factorial** is defined to be $1 \cdot 2 \cdots (n-1) \cdot n$ and is denoted by $n!$.*

Example 2:
By an argument like the one in Example 1, we see that there are $n!$ elements in the set $\mathbb{B}(\{1, 2, 3, \ldots, n\})$.

Example 3:
Consider the two ($2 = 2!$) elements in $\mathbb{B}(\{1, 2\})$. They are:

$$\begin{pmatrix} 12 \\ 12 \end{pmatrix} \quad \text{and} \quad \begin{pmatrix} 12 \\ 21 \end{pmatrix}$$

All possible compositions of these permutations are given by the list:

$$\begin{pmatrix} 12 \\ 12 \end{pmatrix} \circ \begin{pmatrix} 12 \\ 12 \end{pmatrix} = \begin{pmatrix} 12 \\ 12 \end{pmatrix}$$

$$\begin{pmatrix} 12 \\ 12 \end{pmatrix} \circ \begin{pmatrix} 12 \\ 21 \end{pmatrix} = \begin{pmatrix} 12 \\ 21 \end{pmatrix}$$

$$\begin{pmatrix} 12 \\ 21 \end{pmatrix} \circ \begin{pmatrix} 12 \\ 12 \end{pmatrix} = \begin{pmatrix} 12 \\ 21 \end{pmatrix}$$

$$\begin{pmatrix} 12 \\ 21 \end{pmatrix} \circ \begin{pmatrix} 12 \\ 21 \end{pmatrix} = \begin{pmatrix} 12 \\ 12 \end{pmatrix}$$

The list is conveniently displayed in the following table:

$\circ$	$\begin{pmatrix}12\\12\end{pmatrix}$	$\begin{pmatrix}12\\21\end{pmatrix}$
$\begin{pmatrix}12\\12\end{pmatrix}$	$\begin{pmatrix}12\\12\end{pmatrix}$	$\begin{pmatrix}12\\21\end{pmatrix}$
$\begin{pmatrix}12\\21\end{pmatrix}$	$\begin{pmatrix}12\\21\end{pmatrix}$	$\begin{pmatrix}12\\12\end{pmatrix}$

The table entry in the row labeled with a permutation—call it x—and the column labeled with a permutation—say, y—is $x \circ y$.

Exercises 2.6

Practice Exercise

1. Let E denote the set of even integers $\{\ldots, -4, -2, 0, 2, 4, \ldots\}$. Prove that $f: \mathbb{Z} \to E$ given by $f(x) = 2x$ is a bijection.

Problems

2. Let $f: [0, 1] \to [a, b]$ for $a < b$ in $\mathbb{R}$ be defined by $f(x) = a + (b-a)x$. Prove that f is a bijection.
3. Let $X = \{x \in \mathbb{R} \mid x \geq 0\}$. Prove that $f: X \to X$ defined by $f(x) = 3x$ is a bijection.
4. Decide and prove whether or not each of the following functions is a bijection:
   (a) $f: \mathbb{R} \to \mathbb{R}$ defined by $f(x) = 3x + 2$
   (b) $f: \mathbb{Z} \to \mathbb{Z}$ defined by $f(x) = 3x + 2$
   (c) $f: \mathbb{R} \to \mathbb{R}$ defined by $f(x) = x^2$
   (d) $f: \mathbb{R} \to \{x \in \mathbb{R} \mid x \geq 0\}$ defined by $f(x) = x^2$
   (e) $f: \mathbb{R} \to \mathbb{R}$ defined by $f(x) = x^3$
   (f) $f: \mathbb{Z} \to \mathbb{Z}$ defined by $f(x) = x^3$
5. According to Theorem 2.6.1, every bijection f of $\{1, 2, 3\}$ has an inverse f^{-1} that is also a bijection of $\{1, 2, 3\}$. Pair each permutation f in Example 1 with its inverse f^{-1}. Show by the cancellation rule for composition that $f \circ f^{-1} = i$ and $f^{-1} \circ f = i$ in each case.
6. Prove Theorem 2.6.1. Use previous theorems instead of going back to definitions.
7. Let $f: A \to B$ and $g: B \to C$ be bijections. Prove $(g \circ f)^{-1} = f^{-1} \circ g^{-1}$.
8. Example 3 contains a table that exhibits all the compositions of the elements in $\mathbb{B}(\{1, 2\})$. Construct an analogous table for the elements in $\mathbb{B}(\{1, 2, 3\})$.

9. Show that **arctan:** $\mathbb{R} \to (-\frac{\pi}{2}, \frac{\pi}{2})$ and **tan:** $(-\frac{\pi}{2}, \frac{\pi}{2}) \to \mathbb{R}$ are bijections. Use Exercises 2.3.8 (see the footnote) and 2.4.8 and Theorems 2.5.4 and 2.5.5.

2.7 INFINITE SETS

In the preceding section we considered bijections of a set A—one-to-one functions from A onto A. We wish now to consider the following question:

(1) If A is a nonempty set, does there exist a one-to-one function from A to A that is *not* onto?

It is frequently helpful in mathematics to generalize things somewhat and then come back to the original question. So we ask:

(2) When does there exist a one-to-one function $f: A \to B$ that is not onto?

Suppose we have a one-to-one function $f: A \to B$. The first case we wish to consider is that in which we can count, or enumerate, all the elements of A. In this case, let us call a_1 the first element counted, a_2 the second, and so on until the last element a_k has been counted. Then $A = \{a_1, a_2, \ldots, a_k\}$. Since $a_1 \neq a_2$ and $a_1 \neq a_3 \neq a_2, \ldots$, and since f is one-to-one, then $f(a_1) \neq f(a_2)$ and $f(a_1) \neq f(a_3) \neq f(a_2) \neq \cdots$. The elements $f(a_1), f(a_2), \ldots, f(a_k)$ are all distinct and all contained in B. The function f fails to be onto if there are elements in B in addition to these. So by counting $f(a_1)$ as the first element of B, $f(a_2)$ as the second, . . . , up to $f(a_k)$ as the kth, f will fail to be onto if we can extend this to count more elements in B than we count in A. Specializing this to $B = A$, we see that the function f gives us a second counting of the elements of A. There are k elements in A by the first count: $a_1, a_2, \ldots, a_k$. And there are at least k elements, $f(a_1), f(a_2), \ldots, f(a_k)$, by the second count.

When a child who has just learned to count counts the number of figures, say boxes, on a page, she is initially surprised to find that two counts of the boxes in different orders result in the same number. The child does not yet understand the concept of the "number of boxes" on the page. In fact, it is the experience of getting the same count each time that supports the idea of "number of boxes". Once the concept is grasped, the child will confidently assert there are say, five, boxes after only one count. She knows that a second count, in any order, will also yield five—because "there are five boxes" on the page.

According to this concept—probably the first significant mathematical concept for each of us—we know that a second count of the elements of A will give the same number as the first. That is, if by first count we have $a_1, a_2, \ldots, a_k$, then by a second count, the elements $f(a_1), f(a_2), \ldots, f(a_k)$ exhaust the elements of A because "there are k elements" in A. This gives us an intuitive answer to Question (1) in the case where we can count all the elements of A. All one-to-one functions from such an A to A itself must be onto. This is not a formal proof of the fact but an observation that follows from our basic mathematical intuition.

Chapter 2 Functions

Now consider a case where we can't count all the elements of A. For example, consider $A = \mathbb{N}$. Here, there is a one-to-one function from $\mathbb{N}$ to itself that is not onto.

Theorem 2.7.1 *There exists a one-to-one function $g: \mathbb{N} \to \mathbb{N}$ that is not onto.*

Our first task is to define g. Next we must show that the g we have defined is one-to-one. Then we must show that g is not onto.

The proof of this theorem requires that we invent something, namely, g. This is not hard, but we must think of the thing to do. Analyzing proof steps will not suggest it. Theorem 2.7.1 is therefore a small first step toward more creative proofs.

Proof:

Define $g: \mathbb{N} \to \mathbb{N}$ by the rule $g(n) = n + 1$ for all $n \in \mathbb{N}$. First we show that g is one-to-one. Let $g(n_1) = g(n_2)$ for $n_1, n_2 \in \mathbb{N}$. By definition of g, $n_1 + 1 = n_2 + 1$, so that $n_1 = n_2$. Hence g is one-to-one. Next, in order to get a contradiction, assume that g is onto. Then $g(n_0) = 1$ for some $n_0 \in \mathbb{N}$. Thus $n_0 + 1 = 1$, and so $n_0 = 0$, which contradicts $n_0 \in \mathbb{N}$. Hence g is one-to-one but not onto. ∎

The discussion above motivates the following definition:

Definition *A set A is **infinite** if there is a function $f: A \to A$ that is one-to-one but not onto. A set that is not infinite is called **finite**.*

In order that this definition of "infinite" be a good definition, it should conform to our imaginative idea of an infinite set. Finite sets are those that extend our experience with sets of physical things—the set of boxes on a page or marbles in a box. These are the sets we are certain have the property that, if we count all the elements two times, we come up with the same number of elements—the sets for which all one-to-one functions are onto. On the other hand, infinite sets, like $\mathbb{N}$, are those that lack this property. In an infinite set, we can go on counting forever and never exhaust the set, so that we can construct one-to-one functions that are not onto. Theorem 2.7.1 can be reworded using our new definition:

Theorem 2.7.1 $\mathbb{N}$ *is infinite.*

Consider the following theorem:

2.7 Infinite Sets

Theorem 2.7.2 *If A is infinite and $A \subseteq B$, then B is infinite.*

One might think that this theorem is obvious, but this is not so. Even though the word "infinite" conforms to our imaginative idea, a proof of Theorem 2.7.2 must not *depend* on this idea. The idea of infinite might be needed to *find* a proof of Theorem 2.7.2, but the actual proof must not depend on the objects of our imagination. Consider again the definition and theorem:

Definition *A set A is **infinite** if there is a function $f: A \to A$ that is one-to-one but not onto.*

Theorem 2.7.2 *If A is infinite and $A \subseteq B$, then B is infinite.*

Consider also the following definition and theorem:

Definition *A set A is **fragile** if there is a function $f: A \to A$ that is one-to-one but not onto.*

Theorem 2.7.2f *If A is fragile and $A \subseteq B$, then B is fragile.*

It is very important to realize that Theorem 2.7.2 is no more obvious than Theorem 2.7.2f. In fact, from a formal viewpoint the two theorems are exactly the same. Prior to its definition, the word "infinite" has no formal meaning. This is as it should be in a mathematics course. We want "infinite" to mean exactly what its definition says it means without anything extra smuggled in by our imaginations. The same applies to "fragile". It too must mean exactly what its definition says it means. But the definitions of "infinite" and "fragile" are exactly the same. From a formal viewpoint, the two words are merely *identifiers* in the computer-language sense. Either could be replaced by any string of characters.

In any valid proof of Theorem 2.7.2, we could replace the word "infinite" with "fragile" and get a valid proof of Theorem 2.7.2f. It is therefore the syntax in which we find the identifier "infinite" or "fragile" that formally determines its meaning. This point was made by David Hilbert (1862–1943) when (where may be surmised) he asserted that any valid geometry proof about points, lines, and planes must remain valid when "points", "lines", and "planes" are replaced by "tables", "chairs", and "beer mugs".

The proof of Theorem 2.7.2 is for you to attempt (Exercise 6).

Theorem 2.7.2 has an easy corollary:

Corollary 2.7.3 *If B is finite and $A \subseteq B$, then A is finite.*

Proof: By contradiction. ■

A bijection is frequently called a one-to-one correspondence since it pairs elements in its domain with elements in its codomain. If $f: A \to B$ is a bijection, then the sets A and B are said to be in a one-to-one correspondence. In this case, and if A and B are finite, then it is easy to see that A and B must have the same number of elements. This may be easy to see intuitively, but by now you should also be able to see that it can't be proved—simply because we have not defined "number of elements". We can't give a formal proof that A and B have a certain property unless we have defined that property using only previously defined terms. If we can't give a formal proof, then we can't give an informal one either, since an informal proof is just a contraction of a formal one.

Definition *Sets A and B are said to be **equipotent** or have the **same cardinality** if there exists a bijection $f: A \to B$. We write $A \sim B$ to denote that A and B have the same cardinality.*

The idea of cardinality corresponds roughly to the idea of the number of elements in a set. For finite sets, these ideas are the same. We are not yet in a position to define "cardinality", and so we have defined "same cardinality" instead. This is just what we did for sets and functions. "Set" and "function" were not themselves formally defined, and so it was necessary to define "equality" in these two cases. If "function" were defined, then equal functions would simply be functions that were the same under the definition. We will define "function" in the next section. Clearly, however, not everything can be defined using only previously defined terms. We must start somewhere. In most treatments of mathematics, "set" is taken to be a primitive concept, that is, an undefined term.

The following theorem on equipotent sets follows the pattern of the equality laws for sets (page 83):

Theorem 2.7.4 *For all sets A, B, and C:*

(a) $A \sim A$ *(reflexive property)*
(b) *If $A \sim B$, then $B \sim A$.* *(symmetric property)*
(c) *If $A \sim B$ and $B \sim C$, then $A \sim C$.* *(transitive property)*

Proof of (a) and (c): Exercise 1.

Proof of (b):

Assume $A \sim B$. Then there exists a bijection $f: A \to B$. Then $f^{-1}: B \to A$ exists and is also a bijection that shows $B \sim A$. ∎

Theorem 2.7.5 *If A is infinite and $A \sim B$, then B is infinite.*

Proof:

Assume A is infinite and $A \sim B$. We will show B is infinite.

2.7 Infinite Sets

> **Scrap Paper**
> Consider the conclusion. We see we need to show **there exists** $f: B \to B$ **such that f is one-to-one and not onto**. In order to define such an f, we construct it from the functions whose existence is given by the hypotheses:
>
>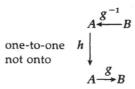
>
> one-to-one h
> not onto

By hypothesis, there exists a bijection $g: A \to B$ and a one-to-one function $h: A \to A$ that is not onto. Then $g^{-1}: B \to A$ is also a bijection. Let $f = g \circ h \circ g^{-1}$. Then f is one-to-one as the composition of one-to-one functions. Assume that f is onto in order to get a contradiction. Then $g^{-1} \circ f \circ g$ is a composition of one-to-one, onto functions, and is therefore one-to-one and onto. But $g^{-1} \circ f \circ g = g^{-1} \circ g \circ h \circ g^{-1} \circ g = h$, which is given as not onto. By contradiction, then, f is not onto. The existence of f shows that B is infinite, by definition. ∎

Definition *For sets A and B, if $A \subseteq B$ but $A \neq B$, then A is called a **proper subset** of B.*

Theorem 2.7.1 tells us that $\mathbb{N}$ is infinite, and since $\mathbb{N} \subseteq \mathbb{Z}$ by Theorem 2.7.2, $\mathbb{Z}$ is therefore infinite. It is possible for an infinite set to have the same cardinality as one of its proper subsets. (Indeed, by definition it must be equipotent with some proper subset.) For example, if E is the set of even integers, the function $f: \mathbb{Z} \to E$ given by $f(x) = 2x$ is a bijection (Exercise 2.6.1), so that $\mathbb{Z} \sim E$.

Theorem 2.7.6 $\mathbb{Z} \sim \mathbb{N}$.

Proof: Exercise 2.

Definition *For sets A and B, we say that A has **smaller cardinality** than B if there exists a one-to-one function $f: A \to B$ but A and B do not have the same cardinality. We write $A < B$ to denote that A has smaller cardinality than B. We write $A \lesssim B$ to mean $A < B$ or $A \sim B$.*

Thus $A \lesssim B$ means that there is a one-to-one function $f: A \to B$, and $A < B$ means that there is a one-to-one function $f: A \to B$ but no function $f: A \to B$ that is one-to-one and onto.

Theorem 2.7.7 If A is infinite and $A \lesssim B$, then B is infinite.

> **Proof:**
> Assume A is infinite and $A \lesssim B$. By the second hypothesis, there exists a one-to-one function $f: A \to B$. Then $f: A \to f(A)$ is one-to-one and onto, so that $A \sim f(A)$. By the first hypothesis and Theorem 2.7.5, $f(A)$ is infinite. But $f(A) \subseteq B$, so by Theorem 2.7.2 B is infinite. ∎

Corollary 2.7.8 If A is infinite and $A < B$, then B is infinite.

Proof: Exercise 3.

Corollary 2.7.9 If B is finite and $A \lesssim B$, then A is finite.

Proof: Exercise 4.

Theorem 2.7.10 Transitivity of $\lesssim$: For all sets A, B, and C: if $A \lesssim B$ and $B \lesssim C$, then $A \lesssim C$.

Proof: Exercise 5.

Additional Proof Ideas

Transitivity of <

Consider the following analog to Theorem 2.7.10:

Theorem T For all sets A, B, and C: if $A < B$ and $B < C$, then $A < C$.

While Theorem T is true, it is very difficult to prove. Consider the following analysis. By the hypotheses, we have one-to-one functions $f: A \to B$ and $g: B \to C$. The composition $g \circ f: A \to C$ shows that $A \lesssim C$. It remains to show that there is no one-to-one function from A onto C. In order to get a contradiction, suppose the contrary. Then there exists $k: A \to C$ that is one-to-one and onto. Then $k^{-1}: C \to A$ is also one-to-one and onto, so that $f \circ k^{-1}: C \to B$ is one-to-one. Thus $C \lesssim B$. We already have $B \lesssim C$ from $B < C$. Suppose we knew the following theorem:

Theorem S-B For sets B and C: if $B \lesssim C$ and $C \lesssim B$, then $B \sim C$.

The conclusion $B \sim C$ would then contradict our hypothesis, $B < C$, so that the proof of Theorem T would follow by contradiction.

2.7 Infinite Sets

Thus Theorem T follows from Theorem S-B. It is not hard to show that Theorem S-B follows from Theorem T, so that the two theorems are equivalent: Assume Theorem T and $B \lesssim C$ and $C \lesssim B$. If $B \sim C$ is not true, then we have $B < C$ and $C < B$, so that by Theorem T, $B < B$. Now, $B < B$ is obviously contradicted by the identity function $i_B : B \to B$. The contradiction shows that $B \sim C$—proving Theorem S-B assuming Theorem T.

Theorem S-B, the Schroeder-Bernstein Theorem, is proved in Chapter 5. Theorem T is given as a corollary.

Formality of Mathematics and Proof

The discussion of the prospective proof of Theorem 2.7.2 has the shocking effect of making one wonder whether mathematics is not some formal game with meaningless symbols. This point of view, known as *formalism*, has always been rejected by most mathematicians. Formalism was dealt a death blow in 1931 by a theorem of Kurt Gödel (1906–1978), which showed that, in any formal system large enough to contain the theory of the natural numbers, there will always remain truths about the natural numbers that cannot be proved. Gödel's proof, of course, is perfectly formal.

While most mathematicians reject a formalistic view of mathematics, they agree on a formalistic view of proof. Mathematics is a thing of the imagination. We want our definitions to faithfully represent the ideas of our imaginations, but these ideas must not be used to validate a proof. A proof must depend on formal definitions only. The fundamental role played by axioms and definitions in a modern proof is the reason for our text's emphasis on developing straightforward proofs from the definitions. This is proof at its basic level.

This idea of a formal proof's depending only on definitions is implicit in texts for graduate courses and advanced undergraduate courses like advanced calculus and abstract algebra. The "rules of the game" in these courses are therefore different from the rules that apply in high-school (Euclidean) geometry, where proofs depend on our imaginative ideas represented in sketches. (This dependence on imagination in high-school geometry is global, not local. We are allowed to use sketches to decide, for example, which side of a line a point falls on but not to prove exact equality of line lengths or angle size.) Hilbert provided an axiom system for geometry in which ideas like "betweenness", for points on a line, are precisely defined and not inferred from sketches.

Exercises 2.7

Problems

1. Prove Theorem 2.7.4a and c.
2. Prove Theorem 2.7.6. Define a function $f: \mathbb{Z} \to \mathbb{N}$ by using different rules for the negative and non-negative integers.

3. Prove Corollary 2.7.8.
4. Prove Corollary 2.7.9.
5. Prove Theorem 2.7.10.

Harder Problem

6. Make an attempt at proving Theorem 2.7.2. This is not just a straightforward problem. It is a step toward more creative proofs (somewhat harder than Theorem 2.7.1). As in the proof of Theorem 2.7.1, you must invent a function. Analyzing proof steps will not suggest it. Analyze the definitions to see what is needed. If you get stuck, do something else and let your subconscious work on the problem. If your subconscious invents the right solution, you will have benefited from the problem and the next one will be a little easier. Chances are you will need to ask your instructor for some kind of hint, but it is a good exercise to see just how far you can go before you ask.

2.8 PRODUCTS, PAIRS, AND DEFINITIONS

The set $\{3, 5\}$ is the same as the set $\{5, 3\}$, whereas the ordered pairs $(3, 5)$ and $(5, 3)$ are different. The two ordered pairs represent different points in the coordinate plane. We would like to define the idea of an ordered pair of either numbers or elements in a set. This definition will be necessary, of course, in order for us to prove facts about ordered pairs.

The critical property we wish to establish from the definition is that the ordered pair (a, b) is the same as the ordered pair (c, d) if and only if $a = c$ and $b = d$. This property can't be considered a definition because it doesn't tell us what an ordered pair *is*. (We don't know formally what a set is either, but the idea in mathematics is to keep the number of undefined things to a minimum.)

Definition Let A and B be sets. For any $a \in A$, $b \in B$, the **ordered pair** (a, b) is the set $\{\{a\}, \{a, b\}\}$.

This unlikely looking candidate for the role of ordered pair will do the job required; that is, with this definition we can prove the following theorem:

Theorem 2.8.1 For $a_1, a_2 \in A$ and $b_1, b_2 \in B$, we have $(a_1, b_1) = (a_2, b_2)$ iff both $a_1 = a_2$ and $b_1 = b_2$.

Proof: Exercise 7.

Theorem 2.8.1 embodies the property we wish to be characteristic of ordered pairs. After we use the definition above to prove Theorem 2.8.1, we will never

2.8 Products, Pairs, and Definitions

have to use this definition again. It serves only to reduce the number of undefined terms. This same sort of trick can be used to define "function" in terms of sets. For this we need the following:

Definition Let X and Y be sets. The **Cartesian product** of X and Y (denoted $X \times Y$) is the set $\{(x, y) \mid x \in X, y \in Y\}$.

Example 1:
If $X = \{(1, 2\}$ and $Y = \{2, 3, 4\}$, we have $X \times Y = \{(1, 2), (1, 3), (1, 4), (2, 2), (2, 3), (2, 4)\}$

Example 2:
Since $\mathbb{R}$ is the set of real numbers, $\mathbb{R} \times \mathbb{R}$ is the set of all ordered pairs of real numbers, represented by the entire coordinate plane.

Example 3:
Consider $[0, 1] = \{x \in \mathbb{R} \mid 0 \leq x \leq 1\}$. Then $[0, 1] \times [0, 1]$ is the set of ordered pairs (x, y) with $0 \leq x \leq 1$ and $0 \leq y \leq 1$, represented by the shaded area in Figure 2.8.1 (following page).

Example 4:
$A \times B = \emptyset$ iff $A = \emptyset$ or $B = \emptyset$.

Proof:
We need to prove if $A \times B = \emptyset$, then $A = \emptyset$ or $B = \emptyset$ and also the converse of this statement. In order to show if $A \times B = \emptyset$, then $A = \emptyset$ or $B = \emptyset$, we will show the contrapositive instead, namely, *if $\neg(A = \emptyset$ or $B = \emptyset)$, then $A \times B \neq \emptyset$*. That is, if $(A \neq \emptyset$ and $B \neq \emptyset)$, then $A \times B \neq \emptyset$. So assume $A \neq \emptyset$ and $B \neq \emptyset$. Then there exists $a \in A$ and $b \in B$, so that $(a, b) \in A \times B \neq \emptyset$.

Also, to show if $A = \emptyset$ or $B = \emptyset$, then $A \times B = \emptyset$, we again use the contrapositive: if $A \times B \neq \emptyset$, then $A \neq \emptyset$ and $B \neq \emptyset$. Assume $A \times B \neq \emptyset$. Then there exists $(a, b) \in A \times B$, so that $a \in A$ and $b \in B$. ∎

Note that the contrapositives of the two implications in Example 4 were useful since they gave us nonempty sets to work with.

As an attempt at a formal definition of "function", one frequently sees the following:

Definition Let A and B be nonempty sets. A **function** f from A to B is a subset of $A \times B$ such that (1) if $(x, y_1) \in f$ and $(x, y_2) \in f$, then $y_1 = y_2$ and (2) if $x \in A$, then $(x, z) \in f$ for some $z \in B$.

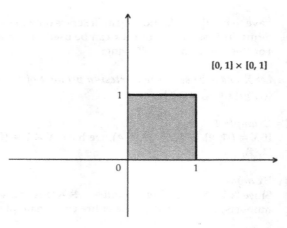

Figure 2.8.1

If $f: A \to B$ is a function according to this definition, the pair (x, y) is in f when f is viewed as mapping x to y. Thus $(x, y) \in f$ and $f(x) = y$ mean the same thing. f must map x to a unique element y in B. Part (1) of the definition assures uniqueness by the usual scheme: we assume different names, y_1 and y_2, for the element in B to which x maps and then require $y_1 = y_2$. Part (2) assures us that the rule for mapping x applies to all elements of A.

The problem with the definition above is that it doesn't quite tell us what a function *is*. A function must be more than just a set of ordered pairs since, from the set of ordered pairs alone, it is not possible to determine the codomain of the function. The domain of the function may be determined as the set of all first coordinates, but if we try to determine the codomain the same way, we get that the function is onto. Using the definition above would mean that all functions were either onto or had an unspecified codomain—where it could not be determined from the definition whether or not they were onto. Since our goal is to reduce the number of undefined terms by defining "function", the definition above will not do. For "function" to be properly defined, *all* properties must follow from the definition.

To this end, we first define the ordered triple (a, b, c) as $((a, b), c)$. With this definition we can prove:

Theorem 2.8.2 *For $a_1, a_2 \in A$, $b_1, b_2 \in B$, $c_1, c_2 \in C$, we have $(a_1, b_1, c_1) = (a_2, b_2, c_2)$ iff $a_1 = a_2$, $b_1 = b_2$, and $c_1 = c_2$.*

Proof: Exercise 8.

This theorem makes possible the formal definition of "function":

2.8 Products, Pairs, and Definitions

Definition *A function $f: A \to B$ is a triple (A, B, f), where f is a subset of $A \times B$ such that:*

(1) *if $(x, y_1) \in f$ and $(x, y_2) \in f$, then $y_1 = y_2$*

and (2) *if $x \in A$, then $(x, z) \in f$ for some $z \in B$*

Here A is called the *domain* of $f: A \to B$ and B is called the *codomain*. From this definition, equal functions have the same codomain (Exercise 11).

Since we now know what a function is by definition, we can no longer define what we mean by "equal" functions. "Equal" must mean "same" according to the definition. Thus our former definition of equal functions ought to be a theorem:

Theorem 2.8.3 *Two functions $f: A \to B$ and $g: A \to B$ are equal iff for all $x \in A$: $f(x) = g(x)$.*

Proof: Exercise 10.

There will be no occasion where we will need to use the definition of function. Instead, we will appeal to Theorem 2.8.3 (or, equivalently, to the definition of equal functions). This parallels the situation for ordered pair, where the useful characterization is given by a theorem instead of the definition.

We can also use the Cartesian product to provide a formal definition of "binary operation". A binary operation on a set A takes two elements from A and obtains a third element of A. For example, addition and multiplication are binary operations on the set $\mathbb{Z}$ of integers: Addition takes two integers, say, a and b, and from them, produces a third integer called the sum of a and b, which is denoted by $a + b$. A binary operation is therefore a rule that tells us how to obtain a third element in a set, given any two elements. As another example, consider the set $\mathbb{B}(A)$ of all bijections of a set A. If f and g are two elements of $\mathbb{B}(A)$, then we obtain a third element $f \circ g \in \mathbb{B}(A)$ by composition. Composition is a binary operation on $\mathbb{B}(A)$.

Just as "function" can be defined formally in terms of ordered pairs, so "binary operation" can be defined formally in terms of functions:

Definition *For any nonempty set A, a **binary operation** on A is a function from $A \times A$ to A.*

Example 5:
The binary operation $+$ on the set $\mathbb{Z}$ is a function $+: \mathbb{Z} \times \mathbb{Z} \to \mathbb{Z}$. If a and b are two integers, then the function $+$ of the pair (a, b) is denoted in functional notation as $+(a, b)$. This is called *prefix* notation. It is customary to denote $+(a, b)$ by $a + b$. This is called *infix* notation. It is customary to denote all binary operations by infix notation.

Example 6:
The commutative property of addition on $\mathbb{Z}$ is given, in infix notation, by **for all** $a, b \in \mathbb{Z}$: $a + b = b + a$. The same property in prefix notation is **for all** $a, b \in \mathbb{Z}$: $+(a, b) = +(b, a)$.

Example 7:
If A is a nonempty set, then function composition ∘ is a binary operation on the set $\mathbf{B}(A)$. ∘ (f, g), given in prefix notation, is denoted by $f \circ g$ in infix notation.

The purpose of this section has been to show how the ideas of function and binary operation can be formally defined. In the remainder of the text, we will continue to think of a function, as most people do, in our former way—as a rule for mapping elements in the domain to elements in the codomain. Similarly, when we get to binary operations in the next chapter, we will consider them as rules for obtaining a third element in a set from two given elements in the set. We will not use the formal definition of binary operation.

Additional Proof Ideas

A function $f: A \rightarrow B$ is thought of as a rule for mapping elements of A to elements in B. "Rule" is undefined. In order to define a particular function $f: A \rightarrow B$, we must know the rule. On the other hand, if we are to define a particular function $f: A \rightarrow B$ using the formal definition of function, we must specify f as a particular subset of $A \times B$. Thus we still need a rule to know which elements of $A \times B$ are in the subset f and which are not. (Recall that although the idea of set is undefined, a particular set is given by a rule for deciding set membership.) Thus, at our level, where sets and functions are given by rules and "rule" is undefined, there is little to be gained by the formal definition of function.

Although most mathematicians in their everyday work think of functions as rules of correspondence, a formal definition of function does shed some light on the attempt to provide a firm foundation for mathematics in terms of set theory and logic. Our rule for determining set membership by a property is made more precise with the idea of a *predicate*. A predicate is a precisely defined (syntactically) formula $\mathcal{P}(x)$ that has the property that every substitution for x from a universal set $\mathcal{U}$ assigns to $\mathcal{P}(x)$ either the value *true* or the value *false*. The predicate $\mathcal{P}(x)$ determines the set $\{x \in \mathcal{U} \mid \mathcal{P}(x)$ is **true**$\}$. We stipulate that all sets are formed in this way. A function $f: A \rightarrow B$ is given by the set $\{(a, b) \in A \times B \mid \mathcal{P}(a, b)\}$ for some predicate $\mathcal{P}$.

Suppose that $f: A \rightarrow B$ is onto. Suppose also for the moment that $B = \{b_1, b_2, b_3\}$. We will use the idea of a predicate and the formal definition of function to obtain a right inverse of the onto function f. From $b_1 \in B$ and the onto statement **for all** $y \in B$: $\mathcal{P}(x, y)$ **for some** $x \in A$, we get $\mathcal{P}(x, b_1)$ **for some** $x \in A$. Therefore, pick $x_1 \in A$ such that $\mathcal{P}(x_1, b_1)$. We also get x_2 and x_3 in A such that $\mathcal{P}(x_2, b_2)$ and $\mathcal{P}(x_3, b_3)$. A predicate defining the function g in the proof of Theorem 2.5.2 is easy to construct: We want g to be of the form $\{(b, a) \in B \times A \mid \mathcal{Q}(b, a)\}$, where $\mathcal{Q}$ can be the statement "$(b = b_1$ **and** $a = x_1)$ **or** $(b = b_2$ **and** $a = x_2)$ **or** $(b = b_3$ **and** $a = x_3)$". If B is an infinite set, however, then the predicate $\mathcal{Q}$ cannot be constructed in this way. (We don't allow infi-

2.8 Products, Pairs, and Definitions

nitely long statements.) We cannot, therefore, construct the right inverse of an onto function in this way. This shows where the Axiom of Choice is needed.

The ordered-pair definition of function makes it possible to define the mathematical idea of formula or rule in terms of the logical idea of predicate. Although the real need for the Axiom of Choice and the ordered-pair definition of function would not ordinarily appear at our level, we have included them since their use has commonly filtered down to everyday informal mathematics.

Exercises 2.8

Practice Exercises

1.
   1. $(a, b) = (c, d)$
   2. * (1; Thm. 2.8.1)
   3. * (1; Thm. 2.8.1)
2. The following properties of $\mathbb{Z}$ are written in the usual infix notation. Write them in prefix notation:
   (a) for all $a, b, c \in \mathbb{Z}$: $a \cdot (b \cdot c) = (a \cdot b) \cdot c$
   (b) for all $a, b, c \in \mathbb{Z}$: $a \cdot (b + c) = (a \cdot b) + (a \cdot c)$

Problems on Cartesian Products

3. Let $A = \{1, 2, 3, 4\}$ and $B = \{x, y, z\}$. Find $A \times B$.
4. Let $A = \{1, 2, 3\}$ and $B = \{1, 2\}$. Find $A \times B$ and $B \times A$. Is $A \times B = B \times A$?
5. Let A and B be nonempty sets. Prove that there exists a one-to-one function from $A \times B$ onto $B \times A$.
6. Let $A, B, C, B_1, \ldots, B_n$ be sets. Prove or disprove:
   (a) $A \times (B \cup C) = A \times B \cup A \times C$
   (b) $A \times (B \cap C) = A \times B \cap A \times C$
   (c) $A \times (B - C) = A \times B - A \times C$
   (d) $A \times \bigcup_{i=1}^{n} B_i = \bigcup_{i=1}^{n} A \times B_i$
   (e) $A \times \bigcap_{i=1}^{n} B_i = \bigcap_{i=1}^{n} A \times B_i$
7. Prove Theorem 2.8.1. The difficulty with this problem is keeping track of all the cases. Use arguments like the following: if $\{x, y\} \subseteq \{a, b\}$, then $x = a$ or $x = b$ by definition of set containment.
8. Prove Theorem 2.8.2.

Problems on the Formal Function Definition

9. Let $A = \{1, 2, 3\}$ and $B = \{1, 2, 3\}$. For each of the following subsets f of $A \times B$, decide whether (A, B, f) is a function and, if so, whether it is one-to-one or onto.

    (a) $f = \{(1, 1), (2, 2), (3, 3)\}$
    (b) $f = \{(1, 1), (1, 2), (2, 3)\}$
    (c) $f = \{(1, 3), (2, 1), (3, 2)\}$
    (d) $f = \{(1, 1), (2, 2)\}$
    (e) $f = \{(1, 2), (2, 2), (3, 2)\}$
    (f) $f = \{(2, 1), (2, 2), (2, 3)\}$

10. To prove Theorem 2.8.3, we need to show $f = g$ iff **for all** $x \in A: f(x) = g(x)$. In order to avoid stumbling over notation, we can rewrite this **for all** statement: **for all** $x \in A$: if $(x, y) \in f$ **and** $(x, z) \in g$, **then** $y = z$. Prove Theorem 2.8.3 by showing this **for all** statement holds iff $f = g$. Note: by definition f and g are sets.

11. Show that equal functions defined by the formal definition have the same codomain.

12. Let $f: A \to C$ and $g: B \to D$. Define a function $f \cup g : A \cup B \to C \cup D$ by the rule $(f \cup g)(x) = f(x)$ if $x \in A$, and $(f \cup g)(x) = g(x)$ if $x \in B$.

    (a) What condition is needed in order to make $f \cup g$ well defined? Assume this condition for parts (b) and (c).
    (b) Show that even if f and g are both one-to-one, $f \cup g$ need not be one-to-one.
    (c) Show that if f and g are one-to-one, $A \cap B = \emptyset$ and $C \cap D = \emptyset$, then $f \cup g : A \cup B \to C \cup D$ is one-to-one.

Chapter 3

Relations, Operations, and the Integers

3.1 INDUCTION

Appendix 1 gives certain properties for $\mathbb{Z}$, called axioms, from which other properties can be derived. Theorems in Appendix 1, derived from the axioms, include the well-known properties useful for working with equations and inequalities, which can be characterized as the algebraic and order properties of $\mathbb{Z}$. They are now introduced early in the school curriculum and repeated frequently. For a completely rigorous treatment of the material in this chapter, Appendix 1 would now need to be covered in order to give formal proofs for these properties. Alternatively, we can merely use any of the axioms or theorems in Appendix 1 as justification for proof steps. This will be our approach in this chapter. The cited justification will be "property of $\mathbb{Z}$" in a manner analogous to our use of "property of $\mathbb{R}$" in examples. Basically, "property of $\mathbb{Z}$" allows you to do any familiar calculation with integers. If in doubt, consult Appendix 1.

Consider the set $\mathbb{N}$ of natural numbers. It is plausible that any nonempty subset S of $\mathbb{N}$ will have a least element: Since S is nonempty, there is some $x_0 \in S$. Either x_0 is the smallest element of S or it is not. If not, then there is an $x_1 < x_0$ in S. Either x_1 is smallest or there is an $x_2 < x_1 < x_0$, and so forth. Note that $x_1 \leq x_0 - 1$ and $x_2 \leq x_0 - 2$. If we continue, we will either find a smallest element in S or get $x_n \leq x_0 - n$ for each $n = 1, 2, 3, \ldots$. For $n = x_0$, we get $x_n \leq x_0 - x_0 = 0$—so that $x_n \notin S$. Therefore we must find a smallest element before we get to $n = x_0$ in this process.

We have not proved that S has a least element, of course, since we don't know formally what "continue", "and so forth", or "process" mean. The property of $\mathbb{N}$ stated above is referred to by saying that $\mathbb{N}$ is *well ordered*. This property is

173

not formally proved but is taken as a foundational property (an *axiom*) from which other properties may be derived. We now list this axiom of the natural numbers:

Well-Ordering Axiom *Every nonempty subset of $\mathbf{N}$ has a smallest element.*

This property will be mentioned explicitly when used as a reason.

Recall that a set A is finite provided that all one-to-one functions from A to itself are onto. This means that if we *can* count all the elements of A, then every time we count them we get the same number. This in turn implies that for a finite set, our concept of "number of elements in a set" is consistent.

Recall that $\{1, 2, 3, \ldots, k\}$ is denoted by $\mathbf{N}_k$. Exercise 8 gives a method, and some hints, for proving the first consequence of the Well-Ordering Axiom, which is:

Theorem 3.1.1 *For each $k \in \mathbf{N}$, the set $\mathbf{N}_k$ is finite.*

Proof: Exercise 8.

This theorem illustrates the relationship of formal mathematics to intuition. It is obvious to our intuitive understanding of the word "finite". In fact, it is just what we want "finite" to mean. A proof of Theorem 3.1.1, however, is a demonstration that it follows from the formal definition of "finite". We may make definitions in any way we wish, but if our formal definition of finite is to be a good one, the fact expressed in Theorem 3.1.1 must be a necessary consequence of the definition.

The next consequence of the Well-Ordering Axiom is one of the most powerful ideas in mathematics:

Theorem 3.1.2 *Induction: Let S be a subset of $\mathbf{N}$ such that*

    (1) $1 \in S$

and   (2)   *for all $n \in \mathbf{N}$: if $n \in S$, then $n + 1 \in S$*

Then $S = \mathbf{N}$.

Proof:

We have $S \subseteq \mathbf{N}$ by hypothesis. To show the reverse inclusion, let $D = \{k \in \mathbf{N} \mid k \notin S\}$. Suppose we can show $D = \emptyset$. Then, to obtain $\mathbf{N} \subseteq S$, let $x \in \mathbf{N}$. If $x \notin S$, then $x \in D$ by definition, which is a contradiction. Therefore $x \in S$. Whence $\mathbf{N} \subseteq S$. Since we then have both inclusions, $\mathbf{N} = S$.

It therefore suffices to show $D = \emptyset$. To do this, suppose the contrary, that is, that D is nonempty. By the well-ordering of $\mathbf{N}$, D contains a smallest element d. Now $d \neq 1$ since $1 \in S$ by hypothesis. Therefore $d - 1$ is a positive integer (property of the integers), and since d is the smallest element

3.1 Induction

in D, $d-1 \notin D$. Therefore $(d-1) \in S$ (by definition of D). By part (2) of the hypothesis, $(d-1)+1 \in S$. But this contradicts the fact that $d \in D$. The contradiction shows $D = \emptyset$. ∎

This theorem is used in doing proofs "by mathematical induction". We illustrate this in our first example:

Example 1:
The following assertion holds for all $n \in \mathbb{N}$:

$$(*) \quad 2+4+6+\cdots+(2n) = n(n+1)$$

If $n \in \mathbb{N}$, then $2n$ is an even natural number. The expression $2+4+6+\cdots+(2n)$ means the sum of all even numbers up to and including $2n$. If $n = 1$, then the expression is taken to mean just 2 (or equivalently, $2n$) since 2 is the only even number up to 2.

Proof of Example 1:

Define S to be the set of all $n \in \mathbb{N}$ for which the assertion (*) is true. We will show:

(1) $1 \in S$

and (2) for all $n \in \mathbb{N}$: if $n \in S$, then $n+1 \in S$

From this it follows by Theorem 3.1.2 that $\mathbb{N} = S$, so that the assertion is true for all $n \in \mathbb{N}$. First, if $n = 1$, then our assertion is that $2 = 2$—which is true. Therefore $1 \in S$.

Second, suppose $n \in S$. Then $2+4+6+\cdots+(2n) = n(n+1)$. By adding $2n+2$ to each side, we get:

$$2+4+6+\cdots+(2n)+2(n+1) = n(n+1)+2(n+1)$$

or $\quad 2+4+6+\cdots+(2n)+2(n+1) = (n+1)(n+2)$

or $\quad 2+4+6+\cdots+(2n)+2(n+1) = (n+1)((n+1)+1)$

This is the same as statement (*) with n replaced by $n+1$. Thus $n+1 \in S$ (recall the definition of S), proving (2). Therefore $\mathbb{N} = S$ by Theorem 3.1.2. ∎

A proof done according to the preceding scheme is said to be a proof "by induction on n". The same scheme will work for the straightforward exercises at the end of this section. Thus a proof by induction consists of two parts: (1) showing that the assertion holds for $n = 1$ and (2) showing that, if the assertion is true for an arbitrary n, then it is true for $n+1$. In proving (2), many people find it convenient to use slightly different notation: we assume the truth of the assertion for n, let $k = n+1$, and then show that the assertion holds for k. Thus the statement to be shown in (2) has exactly the same form as the statement assumed except that k has replaced n. In proving (1), the number 1 is substituted for n. For example, a proof of Example 1 (by induction) would take the following form:

First, show: $\quad 2 = 1(1+1)$

Proof: Property of **N**.

Second, assume: $2+4+6+\cdots+(2n) = n(n+1)$
$\qquad\qquad\qquad k = n+1$
Show: $\qquad 2+4+6+\cdots+(2k) = k(k+1)$

The proof of this last statement follows our proof of Example 1 except that now k is substituted for $n+1$, giving an additional line—the thing to be shown. Either scheme for doing proofs by induction is acceptable.

Example 2:
Suppose that x and y are positive real numbers, with $x < y$. Prove that **for all** $n \in \mathbf{N}: x^n < y^n$.

Proof:

By induction on n:

First, $n = 1$:

$\quad$ Show: $\quad x^1 < y^1$

Proof: By hypothesis.

Next,

$\quad$ Assume: $\quad x^n < y^n$
$\qquad\qquad\quad k = n+1$
$\quad$ Show: $\quad x^k < y^k$

It is frequently helpful to do the first few cases ($n = 2$, 3, or 4) to get an idea of how to proceed in general:

Scrap Paper

$\quad n = 2:\quad x < y$
$\qquad\qquad x^2 < yx,\ xy < y^2,$ so that $x^2 < y^2$

3.1 Induction

> $n = 3$: $x^2 < y^2$, so that $x^3 < y^2 x$
>
> $x < y$, so that $xy^2 < y^3$, so that $x^3 < y^3$

> $x < y$, so that $xy^n < y^{n+1}$ (multiplying both sides by y^n)
>
> $x^n < y^n$, so that $x^{n+1} < y^n x$ (multiplying both sides by x)
>
> therefore $x^{n+1} < y^{n+1}$ (transitivity of $<$)
>
> that is, $x^k < y^k$

In the proof we have assumed $y^n > 0$ since $y > 0$. (See Exercise 2(a).)

Example 3:
Let $S = \{n \in \mathbb{N} \mid n^2 + 2 = 3n\}$. Then $1 \in S$ since $1^2 + 2 = 3 \cdot 1$—so that condition (1) of Theorem 3.1.2 holds. The implication **if** $n \in S$, **then** $n + 1 \in S$ is false when $n = 2$, however, since $n \in S$ is true but $n + 1 \in S$ is false. That is, $2^2 + 2 = 3 \cdot 2$, but $3^2 + 2 \neq 3 \cdot 3$. Thus condition (2) fails. Notice that $3 \notin S$, so that $S \neq \mathbb{N}$.

Example 4:
Let $S = \{n \in \mathbb{N} \mid n = n + 1\}$. The statement $n \in S$ is false for all $n \in \mathbb{N}$, so that the implication **if** $n \in S$, **then** $n + 1 \in S$ is true. Thus condition (2), **for all** $n \in \mathbb{N}$: **if** $n \in S$, **then** $n + 1 \in S$, holds. Condition (1) asserts $1 = 2$ and therefore fails. A mindless proof of condition (2) might follow these lines: let $n \in S$. Then $n = n + 1$. Adding 1 to each side, $n + 1 = (n + 1) + 1$, so that $n + 1 \in S$. Checking condition (1) first may prevent one from proving condition (2) for the empty set.

Examples 3 and 4 show that conditions (1) and (2) in Theorem 3.1.2 are both necessary.

Consider the validity of the inequality $2^n < n!$ for various n:

$n = 1$: $2 < 1$

$n = 2$: $4 < 2$

$n = 3$: $8 < 6$

$n = 4$: $16 < 24$

$n = 5$: $32 < 120$

After a relatively slow start, the factorial grows much faster than the exponential. We believe that $2^n < n!$ is true for all $n \in \{4, 5, 6, \ldots\}$. This can be proved by induction as long as we're not required to start at $n = 1$. In a manner analogous to the proof of Theorem 3.1.2, we could prove the following fact:

Fact   *Suppose S is a subset of $\mathbb{Z}$ with the following two properties:*

(1) $4 \in S$

(2) *for all $n \in \{4, 5, 6, \ldots\}$: if $n \in S$, then $n + 1 \in S$*

Then $\{4, 5, 6, \ldots\} \subseteq S$.

This fact can be used to prove $2^n < n!$ for all $n \geq 4$. It is useful to be able to do induction proofs from starting points other than $n = 1$. In general, we have the following corollary to Theorem 3.1.2:

Corollary 3.1.3   *Suppose $k \in \mathbb{Z}$ and $S \subseteq \{k, k+1, k+2, \ldots\}$ is such that*

(1) $k \in S$

and   (2) *for all $n \in \{k, k+1, k+2, \ldots\}$: if $n \in S$, then $n + 1 \in S$.*

Then $S = \{k, k+1, k+2, \ldots\}$.

Proof:

Let $K = \{k, k+1, k+2, \ldots\}$. Then $S \subseteq K$ by definition. To show the reverse inclusion, let $x \in K$. Then $x - (k - 1) \in \mathbb{N}$. Define $T = \{t \in \mathbb{N} \mid t + (k - 1) \in S\}$. We show $T = \mathbb{N}$ by induction: $1 \in T$: $1 + (k - 1) = k \in S$, so that $1 \in T$ by definition.

For an arbitrary $n \in \mathbb{N}$, assume $n \in T$. Then $n + (k - 1) \in S$, so that $(n + (k - 1)) + 1 \in S$ by hypothesis (2). But $(n + (k - 1)) + 1 = (n + 1) + (k - 1)$, so that $n + 1 \in T$ by definition. By Theorem 3.1.2, $T = \mathbb{N}$. Therefore $x - (k - 1) \in T$, so that $x \in S$. Hence $K \subseteq S$. From both inclusions, $K = S$.   ∎

Example 5:
Prove $2^n < n!$ for all $n \geq 4$.

Proof:

By induction, starting from $n = 4$:

$$n = 4: \quad 2^4 = 16 < 24 = 4!$$

Assume $2^n < n!$ for $n \geq 4$ and $k = n + 1$:

$$2^k = 2^n \cdot 2 < n! \cdot 2 < n! \cdot (n + 1) = (n + 1)! = k!$$

∎

3.1 Induction

In addition to the exercises at the end of this section, problems involving induction appear in future sections. For example, Exercises 3.2.2 and 3.2.3 relate to divisibility, which is defined formally in the next section.

Additional Proof Ideas

Inductive Definitions

An *inductive definition* is a method for defining functions with subsets $\{k, k+1, k+2, \ldots\}$ of $\mathbb{Z}$ as domain. In order to define a function with domain $\mathbb{N}$ inductively, we first define $f(1)$ and then define $f(n+1)$ assuming that we have already defined $f(n)$. Our rule for f is therefore determined as follows: we say what f does to the first natural number and then give a rule for obtaining f of the next natural number. This rule then gives us $f(2)$. Once we know $f(2)$, the rule gives us $f(3)$ and so on.

Example 6:
The function $f(n) = n!$ can be defined inductively by $f(0) = 1$ and $f(n+1) = (n+1) \cdot f(n)$.

Example 7:
Let $a \in \mathbb{R}$. The exponential a^n can be defined inductively on the non-negative integers by $a^0 = 1$ and $a^{n+1} = a \cdot a^n$.

Theorem 3.1.1 states that $\mathbb{N}_k$ is finite. Our next theorem states that any nonempty finite set is equivalent to $\mathbb{N}_k$ for some k:

Theorem 3.1.4 *Let S be a nonempty finite set. Then $S \sim \mathbb{N}_k$ for some $k \in \mathbb{N}$.*

By the definition of finite, every one-to-one function from S to S is onto. We must use this fact to define a one-to-one function f from $\mathbb{N}_k$ onto S (or vice versa). We first define a function $F: \mathbb{N} \to S$ inductively:

Proof:

Define $F: \mathbb{N} \to S$ as follows. Since S is nonempty, pick some element $s_1 \in S$ and define $F(1) = s_1$. If $S - \{s_1\}$ is nonempty, pick some element $s_2 \in S - \{s_1\}$ and define $F(2) = s_2$; otherwise, define $F(2) = F(1)$. Assuming that $F(n)$ has been defined, define $F(n+1)$ as follows: if $S - F(\{1, 2, \ldots, n\})$ is nonempty, pick s_{n+1} in this set and define $F(n+1) = s_{n+1}$; otherwise define $F(n+1) = F(n)$. This defines F inductively.

There are two cases. Case 1: $F(k) = F(k+1)$ for some k. By the Well-Ordering Axiom, we can take k to be the smallest such number. In this case, define $f: \mathbb{N}_k \to S$ by $f(n) = F(n)$. Then f can be shown to be one-to-one and onto (Exercise 12) so that $\mathbb{N}_k \sim S$ (and hence $S \sim \mathbb{N}_k$).

Case 2: $F(k) \neq F(k+1)$ for all $k \in \mathbb{N}$. In this case, F is one-to-one (Exercise 12). Hence $\mathbb{N} \precsim S$. Since $\mathbb{N}$ is infinite, so is S by Corollary 2.7.7, a contradiction.

Thus $S \sim \mathbb{N}_k$ in all cases not leading to a contradiction. ■

Notice that the inductive definition of F depends on making perhaps an infinite number of arbitrary choices and therefore depends on the Axiom of Choice. In fact, Theorem 3.1.4 cannot be proved without the Axiom of Choice. In our discussion of inductive definitions, we have been using our informal idea of function as a rule. If "function" is defined formally, as in Section 2.8, then it is necessary to show that there exists a unique function satisfying the inductive definition. This is done in more advanced texts.

Corollary 3.1.5 A set S is finite iff $S \sim \mathbb{N}_k$ for some natural number k.

Proof:

If S is finite, then $S \sim \mathbb{N}_k$ by Theorem 3.1.4. If $S \sim \mathbb{N}_k$, then S is finite by Corollary 2.7.9 and Theorem 3.1.1. ■

Sigma Notation

The expression "$2 + 4 + 6 + \cdots + (2n)$" in Example 1 can be given in *sigma notation*. In this expression, n is considered to be defined previously, so that $2n$ is the last term in the sum. A *general term* in the sum will have the form $2i$ for i an arbitrary integer from 1 to n. Thus the first term in the sum is $2i$ for $i = 1$. The last is $2i$ for $i = n$. The entire sum is expressed by $\sum_{i=1}^{n} 2i$. The "i" is a local (dummy) variable in this expression and has no meaning outside the expression. The equation of Example 1 can therefore be given as:

$$\sum_{i=1}^{n} 2i = n(n+1)$$

A proof by induction that this equation is true for all $n \in \mathbb{N}$ involves two parts:

(1) (true for $n = 1$) show: $\sum_{i=1}^{1} 2i = 1(1+1)$

3.1 Induction

(2) if $\sum_{i=1}^{n} 2i = n(n+1)$ and $k = n+1$, then $\sum_{i=1}^{k} 2i = k(k+1)$

In part (1) the summation $\sum_{i=1}^{1} 2i$ involves only one term, $2 \cdot 1$, and asserts $2 \cdot 1 = 1(1+1)$, which is true. To prove (2), assume $\sum_{i=1}^{n} 2i = n(n+1)$ and let $k = n+1$. Then $\sum_{i=1}^{k} 2i = \sum_{i=1}^{n} 2i + 2(n+1) = n(n+1) + 2(n+1) = (n+2)(n+1) = (k+1)k$, so that (2) holds.

The Binomial Theorem

In Section 2.8 the ordered triple (a, b, c) was defined in terms of ordered pairs as $((a, b), c)$. We can define an ordered *n-tuple* inductively as (X, z), where X is an ordered $(n-1)$-*tuple*. We will use the notation $(a_1, a_2, \ldots, a_n)$ to denote the ordered n-tuple with ith coordinate given by a_i. With this notation, the ordered n-tuple $(a_1, a_2, \ldots, a_n)$ is equal to the ordered n-tuple $(b_1, b_2, \ldots, b_n)$ iff for all $i = 1, \ldots, n$: $a_i = b_i$ (Exercise 16). A set with n elements is called an *n-set* or sometimes an *unordered n-set*. An ordered n-tuple is sometimes called an *ordered n-set*. Ordered n-tuples are defined so as to distinguish sets of elements on the basis of the order in which the elements are listed. For example, $\{1, 2, 3\} = \{2, 1, 3\}$, but $(1, 2, 3) \neq (2, 1, 3)$.

If $S = \{a_1, a_2, \ldots, a_n\}$ is an *n*-set, there are exactly $n!$ different ordered n-sets that are equal, as unordered sets, to S (Exercise 17). These n distinct ordered n-sets correspond to the $n!$ distinct permutations of the elements $a_1, a_2, \ldots, a_n$. More generally, for $i \leq n$, the number of ordered i-subsets of S is $n \cdot (n-1) \cdot (n-2) \cdot \cdots \cdot (n-i+1)$ (i terms in the product). There are n ways to choose the first term. Having chosen this, there are $n-1$ ways to choose the second, and so on. For a given i-subset A of S, there are $i!$ different permutations of the elements of A. These permutations yield all the different ordered i-subsets of S that are equal to A as sets. Consequently, the number of ordered i-subsets of S is equal to the number of unordered i-subsets times the $i!$ different permutations of each of these. Put the other way around, the number of unordered i-subsets of S is $n \cdot (n-1) \cdot (n-2) \cdot \cdots \cdot (n-i+1)$ divided by $i!$. This is called the *binomial coefficient* "n choose i" and denoted by $\binom{n}{i}$.

Definition
$$\binom{n}{i} = \frac{n \cdot (n-1) \cdot (n-2) \cdot \cdots \cdot (n-i+1)}{i!}$$

The formula for $\binom{n}{i}$ can be put in closed form as $\binom{n}{i} = \frac{n!}{i!(n-i)!}$.

Example 8:
$0!$ is equal to 1 by definition. Let $S = \{s_1, s_2, \ldots, s_n\}$.

$\binom{n}{0} = 1$ (the empty set is the only subset of S with no elements)

$\binom{n}{1} = n$ (there are n 1-subsets of S: $\{s_1\}, \{s_2\}, \ldots, \{s_n\}$)

$\binom{n}{2} = \dfrac{n \cdot (n-1)}{2}$

.
.

$\binom{n}{i} = \dfrac{n!}{i!(n-i)!} = \dfrac{n!}{(n-i)!(n-(n-i))!} = \binom{n}{n-i}$

.
.

$\binom{n}{n} = 1$ (S is the only subset of S with n elements)

Theorem 3.1.6   *Binomial Theorem:* Let $a, b \in \mathbb{R}$, $n \in \mathbb{N}$. Then $(a+b)^n = a^n + \binom{n}{1}a^{n-1}b + \binom{n}{2}a^{n-2}b^2 + \cdots + \binom{n}{1}a \cdot b^{n-1} + b^n$. That is, $(a+b)^n = \sum_{i=0}^{n} \binom{n}{i} a^{n-i} b^i$.

The definition of the binomial coefficient $\binom{n}{i}$ could have been made without the combinatorial discussion preceding it. The Binomial Theorem can also be stated and proved by induction without combinatorial considerations—that is, without it being clear just why the coefficients occur in the theorem. In Exercise 19 you are asked to provide a combinatorial proof. Our purpose here in providing the following proof is to illustrate the use of induction. However, understanding is even more basic to mathematics than is proof. The best proofs provide understanding.

Proof:

By induction. If $n = 1$, the assertion is clear: $a + b = a + b$. Assume the assertion is true for n. Then

$$(a+b)^{n+1} = (a+b) \cdot \sum_{i=0}^{n} \binom{n}{i} a^{n-i} b^i$$

$$= \sum_{i=0}^{n} \binom{n}{i} a^{n-i+1} b^i + \sum_{i=0}^{n} \binom{n}{i} a^{n-i} b^{i+1}$$

3.1 Induction

Let $j = i + 1$ to get:

$$(a+b)^{n+1} = \sum_{i=0}^{n} \binom{n}{i} a^{n-i+1} b^i + \sum_{j=1}^{n+1} \binom{n}{j-1} a^{n-j+1} b^j$$

Then change the dummy variable to get:

$$(a+b)^{n+1} = \sum_{i=0}^{n} \binom{n}{i} a^{n-i+1} b^i + \sum_{i=1}^{n+1} \binom{n}{i-1} a^{n-i+1} b^i$$

$$= \binom{n}{0} a^{n+1} + \sum_{i=1}^{n} \left[\binom{n}{i} + \binom{n}{i-1} \right] a^{n-i+1} b^i + \binom{n}{n} b^{n+1}$$

But $\binom{n}{i} + \binom{n}{i-1} = \binom{n+1}{i}$ (just add by getting a common denominator, Exercise 18), so that:

$$(a+b)^{n+1} = \binom{n}{0} a^{n+1} + \sum_{i=1}^{n} \binom{n+1}{i} a^{n-i+1} b^i + \binom{n}{n} b^{n+1}$$

$$= \binom{n+1}{0} a^{n+1} + \sum_{i=1}^{n} \binom{n+1}{i} a^{n-i+1} b^i + \binom{n+1}{n+1} b^{n+1}$$

Therefore:

$$(a+b)^{n+1} = \sum_{i=0}^{n+1} \binom{n+1}{i} a^{n+1-i} b^i$$

∎

Exercises 3.1

Straightforward Problems

1. Prove the following by induction on n:
   (a) $1 + 2 + 3 + \cdots + n = \dfrac{n(n+1)}{2}$
   (b) $1 + 3 + 5 + \cdots + (2n+1) = (n+1)^2$
   (c) $1 + 3 + 9 + \cdots + 3^n = \dfrac{3^{n+1} - 1}{2}$
   (d) $1^2 + 2^2 + 3^2 + \cdots + n^2 = \dfrac{1}{6} n(n+1)(2n+1)$
   (e) $1^3 + 2^3 + \cdots + n^3 = (1 + 2 + \cdots + n)^2$

2. Suppose $y \in \mathbb{R}$ and $y > 0$. Use induction and Theorem A.1.9 (in Appendix 1) to prove:
   (a) for all $n \in \mathbb{N}$: $y^n > 0$
   (b) for all $n \in \mathbb{N}$: $ny \geq y$

Harder Problems

3. Prove $n < 2^n$ for all $n \in \mathbb{N}$.
4. If $x \geq -1$, prove that $(1+x)^n \geq 1 + nx$ for all $n \in \mathbb{N}$.
5. Prove $x^n - y^n = (x-y)(x^{n-1} + x^{n-2}y + \cdots + xy^{n-2} + y^{n-1})$ for all $n \in \mathbb{N}$.

Challenging Problems

6. Given sets $A_1, A_2, \ldots, A_n$, define:

   $$B_1 = A_1$$
   $$B_2 = A_2 - A_1$$
   $$B_3 = A_3 - (A_1 \cup A_2)$$
   $$\cdot$$
   $$\cdot$$
   $$B_n = A_n - \bigcup_{i=1}^{n-1} A_i$$

   Prove: (a) $\bigcup_{i=1}^{n} B_i = \bigcup_{i=1}^{n} A_i$

   (b) for all $1 \leq i, j \leq n$, $i \neq j$: $B_i \cap B_j = \emptyset$

   Use the well-ordered property of $\mathbb{N}$ or induction to prove this extension of Exercise 1.11.4.

7. Do Exercise 6 for an infinite number of sets $A_1, A_2, \ldots, A_n, \ldots$.

8. Prove Theorem 3.1.1. First prove the following lemma:

   Lemma: Let $k \in \mathbb{N}$. Then $\mathbb{N}_k - \{x\} \sim \mathbb{N}_{k-1}$ for all $x \in \mathbb{N}_k$. (Take $\mathbb{N}_0 = \emptyset$.)

   If $\mathbb{N}_k$ is not finite for all $k \in \mathbb{N}$, then by the Well-Ordering Axiom there is a smallest natural number k such that $\mathbb{N}_k$ is infinite. Use this fact, the lemma, and Theorem 2.3.5 to prove Theorem 3.1.1.

9. Assume Theorem 3.1.2 is true and use it to show the Well-Ordering Axiom. Use all the customary computational properties of the integers but no properties other than Theorem 3.1.2 about *sets* of integers.

10. Consider the following theorem:

    Theorem *Given $a, b \in \mathbb{Z}$ with $b > 0$, there exist unique $q, r \in \mathbb{Z}$ with $0 \leq r < b$ such that $a = qb + r$.*

    This theorem is called the *Division Algorithm*. It states that, if we divide a by b, we get a unique quotient q and remainder r less than b.

    (a) Show the existence part for positive a by induction on a.

3.1 Induction

(b) Show the existence part for negative a and $a = 0$.

(c) Show uniqueness as follows:
Assume $q_1 b + r_1 = q_2 b + r_2$ for $0 \leq r_1, r_2 < b$
Case 1: $r_1 = r_2$
Case 2: (without loss of generality) $0 \leq r_1 < r_2 < b$. Use Exercise 2(b).

Supplementary Problems

11. Let $f: X \to X$ be a function. Let $f^2 = f \circ f$, $f^3 = f \circ f \circ f$, and so forth. Write an inductive definition of f^n for $n \in \mathbb{N}$.

12. Give the details in the proof of Theorem 3.1.4 by showing that in Case 1 f is one-to-one and onto and in case 2 F is one-to-one.

13. Find the fallacy in the following proof that all horses are the same color: Let S be a set of horses. Then all elements of S are the same color by induction on the number of elements in S: If S has one element, then clearly all elements of S are the same color. Now assume that S has n elements and that all sets of horses with $n - 1$ elements are the same color. Remove one horse h_1 from S. Then by the inductive hypothesis all elements of $S_1 = S - \{h_1\}$ are the same color. Remove a horse h_2 from S_1. Horse h_2 is the same color as the rest of the elements of S_1. Add h_1 to the rest of these elements to get a set of size $n - 1$. By the inductive hypothesis, h_1 is the same color as the rest of the elements, so that h_1 is the same color as h_2. Therefore all elements in S are the same color.

14. Theorem 3.1.4 can be stated "Let S be a nonempty set. If S is finite, then $S \sim \mathbb{N}_k$ for some $k \in \mathbb{N}$." Prove Theorem 3.1.4 by showing the contrapositive of the last sentence; that is, "If for all $k \in \mathbb{N}$: $S \nsim \mathbb{N}_k$, then S is infinite". Do this by defining a one-to-one function $f: \mathbb{N} \to S$ inductively.

15. Instead of using induction, show the Division Algorithm directly from the Well-Ordering Axiom for $\mathbb{N}$. Consider $\{a - bq \mid q \in \mathbb{Z}\} \cap \mathbb{N}$.

16. Show $(a_1, a_2, \ldots, a_n) = (b_1, b_2, \ldots, b_n)$ iff for all $i = 1, \ldots, n$: $a_i = b_i$.

17. Let $\{a_1, a_2, \ldots, a_n\}$ be an n-set. Use induction to show that there are $n!$ ordered n-sets with elements $a_1, a_2, \ldots, a_n$.

18. Show $\binom{n}{i} + \binom{n}{i-1} = \binom{n+1}{i}$.

19. Systematically consider all the terms in the product $(a_1 + b_1)(a_2 + b_2) \cdots (a_n + b_n)$. Group the terms according to the number of a_j in each term. Show that there are $\binom{n}{i}$ terms in the group with i of the elements a_j. Let $a = a_1 = a_2 = \cdots = a_n$ and $b = b_1 = b_2 = \cdots = b_n$ to get the Binomial Theorem.

3.2 EQUIVALENCE RELATIONS

Sections 1.1 through 3.1 contain the basic ideas of mathematical proof. In this section we begin a shift of emphasis toward mathematical ideas that are prerequisite for upper-level courses. At the same time, we continue our transition to paragraph-style proofs. At this stage, completely straightforward proofs will be rare. We will still use a straightforward process to set up the form of a proof—to define the variables and give the usual beginning and ending steps (or statements, in a paragraph proof)—but from now on some computation, techniques, or even creative ideas will generally be needed in order to complete proofs. In this chapter, the computation, familiar from previous courses, should not be difficult. As in the preceding section, "property of the integers" is an acceptable reason for any proof step that follows from other steps when the numerical computations familiar from previous mathematics courses are used.

If a is any integer and b is an integer greater than zero, then we may divide a by b and obtain a quotient q and a remainder r, which is either zero or a positive integer less than b. This fact is called the Division Algorithm (somewhat incorrectly since the algorithm is really the process that obtains q and r). It is given, with hints, as a supplementary exercise in the preceding section. We wish to single it out from the other computational "properties of the integers" in order to refer to it specifically in proofs.

Division Algorithm   Let $a, b \in \mathbb{Z}$ with $b > 0$. There exist unique integers q, r such that $a = bq + r$ and $0 \leq r < b$. q is called the quotient and r the remainder upon dividing a by b.

Definition   Let $a, b \in \mathbb{Z}$. We say b divides a iff $a = bc$ for some $c \in \mathbb{Z}$.

The customary notation for denoting "b divides a" is $b \mid a$, which is also expressed as "b is a factor of a" or "a is a multiple of b".

Theorem 3.2.1   Let $a, b, c \in \mathbb{Z}$.

(a) If $b \mid a$ and $b \mid c$, then $b \mid (a + c)$.

(b) If $b \mid a$, then $b \mid ac$.

> **Proof of (a):**
>
> Assume $b \mid a$ and $b \mid c$. We will show $b \mid (a + c)$.

By definition, we need to show there exists $d \in \mathbb{Z}$ such that $a + c = b \cdot d$. According to our existence rule, then, we need to find or define d in the course of our proof. Most of the following steps are straightforward:

3.2 Equivalence Relations

1. $b \mid a$ (hyp.)
2. $a = b \cdot e$ for some $e \in \mathbb{Z}$ (1; def. $\mid$)
3. $b \mid c$ (hyp.)
4. $c = b \cdot f$ for some $f \in \mathbb{Z}$ (3; def. $\mid$)
5. $a + c = b \cdot e + b \cdot f$ (2, 4; prop. $\mathbb{Z}$: + well def.)
6. $a + c = b(e + f)$ (5; prop. $\mathbb{Z}$)
7. Let $d = e + f$.
8. $a + c = b \cdot d$ for some $d \in \mathbb{Z}$ (6, 7; pr. $\exists$)
9. $b \mid (a + c)$ (8; def. $\mid$)

  Step 5 is of course obtained by adding Steps 2 and 4. The justification for doing this is as follows. The significance of the equation in Step 2 is that a and $b \cdot e$ are two different names for exactly the same integer. Similarly, from Step 4, c and $b \cdot f$ are two different names for a second integer. Thus $a + c$ is a name for the sum of these two integers and so is $b \cdot e + b \cdot f$. This sum ought to depend on the integers we're adding and not on the names we give them. This condition is expressed by saying that addition is *well defined* on $\mathbb{Z}$. This is the specific property of the integers, then, that we use as a reason for Step 5. In the future we will use this specific property whenever it applies, instead of the more general "property of the integers".

  In going from Step 8 to Step 9, we used only the definition of "divides", and this is explicitly clear by the way it is presented. We found our needed "d" in Step 7 when we defined it to be equal to the sum of e and f. This shows explicitly how the rule for proving a **for some** statement was used. It is clear, however, right from Step 6, that $b \mid (a + c)$ by definition since $e + f$ is an integer. Further, by the way we arrived at $e + f$ (knowing e and f from Steps 2 and 4), it is clear that the criteria for proving a **for some** statement were met. Thus, in the future, where it is clear what plays the role of, say, d in a **for some** d statement, we will use the rule implicitly. Thus the following is an acceptable version of our proof:

1. $b \mid a$ (hyp.)
2. $a = b \cdot e$ for some $e \in \mathbb{Z}$ (1; def.$\mid$)
3. $b \mid c$ (hyp.)
4. $c = b \cdot f$ for some $f \in \mathbb{Z}$ (3; def.$\mid$)
5. $a + c = b \cdot e + b \cdot f$ (2, 4; + well def.)
6. $a + c = b(e + f)$ (5; prop. $\mathbb{Z}$)
7. $b \mid (a + c)$ (6; def. $\mid$)

Chapter 3 Relations, Operations, and the Integers

The point of the preceding paragraph is that somehow "d" must be clearly identified. The writer and reader of the proof need to see what plays the role of "d". It would not be acceptable to go right from Step 5 to Step 7 since it is not clear just what, in Step 5, plays the role of "d". This is a matter of exposition, not logic.

Proof of (b): Exercise 7.

For any nonempty sets A and B, a *relation R from A to B* is a *rule* of association between the elements of A and those of B. If $a \in A$ is associated under R with $b \in B$, we write $a R b$ (a related to b); otherwise $a \not R b$. If A and B are the same yet, we say that R is a *relation on A*.

We can give a formal definition of "relation" in a manner analogous to the formal definition of "function" in Section 2.7.

Definition *For any nonempty sets A and B, a **relation from A to B** is a triple (A, B, R) where $R \subseteq A \times B$.*

Example 1:
Let $A = \{1, 2, 3\}$ and $B = \{8, 9\}$. Define $R \subseteq A \times B$ by $R = \{(1, 8), (1, 9), (2, 8)\}$. In this example, 1 is related to 8 and 9, and 2 is related to 8. Since 3 is related to nothing and 1 is related to both 8 and 9, this relation is not a function. All functions are relations, but the converse is not true. A relation from A to B is defined by *any* subset of $A \times B$. In our notation, $1 R 8$, $1 R 9$, and $2 R 8$.

Example 2:
The following are relations:
(a) Equality on any set A: $a \in A$ is related to $b \in A$ iff $a = b$.
(b) $\leq$ is a relation on **R**.
(c) A function $f: A \to B$ is a relation from A to B such that each $a \in A$ is related to one and only one element $f(a)$ in B.
(d) Congruence of triangles is a relation on the set of plane triangles: a is related to b iff $a \cong b$.

For our purpose, a particular relation will be defined if we have a rule that tells us whether two given elements are related or not. Thus we will not need to use the formal definition of "relation".

Definition *A relation R on a set A is called an **equivalence relation** iff:*

    (1) for all $a \in A$: $a R a$ *(reflexive property)*

and (2) for all $a, b \in A$: if $a R b$, then $b R a$ *(symmetric property)*

and (3) for all $a, b, c \in R$: if $a R b$ and $b R c$,   *(transitive property)*
            then $a R c$

3.2 Equivalence Relations

Example 3:

(a) Equality on any set A is an equivalence relation—since conditions (1), (2), and (3) in the definition above hold. That is,
  (1) For all $a \in A$: $a = a$.
  (2) For all $a, b \in A$: if $a = b$, then $b = a$.
  (3) For all $a, b, c \in A$: if $a = b$ and $b = c$, then $a = c$.
(b) $\leq$ on $\mathbb{R}$ is *not* an equivalence relation since (1) and (3) hold but (2) does not.
(c) Congruence of triangles is an equivalence relation on the plane. Note that (1), (2), and (3) hold.

Example 4:
Let X be any set. By Theorem 2.7.4, the relation $\sim$ of equipotence, or having the same cardinality, is an equivalence relation on $\mathbb{P}(X)$.

Example 5:
Let $f: A \to B$. Define R_f on A by $a\, R_f\, b$ iff $f(a) = f(b)$. Conditions (1), (2), and (3) are easily seen to hold, so that R_f is an equivalence relation on A.

Example 6:
Define the relation R_Z on $\mathbb{N} \times \mathbb{N}$ by $(a, b)\, R_Z\, (c, d)$ iff $a + d = b + c$. Then R_Z is an equivalence relation:

(1) For all $(a, b) \in \mathbb{N} \times \mathbb{N}$: $(a, b)\, R_Z\, (a, b)$.

Proof:
$a + b = b + a$

(2) For all $(a, b), (c, d) \in \mathbb{N} \times \mathbb{N}$: if $(a, b)\, R_Z\, (c, d)$, then $(c, d)\, R_Z\, (a, b)$.

Proof:
Assume $(a, b)\, R_Z\, (c, d)$. Then $a + d = b + c$, so that $c + b = d + a$, whence $(c, d)\, R_Z\, (a, b)$.

(3) For all $(a, b), (c, d), (e, f) \in \mathbb{N} \times \mathbb{N}$: if $(a, b)\, R_Z\, (c, d)$ and $(c, d)\, R_Z\, (e, f)$, then $(a, b)\, R_Z\, (e, f)$.

Proof:

Assume $(a, b) R_Z (c, d)$ and $(c, d) R_Z (e, f)$. Show $(a, b) R_Z (e, f)$.

1. $a + d = b + c$ (hyp.)
2. $c + f = d + e$ (hyp.)
3. $a + d + c + f = b + c + d + e$ (1, 2; + well def.)
4. $a + f = b + e$ (3; prop. $\mathbb{Z}$)
5. $(a, b) R_Z (e, f)$ (4; def. R_Z) ∎

Note: R_Z is formally defined by addition on $\mathbb{N}$: $(a, b) R_Z (c, d)$ iff $a + d = b + c$. However, notice that the equation $a + d = b + c$ holds in $\mathbb{N}$ iff the equation $a - b = c - d$ holds in $\mathbb{Z}$.

Example 7:
Let $\mathbb{Z}^* = \mathbb{Z} - \{0\}$. Define the relation R_Q on $\mathbb{Z} \times \mathbb{Z}^*$ by $(a, b) R_Q (c, d)$ iff $a \cdot d = b \cdot c$. Then R_Q is an equivalence relation (Exercise 5).

Definition A real number r is called **rational** (from "ratio") if $r = a/b$ for some $a, b \in \mathbb{Z}, b \neq 0$. The set of rational numbers is denoted by $\mathbb{Q}$.

Note: R_Q is formally defined by multiplication on $\mathbb{Z}$: $(a, b) R_Q (c, d)$ iff $a \cdot d = b \cdot c$. But for $b \neq 0 \neq d$, the equation $a \cdot d = b \cdot d$ holds in $\mathbb{Z}$ iff the equation $a/b = c/d$ holds in the rational numbers $\mathbb{Q}$.

Definition Let $n \in \mathbb{N}$. For $a, b \in \mathbb{Z}$, define a to be **congruent to b modulo n** (written $a \equiv b \bmod n$) iff $n | (a - b)$.

Theorem 3.2.2 Congruence modulo n is an equivalence relation.

Proof: Exercise 8.

Our next theorem states that we can add the same number to both sides of a congruence as we can to both sides of an equality. Similarly, we can multiply both sides by the same number. This idea is developed further in the exercises.

Theorem 3.2.3
(a) If $a \equiv b \bmod n$ and $c \in \mathbb{Z}$, then $a + c \equiv b + c \bmod n$.
(b) If $a \equiv b \bmod n$ and $c \in \mathbb{Z}$, then $ac \equiv bc \bmod n$.

Proof of (a):

Suppose $a \equiv b \bmod n$ and $c \in \mathbb{Z}$. By definition of congruence, $n | (a - b)$; that is, there exists $d \in \mathbb{Z}$ such that $n \cdot d = a - b$.

3.2 Equivalence Relations

If we analyze the conclusion, we see that we need $(a+c) - (b+c) = n \cdot f$ **for some** f. The left side of this equation is just $a - b$, however, so that the d we have will do for f. Starting again:

Proof of (a):

Suppose $a \equiv b \bmod n$ and $c \in \mathbb{Z}$. By definition of congruence, $n \mid (a - b)$; that is, there exists $d \in \mathbb{Z}$ such that $n \cdot d = a - b$. Thus $n \cdot d = (a + c) - (b + c)$, whence $a + c \equiv b + c \bmod n$. ∎

Proof of (b): Exercise 9.

Exercises 3.2

Practice Exercise

1. State (1) which properties—reflexive, symmetric, and transitive—hold for the following relations and (2) which of the relations are equivalence relations:

    (a) Congruence on the set of plane triangles
    (b) Similarity on the set of plane triangles
    (c) Parallelism of lines in the plane
    (d) Perpendicularity of lines in the plane
    (e) $=$ on $\mathbb{R}$
    (f) $<$ on $\mathbb{R}$
    (g) $\leq$ on $\mathbb{R}$
    (h) R defined by $x\,R\,y$ for $x, y \in \mathbb{R}$ iff x and y agree to the tenth decimal place
    (i) R defined by $x\,R\,y$ for $x, y \in \mathbb{R}$ iff $|x - y| \leq 2$

Problems

2. Show by induction that for all $n \in \mathbb{N}$: 3 divides $4^n - 1$.
3. Show by induction that for all $n \in \mathbb{N}$: 4 divides $5^n - 1$.
4. Show whether or not the following relations R are equivalence relations:

    (a) On $\mathbb{Z} \times \mathbb{Z}$, define $(a, b)\,R\,(c, d)$ iff $a + b = c + d$
    (b) On $\mathbb{Z} \times \mathbb{Z}$, define $(a, b)\,R\,(c, d)$ iff $a \cdot d = b \cdot c$. Compare with R_Q of Example 7.

5. Show that R_Q in Example 7 is an equivalence relation.

6. The following conjecture continues the idea that we can deal with congruences the way we deal with equations, namely, that we can add or multiply congruences or divide a congruence by a nonzero (not congruent to zero) integer. Prove the following or find a counterexample:

Conjecture

(a) If $a \equiv b \bmod n$ and $c \equiv d \bmod n$, then $a + c \equiv b + d \bmod n$.
(b) If $a \equiv b \bmod n$ and $c \equiv d \bmod n$, then $ac \equiv bd \bmod n$.
(c) If $ac \equiv bc \bmod n$ and $c \not\equiv 0 \bmod n$, then $a \equiv b \bmod n$.

7. Prove Theorem 3.2.1b.
8. Prove Theorem 3.2.2.
9. Prove Theorem 3.2.3b.

Supplementary Problem

10. A trick question can be defined by the property that people usually make an unwarranted assumption when answering. Since mathematicians are not supposed to make any unexamined assumptions, there are no trick questions in mathematics. Prove or find a counterexample to the following conjecture:

Conjecture: Let R be a relation on A such that for all $a, b, c \in A$:

(2) if aRb, then bRa

and (3) if aRb and bRc, then aRc.

Then R is an equivalence relation.

3.3 EQUIVALENCE CLASSES

Definition Let A be a set and R an equivalence relation on A. For each $a \in A$, define $[a] = \{b \in A \mid bRa\}$. $[a]$ is called the **equivalence class of a** defined by R.

The equivalence class of a is, then, the set of all elements that are related to a. Since aRa by definition of equivalence relation, a itself is in this equivalence class.

Theorem 3.3.1 Let R be an equivalence relation on A and $a, b \in A$. Then $[a] = [b]$ iff aRb.

Proof: Exercise 1.

3.3 Equivalence Classes

Corollary 3.3.2 *Let $[a]$ and $[b]$ be two equivalence classes for the equivalence relation of congruence modulo n. Then $[a] = [b]$ iff $a \equiv b \bmod n$.*

Example 1:
Let the function $f: \mathbb{R} \to \mathbb{R}$ be given by $f(x) = x^2$ and let R_f be defined on $\mathbb{R}$ by $x\, R_f\, y$ iff $f(x) = f(y)$. As in Example 3.2.5, R_f is an equivalence relation. The equivalence class determined by $x \in \mathbb{R}$ is $[x] = \{x, -x\}$. For example, $[2] = \{2, -2\}$, $[-2] = \{2, -2\}$, $[0] = \{0\}$.

Example 2:
By Theorem 3.2.2, congruence mod 5 is an equivalence relation on $\mathbb{Z}$. The equivalence classes are given by:

$$[1] = \{\ldots, -19, -14, -9, -4, 1, 6, 11, 16, 21, \ldots\}$$
$$[2] = \{\ldots, -18, -13, -8, -3, 2, 7, 12, 17, 22, \ldots\}$$
$$[3] = \{\ldots, -17, -12, -7, -2, 3, 8, 13, 18, 23, \ldots\}$$
$$[4] = \{\ldots, -16, -11, -6, -1, 4, 9, 14, 19, 24, \ldots\}$$
$$[5] = \{\ldots, -15, -10, -5, 0, 5, 10, 15, 20, 25, \ldots\}$$

Consider $[1]$. It is clear that the difference between 1 and any element in $[1]$ is divisible by 5. This is what defines the set. Furthermore, if two integers are related to 1, they are related to each other (definition of equivalence relation). So any two elements in $[1]$ are congruent mod 5 (related to each other). $[1]$ could have been called $[6]$, $[-4]$, or $[x]$, where x is any element in the equivalence class. Extrapolating from the pattern in listing -19 through 25 in the classes above, we would conjecture that each integer is in one of the classes named and that any two distinct classes have no elements in common.

Example 3:
Let $A = \{1, 2, 3, 4, 5\}$. Let R be defined by

$$R = \{(1, 1), (1, 2), (2, 1), (2, 2), (3, 3), (3, 4), (4, 3), (4, 4), (5, 5)\}$$

The equivalence classes are $[1], [2], [3], [4], [5]$. These are not all distinct since $[1] = [2] = \{1, 2\}$ and $[3] = [4] = \{3, 4\}$.

Example 4:
In Example 3.2.6, an equivalence relation R_Z was defined on $\mathbb{N} \times \mathbb{N}$ by $(a, b)\, R_Z\, (c, d)$ iff $a + d = b + c$. The equation $a + d = b + c$ holds in $\mathbb{N}$ iff the equation $a - b = c - d$ holds in $\mathbb{Z}$. An R_Z equivalence class is a set of pairs, all of whose first coordinate differs from the second coordinate by a fixed integer. For example, $[(6, 3)]$ contains the pairs $(4, 1), (5, 2), (6, 3), (7, 4), (8, 5), \ldots$, where the difference between first and second coordinates is 3. The class $[(4, 5)]$ contains the pairs $(1, 2), (2, 3), (3, 4), \ldots$, where the difference between the first and second coordinates is -1.

Example 5:
In Example 3.2.7, an equivalence relation R_Q was defined on $\mathbb{Z} \times \mathbb{Z}^*$ by $(a, b)\, R_Q\, (c, d)$ iff $a \cdot d = b \cdot c$. The equation $a \cdot d = b \cdot c$ holds in $\mathbb{Z}$ iff the equation $a/b = c/d$ holds in the rationals $\mathbb{Q}$. An R_Q equivalence class is a set of pairs whose coordinates form a fixed ratio. For example, $[(6, 3)]$ contains the pairs $(4, 2)$, $(2, 1)$, $(-6, -3), \ldots$, where the ratio of the first coordinate to the second is 2. The class $[(2, 4)]$ contains the pairs $(1, 2)$, $(-2, -4)$, $(3, 4), \ldots$, where the ratio of the first coordinate to the second is 1/2.

A look at the examples above might suggest that if R is an equivalence relation on a set A, then any two distinct equivalence classes of R have nothing in common. This is the assertion of our next theorem.

Recall that from the statement **if** $\mathscr{P}$, **then** $\mathcal{Q}$ we can form the contrapositive **if** $\sim\mathcal{Q}$, **then** $\sim\mathscr{P}$ by negating $\mathscr{P}$ and $\mathcal{Q}$ and reversing their roles. According to Theorem 2.3.7L, a statement and its contrapositive are equivalent. To show the truth of some statement, it is frequently useful to attempt a proof of the contrapositive. This is illustrated in our next theorem.

Theorem 3.3.3 *Let A be a nonempty set and R an equivalence relation on A. For $a, b \in A$, either $[a] \cap [b] = \varnothing$ or $[a] = [b]$ but not both.*

Our first task in proving this theorem is to express its meaning in terms of our language so that we can start a proof. We are given, in our metalanguage, a statement of the form "either $\mathscr{P}$ or $\mathcal{Q}$ but not both". It seems clear that this ought to mean the same as

(1)   ($\mathscr{P}$ **or** $\mathcal{Q}$) **and** $\sim$($\mathscr{P}$ **and** $\mathcal{Q}$)

There are other statements that are logically equivalent to this, however. One of these is

(2)   (**if** $\mathscr{P}$, **then** $\sim\mathcal{Q}$) **and** (**if** $\sim\mathscr{P}$, **then** $\mathcal{Q}$)

In fact, we can prove that (1) and (2) are equivalent (Exercise 4). Statements, such as (1) and (2), that mean the same thing logically will be equivalent. Two language statements will be equivalent if they are both faithful interpretations of the same metalanguage statement. Recall that the definition of equivalence depends on the existence or nonexistence of a proof.

There is an effective procedure for determining (that is, a way of mechanically deciding) whether two *unquantified* language statements are equivalent. This is the method of truth tables, which are introduced in Appendix 2 and used to verify again that (1) and (2) are equivalent. Truth tables cannot decide the equivalence of quantified statements (**for some** and **for all**), which form a major part of our mathematics. In a classical course in logic, they are used to define the formal language words **and, or, not,** and **if-then.** For us these are defined, along with **for some** and **for all**, by the rules for proving and using statements involving these words.

3.3 Equivalence Classes

Returning now to Theorem 3.3.3, we use formulation (2) in the following proof:

Proof:

Assume: R an equivalence relation on A; $a, b \in A$
Show: (1) if $[a] = [b]$, then $[a] \cap [b] \neq \emptyset$
and (2) if $[a] \neq [b]$, then $[a] \cap [b] = \emptyset$

First we show (1). Let $[a] = [b]$. Then $b \in [a] \cap [b] \neq \emptyset$.
Next we show (2) by showing the contrapositive of (2), namely:

Show: (2') if $[a] \cap [b] \neq \emptyset$, then $[a] = [b]$

1. Suppose $[a] \cap [b] \neq \emptyset$.
2. There exists $c \in [a] \cap [b]$. (1; def. $\emptyset$)
3. $c \in [a]$ (2; def. $\cap$)
4. cRa (3; def. [ ])
5. $c \in [b]$ (2; def. $\cap$)
6. cRb (5; def. [ ])
7. aRc (4; def. eq. rel. (symmetric))
8. aRb (6, 7; def. eq. rel. (transitive))
9. Let $x \in [a]$.
10. xRa (9; def. [ ])
11. xRb (8, 10; def. eq. rel. (transitive))
12. $x \in [b]$ (11; def. [ ])
13. $[a] \subseteq [b]$ (9—12; def. $\subseteq$)
14. $[b] \subseteq [a]$ (1—13; symmetry)
15. $[a] = [b]$ (13, 14; def. =)

In the course of proving Theorem 3.3.3, we have discovered a third, equivalent formulation of "either $\mathcal{P}$ or $\mathcal{Q}$ but not both"—which was obtained by replacing one of the implications in (2) by its contrapositive:

(3) $\mathcal{P}$ iff $\sim \mathcal{Q}$

Our reason for proving (2') rather than (2) is that the statement in (2') gives us elements to work with, using the appropriate definitions. Assuming a set (the intersection) is nonempty gives us an element by definition. Showing two sets to be equal is something with which we are well acquainted. On the other hand, the task in proving (2) would be to show a certain set is empty assuming two other sets are not equal—a rather awkward thing to do. This illustrates how useful the equivalence of a statement and its contrapositive can be.

If R is an equivalence relation on a set A, then Theorem 3.3.3 tells us that A is partitioned by the equivalence classes. We give a formal definition of a partition:

Definition

Let A be a nonempty set and $\{A_i | i \in \mathscr{I}\}$ a family of nonempty subsets of A. The family is called a **partition** of A iff

(1) $A = \bigcup_{i \in \mathscr{I}} A_i$

and (2) $A_i \cap A_j = \emptyset$ for $A_i \neq A_j$

By a "family" we simply mean a set whose elements are themselves sets. Define $\mathscr{S} = \{[a] | a \in A\}$, so that $\mathscr{S}$ is the family of equivalence classes defined by R. Note that A is used as an index set to describe the elements of $\mathscr{S}$.

Example 6:

The relation R on $\{1, 2, 3, 4, 5\}$ defined in Example 3 has the equivalence classes $[1], [2], [3], [4], [5]$. These are not all distinct since $[1] = [2]$ and $[3] = [4]$. We could write the set $\mathscr{S}$ of all equivalence classes in the following ways since it does not matter if an element is listed more than once in a set:

$$\{[i] | 1 \leq i \leq 5\}$$

$$\{[1], [2], [3], [4], [5]\}$$

$$\{[1], [3], [5]\}$$

Theorem 3.3.4

If R is an equivalence relation on A, then the set of equivalence classes defined by R is a partition of A. It is called the partition **induced by** R.

Proof:

Assume: R an equivalence relation on A
Show: (1) $A = \bigcup_{a \in A} [a]$

(2) $[a] \cap [b] = \emptyset$ for $[a] \neq [b]$

That $\bigcup_{a \in A} [a] \subseteq A$ is clear by the definition of equivalence class. To show $A \subseteq \bigcup_{a \in A} [a]$, let $x \in A$. Since $x R x$ (reflexive property), $x \in [x]$, and so $x \in \bigcup_{a \in A} [a]$. Hence $A = \bigcup_{a \in A} [a]$.

Part (2) of the conclusion follows from Theorem 3.3.3. ∎

3.3 Equivalence Classes

Theorem 3.3.5 *For any positive integer n, the set of congruence classes modulo n forms a partition of $\mathbb{Z}$. There are n such classes, given by $[0], [1], [2], \ldots, [n-1]$.*

Proof:
The set of congruence classes forms a partition of $\mathbb{Z}$ by Theorems 3.2.2 and 3.3.4. To prove the second part of Theorem 3.3.5, we need to show:

(1) if $[x]$ is any equivalence class, then $[x] = [j]$ for some $0 \le j \le (n-1)$

and (2) if $0 \le i < j \le (n-1)$, then $[i] \ne [j]$.

To prove (1), let $[x]$ be an equivalence class. Then by the Division Algorithm, $x = nq + r$ for $0 \le r < n$. Thus $n \mid (x - r)$, so that $x \equiv r \bmod n$ and $[x] = [r]$.

To prove (2), suppose $0 \le i < j \le (n-1)$. Assume $[i] = [j]$ in order to get a contradiction. Then $i \in [j]$, so that $i \equiv j \bmod n$. That is, $i - j = nq$ for some q. Then $i = nq + j$, but also $i = n \cdot 0 + i$. Since $0 \le i, j < n$, it follows from the uniqueness part of the Division Algorithm that $i = j$, a contradiction. Therefore $[i] \ne [j]$. ∎

Notice that the metalanguage assertion in Theorem 3.3.5 that there are n equivalence classes is given in our language in statements (1) and (2) above. (1) follows from the existence of a remainder given by the Division Algorithm. (2) follows from the uniqueness of this remainder. It is clear from the proof of Theorem 3.3.5 that a $\equiv$ equivalence class is a set of integers, all of which give the same remainder on division by n.

Example 7:
In Example 3.2.5, given a function $f: A \to B$, the equivalence relation R_f is defined on A by $a R_f b$ iff $f(a) = f(b)$. If f is onto, the family $\{f^{-1}(b) \mid b \in B\}$ is a partition of the domain A (Exercise 2).

Additional Ideas

Theorem 3.3.6 *Let $\{A_i \mid i \in \mathcal{I}\}$ be any partition of a set A. Define the relation R on A by: for $a, b \in A$, $a R b$ iff a and b are in the same set A_i. Then R is an equivalence relation on A. It is called the relation induced by the partition $\{A_i \mid i \in \mathcal{I}\}$.*

Proof: Exercise 6.

By Theorem 3.3.4, the set of equivalence classes induced by an equivalence relation R on a set A is a partition of A. You should convince yourself that applying Theorem 3.3.6 to this partition gives back R. Also, starting with any partition of a set A, Theorem 3.3.6 gives an equivalence relation R. Convince yourself that applying Theorem 3.3.4 to R gives the original partition.

Example 8:
Let $\{\{1\}, \{2, 3\}, \{4, 5\}\}$ be a partition of the set $S = \{1, 2, 3, 4, 5\}$. By Theorem 3.3.6, the relation R on S defined by $1R1, 2R2, 2R3, 3R3, 3R2, 4R4, 4R5, 5R5, 5R4$ is an equivalence relation. This relation induces a partition of S by Theorem 3.3.4—just the partition we started with.

For any set A, let P be the set of all partitions of A and let E be the set of equivalence relations on A. A function $f: E \to P$ is defined by Theorem 3.3.4; that is, for any equivalence relation $R \in E$, $f(R)$ is the partition induced by R. A function $g: P \to E$ is given by Theorem 3.3.6; that is, for any partition $\mathscr{S} \in P$, $g(\mathscr{S})$ is the relation induced by $\mathscr{S}$. The functions f and g are bijections—one-to-one correspondences—between the equivalence relations on A and the partitions of A. This can be seen by the relationships $f \circ g = i_P$ and $g \circ f = i_E$ that follow from your observations in the preceding paragraph.

Exercises 3.3

Problems

1. Prove Theorem 3.3.1.
2. Prove that if $f: A \to B$ is onto, then $\{f^{-1}(b) \mid b \in B\}$ is the partition induced by R_f (see Example 7).
3. Find the equivalence classes defined by the equivalence relations of Exercise 3.2.4.

Supplementary Problems

4. Using formal proof steps, show that the statements:
   (1) ($\mathscr{P}$ or $\mathscr{Q}$) and $\sim(\mathscr{P}$ and $\mathscr{Q})$
   (2) (if $\mathscr{P}$, then $\sim\mathscr{Q}$) and (if $\sim\mathscr{P}$, then $\mathscr{Q}$)
   are equivalent.
5. Show that the statement
   (3) $\mathscr{P}$ iff $\sim\mathscr{Q}$
   is equivalent to either (1) or (2) (and hence both) in Exercise 4.
6. Prove Theorem 3.3.6.
7. Find the equivalence relation defined by Theorem 3.3.6 from the partition $\{\{1, 2\}, \{3\}, \{4, 5, 6\}\}$ of $\{1, 2, 3, 4, 5, 6\}$.

3.4 WELL-DEFINED OPERATIONS

Given a positive integer n, there are n equivalence classes defined by the equivalence relation of congruence modulo n. These congruence classes can be represented by [0], [1], [2], ..., [n − 1]. If z is any integer, then dividing z by n gives (by the Division Algorithm) $z = qn + r$, where $0 \le r < n$. Thus $z \equiv r \bmod n$, so that z and r are in the same congruence class. Hence [z] = [r]. The set of possible remainders we can get upon division by n then gives us a complete set of representatives for the set of congruence classes modulo n. We could of course choose another set of representatives, such as [n], [n + 1], ..., [2n − 1] or even [n], [n + 1], [2n + 2], [3n + 3], ..., [n^2 − 1], but it is standard practice to use the integers $0, \ldots, n - 1$ as representatives.

For a given positive integer n, the set {[0], [1], ..., [n − 1]} of congruence classes modulo n is denoted by $\mathbb{Z}_n$. We would like to define an addition on this set of equivalence classes. The question is whether this can be done and, if so, how? For an example, consider $\mathbb{Z}_6 = \{[0], [1], [2], [3], [4], [5]\}$. We would like to define the sum of [4] and [5] to be the class [9]. In general, we would like [a] + [b] to be the class [a + b] containing $a + b$. The problem here is, for example, that the class [4] is the same as the class [10], and [5] is the same as [11]. Our rule for adding would give [10] + [11] = [21] but also [4] + [5] = [9]. Is this all right? What we want is to add two classes to get a third class. This third class should depend only on the two classes we're adding and not on the particular way we represent these two classes.

Idea of Being Well Defined

Whenever an operation is defined in terms of representations for the elements being operated on, we must show that the rule specifying the operation gives the same value regardless of the different ways in which each element is represented. We say in this case that the operation is well defined.

Example 1:
The idea that addition is well defined on $\mathbb{Z}$ is used in the discussion that follows the proof of Theorem 3.2.1a.

Example 2:
Consider our discussion of $\mathbb{Z}_6$ above. [10] and [4] are the same element, call it a, of $\mathbb{Z}_6$—represented by 10 and by 4. Call [10] representation 1 and [4] representation 2 for a. Similarly, call [11] representation 1 and [5] representation 2 for element b. Then by representation 1 $a + b = $ [21] and by representation 2 $a + b = $ [9]. If addition is to be well defined on $\mathbb{Z}_6$, then [21] and [9] had better be the same element of $\mathbb{Z}_6$. This indeed is the case.

We now define addition and multiplication on $\mathbb{Z}_n$ in general. After this definition is made, we show that it is an acceptable one.

Definition For any $n \in \mathbb{N}$, the set of congruence classes modulo n is denoted by $\mathbb{Z}_n$. Addition is defined on $\mathbb{Z}_n$ by $[a] + [b] = [a+b]$, multiplication by $[a] \cdot [b] = [ab]$.

Fact 3.4.1
(a) Addition on $\mathbb{Z}_n$ is well defined.
(b) Multiplication on $\mathbb{Z}_n$ is well defined.

Proof of (a): Exercise 1.

In order to see what is involved in showing (b), let A and B be two elements of $\mathbb{Z}_n$ (congruence classes). Suppose, by our rule, that $A \cdot B = C$. We need to show that if $A = [a_1]$ and $A = [a_2]$ (that is, if we have two representations of A) and if $B = [b_1]$ and $B = [b_2]$, then $C = [a_1 b_1] = [a_2 b_2]$. Note that by our rule $C = [a_1 b_1]$. Also by our rule, $C = [a_2 b_2]$. Our rule is no good, therefore, unless $[a_1 b_1] = [a_2 b_2]$. This illustrates what is needed in order to show multiplication is well defined.

Proof of (b):

Assume: $[a_1] = [a_2]$
$[b_1] = [b_2]$
Show: $[a_1 b_1] = [a_2 b_2]$

1. $[a_1] = [a_2]$ (hyp.)
2. $a_1 \equiv a_2 \bmod n$ (1; Cor. 3.3.2)
3. $n \mid (a_1 - a_2)$ (2; def. $\equiv$)
4. $(a_1 - a_2) = qn$ for some $q \in \mathbb{Z}$ (3; def. $\mid$)
5. $(b_1 - b_2) = pn$ for some $p \in \mathbb{Z}$ (hyp., 1—4; symmetry)

   .
   .

k. $(a_1 b_1 - a_2 b_2) = rn$ for some $r \in \mathbb{Z}$
k+1. $n \mid (a_1 b_1 - a_2 b_2)$ (k; def. $\mid$)
k+2. $a_1 b_1 \equiv a_2 b_2 \bmod n$ (k+1; def. $\equiv$)
k+3. $[a_1 b_1] = [a_2 b_2]$ (k+2; Cor. 3.3.2)

Straightforward analysis of the hypotheses and conclusion leads to the steps above. We need to prove a **for some** statement (Step k); that is, we need to find r in the course of our proof.

Scrap Paper:

We have: (1) $a_1 - a_2 = qn$
and (2) $b_1 - b_2 = pn$

3.4 Well-Defined Operations

> Nonstraightforward insight from computational experience
>
> We want: $a_1 b_1 - a_2 b_2 = rn$
>
> Multiply (1) by b_1 and (2) by a_2:
>
> $a_1 b_1 - a_2 b_1 = b_1 q n$
>
> $a_2 b_1 - a_2 b_2 = a_2 p n$
>
> Add: $a_1 b_1 - a_2 b_2 = (b_1 q + a_2 p) n$
>
> We have now found r, namely, $b_1 q + a_2 p$.

Proof of (b):

In order to show multiplication is well defined, assume

$$[a_1] = [a_2] \text{ and } [b_1] = [b_2]$$

We will show $[a_1 a_2] = [b_1 b_2]$. From our assumptions we get:

$a_1 - a_2 = qn$ for some $q \in \mathbb{Z}$

and $b_1 - b_2 = pn$ for some $p \in \mathbb{Z}$

Whence $a_1 b_1 - a_2 b_1 = b_1 q n$

and $a_2 b_1 - a_2 b_2 = a_2 p n$

Adding: $a_1 b_1 - a_2 b_2 = (b_1 q + a_2 p) n$

Therefore $a_1 b_1 \equiv a_2 b_2 \bmod n$, so that $[a_1 b_1] = [a_2 b_2]$. ∎

Proving that operations (or functions, in Section 3.7) are well defined is a technical necessity whenever we make definitions based on representations, but the statement that the operations are well defined does not rate the status of a theorem. We have therefore called this statement a *fact*. In fact, people usually don't bother to make a formal statement at all—but they do check.

The set $\mathbb{Z}_n$, on which we have just defined an addition, has the following properties (recall Theorem 2.6.1 and note the similarity of properties):

Theorem 3.4.2
(a) For all $x, y \in \mathbb{Z}_n$: $x + y \in \mathbb{Z}_n$.
(b) For all $x, y, z \in \mathbb{Z}_n$: $x + (y + z) = (x + y) + z$.
(c) There exists $i \in \mathbb{Z}_n$ such that for all $x \in \mathbb{Z}_n$: $i + x = x$ and $x + i = x$.
(d) For each $x \in \mathbb{Z}_n$, there exists $y \in \mathbb{Z}_n$ such that $x + y = i$ and $y + x = i$.
(e) For all $x, y \in \mathbb{Z}_n$: $x + y = y + x$.

Proof: Exercise 3.

Chapter 3   Relations, Operations, and the Integers

The multiplication we have defined on $\mathbb{Z}_n$ has the following properties:

Theorem 3.4.3

(a) For all $x, y \in \mathbb{Z}_n$: $x \cdot y \in \mathbb{Z}_n$.
(b) For all $x, y, z \in \mathbb{Z}_n$: $x \cdot (y \cdot z) = (x \cdot y) \cdot z$.
(c) For all $x, y, z \in \mathbb{Z}_n$: $x \cdot (y + z) = x \cdot y + x \cdot z$ and $(y + z) \cdot x = y \cdot x + z \cdot x$.
(d) For all $x, y \in \mathbb{Z}_n$: $x \cdot y = y \cdot x$.

Proof: Exercise 4.

Example 3:
The equivalence relation R_Z on $\mathbb{N} \times \mathbb{N}$ is defined in Example 3.2.6. Example 3.3.4 considers the equivalence classes defined by R_Z. The set of all equivalence classes is a partition, denoted by P_Z, of $\mathbb{N} \times \mathbb{N}$. We define addition and multiplication on P_Z by the following rules:

$$[(a, b)] + [(c, d)] = [(a + c, b + d)]$$

and $[(a, b)] \cdot [(c, d)] = [(ac + bd, ad + bc)]$

Since addition and multiplication on P_Z are defined in terms of representatives of the equivalence classes, we must show that they are well defined. Following the procedure after Fact 3.4.1, we have for multiplication:

Assume: $[(a_1, b_1)] = [(a_2, b_2)]$
$[(c_1, d_1)] = [(c_2, d_2)]$
Show: $[(a_1 c_1 + b_1 d_1, a_1 d_1 + b_1 c_1)] =$
$[(a_2 c_2 + b_2 d_2, a_2 d_2 + b_2 c_2)]$

1. $(a_1, b_1) \, R_Z \, (a_2, b_2)$ (hyp.; Thm. 3.3.1)
2. $(c_1, d_1) \, R_Z \, (c_2, d_2)$ (hyp.; Thm. 3.3.1)
3. $a_1 + b_2 = b_1 + a_2$ (1; def. R_Z)
4. $c_1 + d_2 = d_1 + c_2$ (2; def. R_Z)
5. $a_1 - b_1 = a_2 - b_2$ (3; prop. $\mathbb{Z}$)
6. $c_1 - d_1 = c_2 - d_2$ (4; prop. $\mathbb{Z}$)
7. $(a_1 - b_1)(c_1 - d_1) =$ (5, 6; $\cdot$ well def. on $\mathbb{Z}$)
$(a_2 - b_2)(c_2 - d_2)$
8. $a_1 c_1 + b_1 d_1 - b_1 c_1 - a_1 d_1 =$ (7; prop. $\mathbb{Z}$)
$a_2 c_2 + b_2 d_2 - b_2 c_2 - a_2 d_2$
9. $a_1 c_1 + b_1 d_1 + b_2 c_2 + a_2 d_2 =$ (8; prop. $\mathbb{Z}$)
$b_1 c_1 + a_1 d_1 + a_2 c_2 + b_2 d_2$
10. $(a_1 c_1 + b_1 d_1, a_1 d_1 + b_1 c_1) \, R_Z$ (9; def. R_Z)
$(a_2 c_2 + b_2 d_2, a_2 d_2 + b_2 c_2)$

3.4 Well-Defined Operations

11. $[(a_1c_1 + b_1d_1, a_1d_1 + b_1c_1)] =$ (10; Thm. 3.3.1)
    $[(a_2c_2 + b_2d_2, a_2d_2 + b_2c_2)]$

Exercise 5 asks you to show that addition is well defined on P_Z.

Example 4:
The equivalence relation R_Q on $\mathbb{Z} \times \mathbb{Z}^*$ is defined in Example 3.2.7. Example 3.3.5 considers the equivalence classes defined by R_Q. The set of all equivalence classes is a partition, denoted by P_Q, of $\mathbb{Z} \times \mathbb{Z}^*$. We define addition and multiplication on P_Q by the following rules:

$$[(a, b)] + [(c, d)] = [(ad + bc, bd)]$$
and $$[(a, b)] \cdot [(c, d)] = [(ac, bd)]$$

Exercises 6 and 7 ask you to show that addition and multiplication are well defined on P_Q.

Additional Ideas

Theorems 3.4.2 and 3.4.3, which hold for $\mathbb{Z}_n$, also hold for the sets P_Z and P_Q with operations defined in Examples 3 and 4.

Theorem 3.4.4
(a) For all $x, y \in P_Z$: $x + y \in P_Z$.
(b) For all $x, y, z \in P_Z$: $x + (y + z) = (x + y) + z$.
(c) There exists $i \in P_Z$ such that for all $x \in P_Z$: $i + x = x$ and $x + i = x$.
(d) For each $x \in P_Z$, there exists $y \in P_Z$ such that $x + y = i$ and $y + x = i$.
(e) For all $x, y \in P_Z$: $x + y = y + x$.

Proof: Exercise 9.

Theorem 3.4.5
(a) For all $x, y \in P_Z$: $x \cdot y \in P_Z$.
(b) For all $x, y, z \in P_Z$: $x \cdot (y \cdot z) = (x \cdot y) \cdot z$.
(c) For all $x, y, z \in P_Z$: $x \cdot (y + z) = x \cdot y + x \cdot z$ and $(y + z) \cdot x = y \cdot x + z \cdot x$.
(d) For all $x, y \in P_Z$: $x \cdot y = y \cdot x$.

Proof: Exercise 10.

Theorem 3.4.6
(a) For all $x, y \in P_Q$: $x + y \in P_Q$.
(b) For all $x, y, z \in P_Q$: $x + (y + z) = (x + y) + z$.

(c) There exists $i \in P_Q$ such that for all $x \in P_Q$: $i + x = x$ and $x + i = x$.
(d) For each $x \in P_Q$, there exists $y \in P_Q$ such that $x + y = i$ and $y + x = i$.
(e) For all $x, y \in P_Q$: $x + y = y + x$.

Proof: Exercise 11.

Theorem 3.4.7
(a) For all $x, y \in P_Q$: $x \cdot y \in P_Q$.
(b) For all $x, y, z \in P_Q$: $x \cdot (y \cdot z) = (x \cdot y) \cdot z$.
(c) For all $x, y, z \in P_Q$: $x \cdot (y + z) = x \cdot y + x \cdot z$ and $(y + z) \cdot x = y \cdot x + z \cdot x$.
(d) For all $x, y \in P_Q$: $x \cdot y = y \cdot x$.

Proof: Exercise 12.

Exercises 3.4

Problems

1. Prove Fact 3.4.1a.
2. Is it possible to define subtraction (the inverse of addition) on Z_n? Can you give a rule for subtraction in terms of representatives? (We will investigate division later.)
3. Prove Theorem 3.4.2.
4. Prove Theorem 3.4.3.
5. Show that addition is well defined on P_Z (see Example 3).
6. Show that addition is well defined on P_Q (see Example 4).
7. Show that multiplication is well defined on P_Q (see Example 4).
8. For any set X, the relation $\sim$ on $P(X)$ is an equivalence relation (Example 3.2.4). Let P denote the set of $\sim$ equivalence classes of $P(X)$. Define addition and multiplication on P by the rules:

$$[S] + [T] = [S \cup T]$$
$$[S] \cdot [T] = [S \cap T]$$

Show by a counterexample that these operations are not well defined.

Supplementary Problems

9. Prove Theorem 3.4.4.
10. Prove Theorem 3.4.5.
11. Prove Theorem 3.4.6.
12. Prove Theorem 3.4.7.

3.5 GROUPS AND RINGS

Recall from Section 2.8 that a binary operation on a set S is a rule that assigns to any pair of elements of S a third element. For example, addition is a binary operation on the set of real numbers: to any pair x, y of real numbers, we assign a third number $x + y$ called the sum of x and y. Composition is a binary operation on the set of all functions $f: A \to A$ from A to A: to any pair f, g of functions, we assign a third function $f \circ g$ called the composition of f and g.

Theorem 3.4.2a (for all $x, y \in \mathbb{Z}_n: x + y \in \mathbb{Z}_n$) states that $+$ is a binary operation on the set $\mathbb{Z}_n$. Conversely, if we say that $+$ is a binary operation on $\mathbb{Z}_n$, then there is no need to assert Theorem 3.4.2a also, because the two statements mean the same thing. Similarly, Theorem 2.6.1a (for all $f, g \in \mathbb{B}(A): f \circ g \in \mathbb{B}(A)$) is equivalent to the assertion that $\circ$ is a binary operation on $\mathbb{B}(A)$. The properties common to $\mathbb{Z}_n$ (given in Theorem 3.4.2) and $\mathbb{B}(A)$ (given in Theorem 2.6.1) are now abstracted. Sets on which there is an operation satisfying these properties are called *groups*. Thus $\mathbb{B}(A)$ under the operation of composition is a group, and $\mathbb{Z}_n$ under the operation of addition is a group. A group can be any set whatever with an operation that satisfies the four properties. The properties therefore define what a group is. They are called the *axioms* for a group.

Definition A nonempty set G with a binary operation $\cdot$ is called a ***group*** provided that:

(1) For all $x, y, z \in G: x \cdot (y \cdot z) = (x \cdot y) \cdot z$.

(2) There exists $i \in G$ such that for all $x \in G: i \cdot x = x$ and $x \cdot i = x$.

and (3) For each $x \in G$, there exists $y \in G$ such that $x \cdot y = i$ and $y \cdot x = i$.

Definition A group G is called ***Abelian*** provided that

(4) $x \cdot y = y \cdot x$ for all $x, y \in G$.

Formally, a group is a pair $(G, \cdot)$ with the set G as first coordinate and the binary operation $\cdot$ on G as the second coordinate, subject to the axioms above. The definition above expresses this idea in a slightly less formal way. If we do not wish to emphasize the binary operation, the group is simply called G.

Example 1:
For any $n \in \mathbb{N}$, $(\mathbb{Z}_n, +)$ is an Abelian group (Theorem 3.4.2).

Example 2:
For any set A, $(\mathbb{B}(A), \circ)$ is a group (Theorem 2.6.1). If $A = \{1, 2, 3\}$, then the calculations in Example 2.6.1 show that $\mathbb{B}(A)$ is not Abelian.

Example 3:
$(P_Z, +)$ of Example 3.4.3 is an Abelian group (Theorem 3.4.4).

Example 4:
$(P_Q, +)$ of Example 3.4.4 is an Abelian group (Theorem 3.4.6).

The well-known properties of the number systems assure us that $(\mathbb{Z}, +)$, $(\mathbb{Q}, +)$ and $(\mathbb{R}, +)$ are Abelian groups. The sets $\mathbb{Q}^+$ of positive rational numbers and $\mathbb{R}^+$ of positive real numbers are Abelian groups under the binary operation of multiplication.

The logical consequences of properties 1, 2, and 3 in the definition of a group—the group axioms—must be true for all groups, and the study of these consequences is called *group theory*. Particular examples of groups abound. The set of all symmetries of some geometric object, the set of all nonsingular $n \times n$ matrices under multiplication, the set of k-dimensional vectors under vector addition are all groups, as are the several examples we will consider. Any theorem on groups, then, must be true about all of these examples—and this is one reason for studying groups. Any understanding we obtain by studying groups will apply to each example of a group. This can be put another way. In order to study, say, the set of all nonsingular $n \times n$ matrices, we can deny ourselves all information about the matrices except certain properties and then look to see what properties the matrices have by virtue only of the starting properties and logic. In this way it is possible to see the precise relationship between properties.

The particular examples of groups we have at hand motivate the questions we ask about groups in general. For example, the element i in group axiom 2 is called an *identity element*. There is only one identity element in each of the examples of groups with which we are familiar. In $\mathbb{Z}_n$, it is the class $[0]$; in $B(A)$, it is the function i_A; in $\mathbb{Z}$ under addition, the number 0; in $\mathbb{R}^+$ under multiplication, the number 1. Is this true for all groups? That is, does it follow from the axioms?

Definition Suppose $(G, \cdot)$ is a group. An element $i \in G$ that has the property $i \cdot x = x$ and $x \cdot i = x$ for all $x \in G$ is called an **identity element**.

Theorem 3.5.1 A group has a unique identity element.

Proof:

Assume that G is a group. G has an identity by property 2. Now assume that i and j are identity elements of G. We will show that $i = j$:

$$i = i \cdot j \quad (j \text{ is an identity})$$
$$i \cdot j = j \quad (i \text{ is an identity})$$
$$i = j \quad (\text{sub.})$$

∎

3.5 Groups and Rings

Theorem 3.5.1 allows us to refer to *the* identity element of a group. By 3.5.1, any element y in a group G that satisfies $x \cdot y = x$ and $y \cdot x = x$ for all $x \in G$ must be the identity i. We can show a much stronger result:

Theorem 3.5.2 *Let G be a group with identity i, and let $e \in G$. If $x \cdot e = x$ or $e \cdot x = x$ for some $x \in G$, then $e = i$.*

> **Proof:**
>
> For Case 1 we assume $x \cdot e = x$ for some $x \in G$ and show $e = i$:
>
> 1. $x \cdot e = x$ (hyp.)
> 2. There exists $y \in G$ such that $y \cdot x = i$.   (def. group (ax. 3))
> 3. $y \cdot (x \cdot e) = y \cdot x$ (1; $\cdot$ well def.)
> 4. $(y \cdot x) \cdot e = y \cdot x$ (3; def. group (ax. 1))
> 5. $i \cdot e = i$ (2, 4; sub.)
> 6. $e = i$ (5; def. i)
>
> The case where $e \cdot x = x$ is similar (Exercise 1). ∎

Note that it follows from Theorem 3.5.2 that if $y \cdot y = y$ for some $y \in G$, then $y = i$.

Theorem 3.5.3 *Let G be a group with identity i. For each element $x \in G$, there is a unique element $y \in G$ such that $x \cdot y = i$ and $y \cdot x = i$.*

Proof: Exercise 2.

For a given element x in a group G, the unique element given by Theorem 3.5.3 is called the *inverse* of x. The notation used to denote the inverse of an element in a group depends on the notation used for the binary operation. As in the case of multiplication of real numbers, the symbol "$\cdot$" for the abstract group operation is frequently omitted and juxtaposition alone used to denote the group operation. Thus $x \cdot y$ is written xy. With this convention, we are using "multiplicative" notation for the group operation (which is then also called a *product*). It is customary to use $+$, "additive" notation, as the symbol for the operation of Abelian groups and to use juxtaposition for groups that are not necessarily Abelian.

The inverse of the element x is denoted by x^{-1} in multiplicative notation and by $-x$ in additive notation. Since the groups we will be studying are not necessarily Abelian, we will use multiplicative notation in our theorems and general discussion. In applications to the groups $\mathbb{Z}$ and $\mathbb{Z}_n$, for example, we will translate into additive notation.

Without further reference, i will be used to denote the identity in a group. Thus for all $x \in G$:

$$xx^{-1} = i \quad \text{and} \quad x^{-1}x = i$$

Theorem 3.5.3 asserts that for any $x \in G$ there is only one element y that satisfies both $xy = i$ and $yx = i$, namely, the element we have called the inverse of x. This result can be sharpened; that is, there is a weaker hypothesis that will identify the inverse of x:

Theorem 3.5.4 *If x and y are elements of a group and if $xy = i$, then $y = x^{-1}$ and $x = y^{-1}$.*

Proof: Exercise 3.

Theorem 3.5.4 asserts that, in a group, the idea of left or right inverse is no more general than the idea of a two-sided inverse: any left inverse is a two-sided inverse and any right inverse is a two-sided inverse.

Theorem 3.5.5 *Let x and y be elements of a group G. Then*

(a) $(x^{-1})^{-1} = x$

(b) $(xy)^{-1} = y^{-1}x^{-1}$

Proof: Exercise 4.

Consider again Example 2.6.3, which we now renumber as Example 5:

Example 5:
The two elements in $\mathbb{B}(\{1, 2\})$ are

$$\begin{pmatrix} 1 & 2 \\ 1 & 2 \end{pmatrix} \quad \text{and} \quad \begin{pmatrix} 1 & 2 \\ 2 & 1 \end{pmatrix}$$

All possible compositions of these permutations are:

$\circ$	$\begin{pmatrix} 1 & 2 \\ 1 & 2 \end{pmatrix}$	$\begin{pmatrix} 1 & 2 \\ 2 & 1 \end{pmatrix}$
$\begin{pmatrix} 1 & 2 \\ 1 & 2 \end{pmatrix}$	$\begin{pmatrix} 1 & 2 \\ 1 & 2 \end{pmatrix}$	$\begin{pmatrix} 1 & 2 \\ 2 & 1 \end{pmatrix}$
$\begin{pmatrix} 1 & 2 \\ 2 & 1 \end{pmatrix}$	$\begin{pmatrix} 1 & 2 \\ 2 & 1 \end{pmatrix}$	$\begin{pmatrix} 1 & 2 \\ 1 & 2 \end{pmatrix}$

The table in Example 5 gives the group products (that is, the composition) for all elements in the group $\mathbb{B}(\{1, 2\})$. In the context of group theory, it is called a *group table*.

3.5 Groups and Rings

Example 6:
The group table for Z_3 is

$$Z_3$$

+	[0]	[1]	[2]
[0]	[0]	[1]	[2]
[1]	[1]	[2]	[0]
[2]	[2]	[0]	[1]

The table in Example 6 consists of two parts. (1) The elements of the group Z_3 are listed to the right of and below the group operation +. This part is considered to exhibit the labels for the table proper. (2) The rest of the table lists the sums of the elements labeling the rows and columns. It is this part that is considered "the group table" when mathematical properties of the table are considered—for example, the fact that each element of the group is listed precisely once in each row and column (see Exercise 5).

Example 7:
The group table for Z_2 is

$$Z_2$$

+	[0]	[1]
[0]	[0]	[1]
[1]	[1]	[0]

Consider the tables for Z_2 and $B(\{1, 2\})$:

$$Z_2 \qquad\qquad B(\{1, 2\})$$

+	[0]	[1]
[0]	[0]	[1]
[1]	[1]	[0]

∘	$\binom{1\,2}{1\,2}$	$\binom{1\,2}{2\,1}$
$\binom{1\,2}{1\,2}$	$\binom{1\,2}{1\,2}$	$\binom{1\,2}{2\,1}$
$\binom{1\,2}{2\,1}$	$\binom{1\,2}{2\,1}$	$\binom{1\,2}{1\,2}$

It is easy to see that these two tables are the same except for the names of the operations and elements in the two sets. The two groups are said to be *isomorphic*. This idea can be made precise by using the function f from Z_2 to $B(\{1, 2\})$ given by [0] → $\binom{1\,2}{1\,2}$ and [1] → $\binom{1\,2}{2\,1}$. If $x + y = z$ in Z_2, then the entry in row x and column

y in that table is z. Consider the entry in row $f(x)$ and column $f(y)$ in the **B** table. If $f(x)$ is the new name for x in $\mathbf{B}(\{1, 2\})$ and $f(y)$ is the new name for y, then the fact that the tables are the same except for element names dictates that the entry in row $f(x)$ and column $f(y)$ in the **B** table is just $f(z)$—the new name for z. Hence the "sameness" of the tables is expressed by the following condition:

If $x + y = z$, then $f(x) \circ f(y) = f(z)$.

Substituting $x + y$ for z, we have:

$$f(x) \circ f(y) = f(x + y)$$

This leads to our formal definition:

Definition   *Let $(G, \oplus)$ and $(H, \otimes)$ be groups. A function $f: G \to H$ is a **group isomorphism** provided that:*

     (1) f is one-to-one and onto.

and   *(2) $f(x \oplus y) = f(x) \otimes f(y)$ for all $x, y \in G$.*

An isomorphism is therefore a renaming of the elements, not only in a group but in a group table. If the operation symbols $\oplus$ and $\otimes$ are both replaced by juxtaposition, condition 2 becomes $f(xy) = f(x)f(y)$.

Definition   *Groups G and H are called **isomorphic** (denoted by $G \cong H$) if there is an isomorphism $f: G \to H$.*

The condition that an isomorphism be one-to-one and onto can be relaxed to give a more general function that still respects the group operations. Using juxtaposition for the operations:

Definition   *A function f from a group G to a group H is called a **group homomorphism** provided that $f(xy) = f(x)f(y)$ for all $x, y \in G$.*

Problems involving homomorphisms are given in Section 3.6, where they are first considered in the simple context of the group $(\mathbb{Z}, +)$.

The sets $\mathbb{Z}_n$, $P_\mathbb{Z}$, $P_\mathbb{Q}$, $\mathbb{R}$, $\mathbb{Q}$, and $\mathbb{Z}$, which are Abelian groups under the operation of "addition", all have defined on them a second binary operation, "multiplication", which satisfies the properties listed in Theorems 3.4.3, 3.4.5, and 3.4.7. These sets, together with their two operations, are examples of *rings*.

Definition   *A set R with binary operations $+$ and $\cdot$ is called a **ring** provided that:*

(1) $(R, +)$ is an Abelian group.

(2) For all $x, y, z \in R$: $x \cdot (y \cdot z) = (x \cdot y) \cdot z$.

(3) For all $x, y, z \in R$: $x \cdot (y + z) = x \cdot y + x \cdot z$ and $(y + z) \cdot x = y \cdot x + z \cdot x$.

3.5 Groups and Rings

R is called **commutative** if, in addition:

(4) For all $x, y \in R: x \cdot y = y \cdot x$.

A ring R with operations $+$ and $\cdot$ is formally defined to be the triple $(R, +, \cdot)$, subject to the conditions above. Our definition uses the customary less formal wording. The examples above, which we will consider, are all commutative rings. The set of all $n \times n$ matrices under matrix addition and multiplication is a noteworthy example of a noncommutative ring.

Definition Let $(R, +, \cdot)$ and $(S, +, \cdot)$ be rings. A function $f: R \to S$ is a **ring homomorphism** provided that for all $x, y \in R$:

(1) $f(x + y) = f(x) + f(y)$

and (2) $f(x \cdot y) = f(x) \cdot f(y)$

A ring **isomorphism** is a one-to-one, onto ring homomorphism.

Examples of ring isomorphisms are given in Section 3.7. In Section 3.8 we introduce some ideas of "ring theory" in the context of the specific ring $(\mathbb{Z}, +, \cdot)$.

Exercises 3.5

Problems

1. Finish the proof of Theorem 3.5.2.
2. Prove Theorem 3.5.3.
3. Prove Theorem 3.5.4.
4. Prove Theorem 3.5.5.
5. Let x, y, z be elements of a group G. Prove the following cancellation properties:
   (a) If $xy = xz$, then $y = z$.
   (b) If $yx = zx$, then $y = z$.

Note: Exercise 5 shows that in a group table (proper), each element of the group appears exactly once in each row and each column. Such a table is called a *Latin square*.

Supplementary Problem

6. There are just two ways the elements in the product $a_1 \cdot a_2 \cdot a_3$ can be grouped: $a_1 \cdot (a_2 \cdot a_3)$ and $(a_1 \cdot a_2) \cdot a_3$. The associative property asserts that these two ways are equal, so that grouping is irrelevant. A generalized associative property states that grouping of n-fold products is irrelevant. Prove this by induction. How many distinct groupings of $a_1 \cdot a_2 \cdot a_3 \cdot \cdots \cdot a_n$ are there?

3.6 HOMOMORPHISMS AND CLOSED SUBSETS OF $\mathbb{Z}$

Definition   A function $f: \mathbb{Z} \to \mathbb{Z}$ is called an **additive homomorphism** iff for all $a, b \in \mathbb{Z}$: $f(a+b) = f(a) + f(b)$. f is called a **multiplicative homomorphism** iff for all $a, b \in \mathbb{Z}$: $f(a \cdot b) = f(a) \cdot f(b)$.

Additive homomorphisms are, then, just the group homomorphisms of the Abelian group $(\mathbb{Z}, +)$. Functions that are both additive and multiplicative homomorphisms are ring homomorphisms of $(\mathbb{Z}, +, \cdot)$.

For a while we will focus solely on additive homomorphisms. These are the functions from $\mathbb{Z}$ to $\mathbb{Z}$ whose action "commutes" with addition. Thus in the left-hand side of the equation in our definition, we first add a and b and then apply f. In the right-hand side, we first apply f to a and b and then add the functional images (or "values"). We say that the image of a sum is the sum of the images.

Example 1:
Define $f: \mathbb{Z} \to \mathbb{Z}$ by $f(x) = 3x$. Then f is an additive homomorphism since for all $x, y \in \mathbb{Z}$:

$$f(x+y) = 3(x+y) = 3x + 3y = f(x) + f(y)$$

f is not a multiplicative homomorphism since:

$$f(1 \cdot 1) = f(1) = 3 \cdot 1 = 3 \neq 3 \cdot 3 = f(1) \cdot f(1)$$

The following theorems can all be proved using ideas we have already seen.

Theorem 3.6.1   If $f: \mathbb{Z} \to \mathbb{Z}$ and $g: \mathbb{Z} \to \mathbb{Z}$ are additive homomorphisms, then $g \circ f: \mathbb{Z} \to \mathbb{Z}$ is an additive homomorphism.

Proof: Exercise 1.

Theorem 3.6.2   Let $f: \mathbb{Z} \to \mathbb{Z}$ be an additive homomorphism.
(a) $f(0) = 0$
(b) $f(-a) = -f(a)$ for all $a \in \mathbb{Z}$
(c) $f(a-b) = f(a) - f(b)$ for all $a, b \in \mathbb{Z}$

Proof: Exercise 2.

Definition   The function $\zeta: \mathbb{Z} \to \mathbb{Z}$ defined by $\zeta(x) = 0$ for all $x \in \mathbb{Z}$ is called the **zero-function**.

3.6 Homomorphisms and Closed Subsets of Z

Theorem 3.6.3 $\zeta: \mathbb{Z} \to \mathbb{Z}$ *is an additive and multiplicative homomorphism.*

Proof: Exercise 3.

Definition *For any nonempty subset M of $\mathbb{Z}$, a function $f: \mathbb{Z} \to \mathbb{Z}$* **respects multiplication by** *M provided that for all $z \in \mathbb{Z}$, $m \in M$: $f(mz) = mf(z)$.*

Theorem 3.6.4 *If $f: \mathbb{Z} \to \mathbb{Z}$ is an additive homomorphism, then f respects multiplication by $\mathbb{N}$.*

Proof: Exercise 4.

Corollary 3.6.5 *If $f: \mathbb{Z} \to \mathbb{Z}$ is an additive homomorphism, then f respects multiplication by $\mathbb{Z}$.*

Proof: Exercise 5.

Corollary 3.6.6 *The only functions $f: \mathbb{Z} \to \mathbb{Z}$ that are both additive and multiplicative homomorphisms are the identity $i_\mathbb{Z}$ and the zero-function.*

Proof: Exercise 6.

Definition *A subset $X \subseteq \mathbb{Z}$ is called* **closed under addition** *if for all $x, y \in X$ we have $x + y \in X$.*

Theorem 3.6.7 *If X and Y are closed under addition, then so is $X \cap Y$.*

Proof: Exercise 7.

Theorem 3.6.8 *If $f: \mathbb{Z} \to \mathbb{Z}$ is an additive homomorphism and X is closed under addition, then*

(a) *$f(X)$ is closed under addition.*
(b) *$f^{-1}(X)$ is closed under addition.*

Proof: Exercise 8.

Exercises 3.6

Straightforward Problems

1. Prove Theorem 3.6.1.
2. Prove Theorem 3.6.2.
3. Prove Theorem 3.6.3.

4. Use induction to prove Theorem 3.6.4.
5. Use Theorems 3.6.2 and 3.6.4 and cases to prove Corollary 3.6.5.
6. Prove Corollary 3.6.6.
7. Prove Theorem 3.6.7.
8. Prove Theorem 3.6.8.
9. If $f: \mathbb{Z} \to \mathbb{Z}$ is an additive homomorphism, then $f^{-1}(0)$ is called the **kernel** of f. Prove:
   (a) $0 \in f^{-1}(0)$
   (b) $f^{-1}(0) = \mathbb{Z}$ iff $f(1) = 0$ iff $f^{-1}(0) \neq \{0\}$
10. Let X_i for $i \in \mathcal{I}$ be a family of subsets of $\mathbb{Z}$ that are closed under addition. Prove $\bigcap_{i \in \mathcal{I}} X_i$ is closed under addition.

Definition *Let A and M be subsets of $\mathbb{Z}$. A is called* closed under multiplication by M *provided for all $a \in A$, $m \in M$: $ma \in A$.*

11. Prove that if A and B are closed under multiplication by M, then so is $A \cap B$.
12. Prove that if $M \subseteq N$ and A is closed under multiplication by N, then it is closed under multiplication by M.
13. Prove by induction that if A is closed under addition, then it is closed under multiplication by $\mathbb{N}$.
14. Let $f: \mathbb{Z} \to \mathbb{Z}$ be defined by $f(x) = 3x$ for all $x \in \mathbb{Z}$. Prove that f respects multiplication by $\mathbb{Z}$.
15. Prove that if $f: \mathbb{Z} \to \mathbb{Z}$ and $g: \mathbb{Z} \to \mathbb{Z}$ respect multiplication by M, then so does $g \circ f$.
16. Let $f: \mathbb{Z} \to \mathbb{Z}$ respect multiplication by a subset M of $\mathbb{Z}$ and let A be a subset of $\mathbb{Z}$ that is closed under multiplication by M.
    (a) Prove that $f^{-1}(A)$ is closed under multiplication by M.
    (b) Prove that $f(A)$ is closed under multiplication by M.

Harder Problems

17. Is every function $f: \mathbb{Z} \to \mathbb{Z}$ that respects multiplication by $\mathbb{N}$ an additive homomorphism?
18. Is every function $f: \mathbb{Z} \to \mathbb{Z}$ that respects multiplication by $\mathbb{Z}$ an additive homomorphism?
19. Prove that if A is a nonempty subset of $\mathbb{Z}$ closed under addition and under multiplication by $\{-1\}$, then A is a group under addition.

3.7 WELL-DEFINED FUNCTIONS

Idea of Being Well Defined

Whenever a function f is defined in terms of representations for the elements in its domain, we must show that the rule specifying the function gives the same value regardless of the different ways in which we represent each element. That is, if a and b are different representations for the same element, then we must show $f(a) = f(b)$. We say in this case that the function f is **well defined**.

> *Example 1:*
> In the proof of Conjecture 2.3.3a, the function g was applied to an element in its domain represented by $f(a_1)$ and also by $f(a_2)$. That is, g was applied to the equation $f(a_1) = f(a_2)$, and we obtained $g(f(a_1)) = g(f(a_2))$. In this case we knew that g was a function. There is no such thing as a function that is not well defined. The words "well defined" merely focus on the necessity of satisfying the requirements for defining a function. When a function is defined in terms of a representation, this is in doubt.

In this section we consider functions with domain $\mathbb{Z}_n$. A function is a rule that tells us where to send each element in its domain. Thus if $g: \mathbb{Z}_n \to \mathbb{Z}_n$ is to be a function, it must tell us where to send a given element of $\mathbb{Z}_n$. Since the elements of $\mathbb{Z}_n$ are equivalence classes and thus may have different representations, we must be careful—as when defining an *operation* on $\mathbb{Z}_n$—that the function g is well defined.

The next theorem, 3.7.1, states that a given homomorphism of $\mathbb{Z}$, by acting on the representatives of equivalence classes, determines a homomorphism of $\mathbb{Z}_n$. The homomorphism of $\mathbb{Z}_n$ is said to be *induced* by the homomorphism of $\mathbb{Z}$. Theorem 3.7.1 shows why the computations in Example 2 are valid in general.

> *Example 2:*
> According to Example 3.6.1, the function $f: \mathbb{Z} \to \mathbb{Z}$ defined by $f(x) = 3x$ for all $x \in \mathbb{Z}$ is a homomorphism. Define the induced function $\bar{f}: \mathbb{Z}_5 \to \mathbb{Z}_5$ by $\bar{f}([x]) = [f(x)]$; that is, $\bar{f}([x]) = [3x]$. Then, for example,
>
> $[2] = [7]$ and $\left. \begin{array}{l} \bar{f}([2]) = [6] = [1] \\ \text{also,} \quad \bar{f}([7]) = [21] = [1] \end{array} \right\}$ idea of being well defined
>
> Further,
>
> $\left. \begin{array}{l} \bar{f}([2] + [4]) = \bar{f}([2+4]) = \bar{f}([6]) = \bar{f}([1]) = [3] \\ \text{and} \quad \bar{f}([2]) + \bar{f}([4]) = [6] + [12] = [18] = [3] \end{array} \right\}$ idea of homomorphism

Theorem 3.7.1

Let $n \in \mathbb{N}$ be given and $f: \mathbb{Z} \to \mathbb{Z}$ be an (additive) homomorphism. Define the function $\bar{f}: \mathbb{Z}_n \to \mathbb{Z}_n$ by $\bar{f}([a]) = [f(a)]$. Then

(a) $\bar{f}$ is well defined.

(b) $\bar{f}$ is a homomorphism (that is, for all $a, b \in \mathbb{Z}$: $\bar{f}([a] + [b]) = \bar{f}([a]) + \bar{f}([b])$).

Proof of (a):

Suppose we have two representations for the same equivalence class in $\mathbb{Z}_n$, that is, suppose $[a] = [b]$ for $a, b \in \mathbb{Z}$. We need to show $\bar{f}([a]) = \bar{f}([b])$. By definition of $\bar{f}$, the last equation is the same as $[f(a)] = [f(b)]$, and so by Corollary 3.3.2 we need to show $f(a) \equiv f(b) \bmod n$.

From $[a] = [b]$, we get $a \equiv b \bmod n$, that is, $n \mid (a - b)$, so that $a - b = nc$ for some $c \in \mathbb{Z}$. Then $a = nc + b$ and $f(a) = f(nc + b)$. Since f is a homomorphism, $f(a) = f(nc) + f(b)$ and, by Theorem 3.6.4, $f(a) = nf(c) + f(b)$, so that $f(a) - f(b) = nf(c)$. Thus $n \mid (f(a) - f(b))$ and $f(a) \equiv f(b) \bmod n$. ∎

Proof of (b): Exercise 1.

Consider the equivalence relation R_Z on $\mathbb{N} \times \mathbb{N}$ and the set P_Z of R_Z equivalence classes given in Examples 3.2.6, 3.3.4, and 3.4.3:

$$(a, b) \, R_Z \, (c, d) \text{ iff } a + d = b + c$$

Addition and multiplication on P_Z are defined by the rules:

$$[(a, b)] + [(c, d)] = [(a + c, b + d)]$$

$$[(a, b)] \cdot [(c, d)] = [(ac + bd, ad + bc)]$$

Define a function $f: P_Z \to \mathbb{Z}$ by the rule $f([(a, b)]) = a - b$. To show f is well defined, we:

Assume: $[(a_1, b_1)] = [(a_2, b_2)]$
Show: $f([(a_1, b_1)]) = f([(a_2, b_2)])$

1. $(a_1, b_1) \, R_Z \, (a_2, b_2)$ (hyp.; Thm. 3.3.1)
2. $a_1 + b_2 = b_1 + a_2$ (1; def. R_Z)
3. $a_1 - b_1 = a_2 - b_2$ (2; prop. $\mathbb{Z}$)
4. $f([(a_1, b_1)]) = f([(a_2, b_2)])$ (3; def. f) ∎

3.7 Well-Defined Functions

Theorem 3.7.2 Let $f: P_Z \to Z$ be defined by $f([(a, b)]) = a - b$. Then f is a ring isomorphism.

Proof: Exercise 2.

Consider the equivalence relation R_Q on $Z \times Z^*$ and the set P_Q of R_Q equivalence classes given by Examples 3.2.7, 3.3.5, and 3.4.4:

$$(a, b)\ R_Q\ (c, d) \text{ iff } ad = bc$$

Addition and multiplication on P_Q are defined by the rules:

$$[(a, b)] + [(c, d)] = [(ad + bc, bd)]$$

and $\quad [(a, b)] \cdot [(c, d)] = [(ac, bd)]$

Let Q denote the set of rational numbers. Define the function $f: P_Q \to Q$ by the rule $f([(a, b)]) = a/b$.

Theorem 3.7.3 Let $f: P_Q \to Q$ be defined by $f([(a, b)]) = a/b$. Then f is a ring isomorphism.

Proof: Exercise 3.

Exercises 3.7

Problems

1. Prove Theorem 3.7.1b.
2. Prove Theorem 3.7.2.
3. Prove Theorem 3.7.3. Show first that f is well defined.
4. Let $n > 0$ and $k > 0$ be given and let $f: Z \to Z_k$ be an (additive) homomorphism such that $f(n) = [0]_k$. Define the function $\tilde{f}: Z_n \to Z_k$ by $\tilde{f}([a]_n) = f(a)$. Prove that $\tilde{f}$ is well defined. Here [ ]$_n$ and [ ]$_k$ are classes in Z_n and Z_k, respectively.

Challenging Problems

5. Let $f: A \to B$ be a function onto B and let R_f be the equivalence relation defined by $a_1 R_f a_2$ iff $f(a_1) = f(a_2)$ (Example 3.3.7). For each $b \in B$, let $A_b = f^{-1}(b)$. Let $\mathscr{A} = \{A_b | b \in B\}$ be the partition of A induced by R_f (Theorem 3.3.6). The elements of $\mathscr{A}$ are equivalence classes. The equivalence class A_b is the set of all $a \in A$ such that $f(a) = b$ and can be represented by $[a]$ for any $a \in A_b$.

    Define the function $F: \mathscr{A} \to B$ by $F([a]) = f(a)$. Prove:
    (a) F is well defined.
    (b) F is one-to-one.
    (c) F is onto.

6. Let $f: A \to B$ be onto. With $\mathscr{A}$ as in Exercise 5, define the function $g: A \to \mathscr{A}$ by $g(a) = [a]$ for all $a \in A$. Prove:
   (a) g is onto.
   (b) There exists a unique function $h: \mathscr{A} \to B$ such that $h \circ g = f$.

3.8 IDEALS OF $\mathbb{Z}$

In this section we consider, in a simple context, a basic idea in the study of that branch of abstract algebra called *ring theory*. An *ideal* is defined for any abstract ring, but we won't consider the idea at this level of generality. Instead, we consider the ideals of a concrete example—$\mathbb{Z}$. This study will enable us later to obtain some results about the natural numbers.

Definition   A subset H of $\mathbb{Z}$ is called an **ideal** of $\mathbb{Z}$ provided that for all $a, b \in \mathbb{Z}$:
   (1) if $a, b \in H$, then $a + b \in H$
   and (2) if $z \in \mathbb{Z}$ and $a \in H$, then $za \in H$.

According to the definitions in Section 3.6, ideals are just those subsets of $\mathbb{Z}$ that are (1) closed under addition and (2) closed under multiplication by $\mathbb{Z}$. In particular, $\{0\}$ is an ideal of $\mathbb{Z}$.

Definition
   (1) For $a \in \mathbb{Z}$, define $\mathbb{Z}a = \{za \mid z \in \mathbb{Z}\}$.
   (2) For $A, B \subseteq \mathbb{Z}$, define $A + B = \{a + b \mid a \in A, b \in B\}$.

Theorem 3.8.1
   (a) If H and J are ideals of $\mathbb{Z}$, then so is $H \cap J$.
   (b) For any $a \in \mathbb{Z}$, $\mathbb{Z}a$ is an ideal of $\mathbb{Z}$.
   (c) For all $a, b \in \mathbb{Z}$: $\mathbb{Z}a \subseteq \mathbb{Z}b$ iff $b \mid a$.
   (d) If H and J are ideals of $\mathbb{Z}$, then so is $H + J$.
   (e) For all $a, b \in \mathbb{Z}$: $\mathbb{Z}a + \mathbb{Z}b$ is an ideal of $\mathbb{Z}$.

Proof of (e): Immediate from (b) and (d).

Proof of (a), (c), and (d): Exercise 1.

Proof of (b):
Let $a \in \mathbb{Z}$. According to the definition of "ideal," we need to show
   (1) if $c, d \in \mathbb{Z}a$, then $c + d \in \mathbb{Z}a$
   and (2) if $z \in \mathbb{Z}$ and $c \in \mathbb{Z}a$, then $zc \in \mathbb{Z}a$

3.8 Ideals of $\mathbb{Z}$ 219

> To show (1), suppose $c, d \in \mathbb{Z}a$. Then since $c \in \mathbb{Z}a$, there exists $z_1 \in \mathbb{Z}$ such that $c = z_1 a$ (def. $\mathbb{Z}a$). Similarly, there exists $z_2 \in \mathbb{Z}$ such that $d = z_2 a$. Hence $c + d = (z_1 + z_2)a \in \mathbb{Z}a$ (def. $\mathbb{Z}a$), showing (1).
> To show (2), suppose $z \in \mathbb{Z}$ and $c \in \mathbb{Z}a$. Then $c = z_1 a$ for some $z_1 \in \mathbb{Z}$. Hence $zc = z(z_1 a) = (zz_1)a \in \mathbb{Z}a$ by definition.

$\mathbb{Z}a$ is called the *ideal of $\mathbb{Z}$ generated by a*. It is the smallest ideal of $\mathbb{Z}$ containing a (Exercise 2).

Theorem 3.8.2 For any $a, b \in \mathbb{Z}$, there exists $c \geq 0$ such that $\mathbb{Z}a + \mathbb{Z}b = \mathbb{Z}c$.

Proof: Exercise 3.

Theorem 3.8.3 For any $a, b \in \mathbb{Z}$, there exists $c \geq 0$ such that $\mathbb{Z}a + \mathbb{Z}b = \mathbb{Z}c$.

Proof: Exercise 4.

Exercises 3.8

Problems

1. Prove parts a, c, and d of Theorem 3.8.1.
2. Let $a \in \mathbb{Z}$ and let S be an ideal of $\mathbb{Z}$ such that $a \in S$. Prove $\mathbb{Z}a \subseteq S$. In this sense, $\mathbb{Z}a$ is the smallest ideal of $\mathbb{Z}$ containing a.
3. Prove Theorem 3.8.2. as follows:
   (1) Show (Case 1) if $H = \{0\}$, then $H = \mathbb{Z} \cdot 0$.
   Then assume (Case 2) $H \neq \{0\}$. For Case 2,
   (2) Show H contains a positive integer.
   (3) Show H contains a least positive integer. Define a to be this element.
   (4) In order to show $H \subseteq \mathbb{Z}a$, let $x \in H$ and then use the Division Algorithm to divide x by a.
   (5) Show $\mathbb{Z}a \subseteq H$.
4. Prove Theorem 3.8.3.

3.9 PRIMES

Definition   Let $a, b \in \mathbb{Z}$. An element $c \in \mathbb{N}$ is called a **greatest common divisor** of a and b provided that

        (1) $c \mid a$ and $c \mid b$

and   (2)   for all $d \in \mathbb{Z}$: if $d \mid a$ and $d \mid b$, then $d \mid c$.

Theorem 3.9.1   If a and b have a greatest common divisor, then it is unique.

     Proof: Exercise 1.

     The unique greatest common divisor of a and b, if it exists, is denoted by **gcd(a, b)**.

Theorem 3.9.2   For any $a, b \in \mathbb{Z}$ (not both zero), gcd(a, b) exists and has the form gcd(a, b) = $ma + nb$ for some $m, n \in \mathbb{Z}$.

     Proof: Exercise 2.

Definition   The numbers $a, b \in \mathbb{Z}$ are called **relatively prime** if $1 = $ gcd(a, b).

Definition   Let $p \in \mathbb{N}$, $p \neq 1$. Then p is called **prime** provided that for $n \in \mathbb{Z}$: if $n \mid p$, then $n = \pm 1$ or $n = \pm p$.

Theorem 3.9.3   Let p be prime and $n \in \mathbb{Z}$. If p does not divide n, then p and n are relatively prime.

     Proof: Exercise 3.

Theorem 3.9.4

    (a)   If gcd(a, b) = 1 and $a \mid bc$, then $a \mid c$.
    (b)   If p is prime and $p \mid bc$, then $p \mid b$ or $p \mid c$.
    (c)   If p is prime and $p \mid b_1 b_2 \cdots b_k$, then $p \mid b_i$ for some $1 \leq i \leq k$.

     Proof: Exercise 4.

     Recall that in Exercise 3.2.6, conjecture c was false. It is not always possible to divide a congruence by any $c \not\equiv 0 \bmod n$. We now give a correct version of this conjecture.

Corollary 3.9.5   If $ac \equiv bc \bmod n$ and if c and n are relatively prime, then $a \equiv b \bmod n$.

     Proof: Exercise 5.

     Theorem 3.9.7 asserts the uniqueness of factorization in the integers. In order to prove this theorem, it is useful to have a form of induction that differs from the form given by Theorem 3.1.2.

3.9 Primes

Theorem 3.9.6 *Principle of Complete Induction: Let $S \subseteq \mathbb{N}$ with the following property:* **for all** $n \in \mathbb{N}$: *if* $\{1, \ldots, n-1\} \subseteq S$, *then* $n \in S$. *Then* $S = \mathbb{N}$.

Proof:
Suppose S has the property indicated. Let $n = 1$. Then $\{1, \ldots, n-1\} = \emptyset$. Since $\emptyset \subseteq S$, we have $1 \in S$ by the property. Define $D = \{n \in \mathbb{N} \mid \{1, \ldots, n\} \subseteq S\}$. Thus $1 \in D$. Assume $n \in D$. By definition of D, $\{1, \ldots, n\} \subseteq S$, so that by our hypothesis on S, $n + 1 \in S$. Therefore $\{1, \ldots, n, n+1\} \subseteq S$, so that $n + 1 \in D$. Thus D satisfies the condition of Theorem 3.1.2. Therefore $D = \mathbb{N}$. It follows that $S = \mathbb{N}$. ∎

Theorem 3.9.7 *Fundamental Theorem of Arithmetic: Any integer a greater than 1 has a unique factorization $a = p_1^{n_1} p_2^{n_2} \cdots p_k^{n_k}$ for primes $p_1 < p_2 < \cdots < p_k$ and $n_i \in \mathbb{N}$.*

Proof:
Let S be the set containing 1 plus the integers greater than 1 that have such unique factorizations. We will show that S has the property given in Theorem 3.9.6 and hence that $S = \mathbb{N}$. Assume $\{1, \ldots, n-1\} \subseteq S$ for $n > 1$. If the only positive divisors of n are 1 and n itself, then n is prime, say, p, and hence has the required unique factorization, $n = p$. If n has a divisor b other than 1 and itself, then $n = b \cdot c$ for $1 < b < n$ and some c. Therefore $1 < c < n$ also, so that b and c are both in S. Therefore n is the product of the prime factors of b and those of c, so that n has a factorization.

Now let us show that the factorization is unique, that is, suppose

(1) $n = p_1^{n_1} \cdot p_2^{n_2} \cdots p_k^{n_k}$

and (2) $n = q_1^{m_1} \cdot q_2^{m_2} \cdots q_l^{m_l}$.

We need to show $k = l$ and for $i = 1, \ldots, k$: $p_i = q_i$ and $m_i = n_i$. From (1) and (2) we can write:

(3) $p_1^{n_1} \cdot p_2^{n_2} \cdots p_k^{n_k} = q_1^{m_1} \cdot q_2^{m_2} \cdots q_l^{m_l}$.

For any given p_i, since p_i divides $q_1^{m_1} \cdots q_l^{m_l}$, p_i divides some q_j by Theorem 3.9.4c. Since p_i and q_j are prime, $p_i = q_j$. Dividing both sides of (3) by p_i $(= q_j)$ gives:

(4) $p_1^{n_1} \cdots p_i^{n_i - 1} \cdots p_k^{n_k} = q_1^{m_1} \cdots q_j^{m_j - 1} \cdots q_l^{m_l}$.

This number is less than n and therefore in S. By our inductive hypothesis, $k = l$ (if $i = k$ and $n_k = 1$, we would get $k - 1 = l - 1$), $n_i - 1 = m_j - 1$, and $n_t = m_t$ for $t \neq i$. The result follows. ∎

Exercises 3.9

Problems

1. Prove Theorem 3.9.1. Use the following property of the integers: for $x, y \in \mathbb{N}$: if $x|y$ and $y|x$, then $x = y$.
2. Prove Theorem 3.9.2. Consider $\mathbb{Z}a + \mathbb{Z}b$ and use results from Section 3.8.
3. Prove Theorem 3.9.3.
4. Prove Theorem 3.9.4. Use induction for part (c) and Theorem 3.9.2 for part (a).
5. Prove Corollary 3.9.5.

Supplementary Problems

Definition *An ideal P of $\mathbb{Z}$ is* **prime** *provided for all $x, y \in \mathbb{Z}$: if $xy \in P$, then $x \in P$ or $y \in P$.*

6. For $a \in \mathbb{N}$, prove that $\mathbb{Z}a$ is a prime ideal iff a is a prime number.
7. A commutative ring in which each nonzero element has a multiplicative inverse is called a *field* (page 306). Prove that if p is prime, then $\mathbb{Z}_p$ is a field. Use Theorems 3.4.2 and 3.4.3.

3.10 PARTIALLY ORDERED SETS

Definition A relation $\leq$ on a set P is called a **partial order** on P provided that for all $x, y, z \in P$:

(1) $x \leq x$ *(reflexive property)*

(2) If $x \leq y$ and $y \leq x$, then $x = y$.   *(antisymmetric property)*

(3) If $x \leq y$ and $y \leq z$, then $x \leq z$.   *(transitive property)*

If $\leq$ is a partial order on P, then P is called a **partially ordered set** (or **poset**) under $\leq$. If $x \leq y$ or $y \leq x$, then x and y are said to be **comparable**.

Example 1:
Let X be a set. The relation $\subseteq$ between subsets of X is a partial order on the set $P(X)$. Note that if $A, B \in P(X)$, it need not be the case that either $A \subseteq B$ or $B \subseteq A$; that is, A and B need not be comparable.

Example 2:
Let $\leq$ be the usual relation of **less than or equal to** on $\mathbb{R}$. For any two elements $x, y \in \mathbb{R}$, either $x \leq y$ or $y \leq x$; that is, any two elements of $\mathbb{R}$ are comparable.

3.10 Partially Ordered Sets

Definition   A partial order $\leq$ on a set P is called a **total order** (or **linear order** or **chain**) if every two elements of P are comparable.

Example 3
Let $\mid$ be the relation of **divides** on the set $\mathbb{N}$. Then $\mid$ is a partial order on $\mathbb{N}$ but not a total order.

Definition   Let P be a poset under $\leq$ and let $S \subseteq P$. An element $m \in P$ is called an **upper bound** for S provided that *for all* $x \in S: x \leq m$. An element $l \in P$ is called a **least upper bound** for S provided that

    (1) l is an upper bound for S

and   (2)   for all m such that m is an upper bound for $S: l \leq m$.

Example 4
Let $P = \mathbf{P}(X)$ for a set X and consider the poset P under $\subseteq$. For any $A, B \in P$, $A \cup B$ is a least upper bound for $\{A, B\}$. Proof: Exercise 1.

Theorem 3.10.1   Let P be a poset under $\leq$ and let $S \subseteq P$. If x and y are least upper bounds for S, then $x = y$.

Proof: Exercise 2.

Theorem 3.10.1 states that the least upper bound of a set S, if it exists, is unique. We can therefore give it a name: **lub(S)**.

As you read the following definition, consider the definitions of upper bound and least upper bound.

Definition   Let P be a poset under $\leq$ and let $S \subseteq P$. An element $m \in P$ is called a **lower bound** for S provided that *for all* $x \in S: m \leq x$. An element $g \in P$ is called a **greatest lower bound** for S provided that

    (1) g is a lower bound for S

and   (2)   for all m such that m is a lower bound for $S: m \leq g$.

The definitions of lower bound and greatest lower bound above are said to be *dual* to the definitions of upper bound and least upper bound. Duality is a kind of symmetry. The second definition can be obtained from the first by replacing the word "upper" with "lower" and the word "least" with "greatest" and writing "$a \leq b$" wherever "$b \leq a$" appears (for any a, b). Theorem 3.10.1 also has a dual:

Theorem 3.10.2   Let P be a poset under $\leq$ and let $S \subseteq P$. If x and y are greatest lower bounds for S, then $x = y$.

> **Proof:**
>
> A proof of Theorem 3.10.2 can be obtained from a proof of Theorem 3.10.1 by replacing the word "upper" with "lower" and the word "least" with "greatest" and writing "$a \leq b$" wherever "$b \leq a$" appears (for any a, b). (Verify this as Exercise 3.) ∎

The proof of Theorem 3.10.2 is said to be *dual* to the proof of Theorem 3.10.1.

The greatest lower bound of a set S (if it exists) is denoted **glb**(S). Example 4 has a dual:

Example 5:
Let $P = \mathbf{P}(X)$ for a set X and consider the poset P under $\subseteq$. For any $A, B \in P$, $A \cap B$ is a greatest lower bound for $\{A, B\}$. Proof: Exercise 4.

Example 6:
Consider $\mathbf{N}$ under $|$. For any $x, y \in \mathbf{N}$, $\gcd(x, y)$ is a greatest lower bound for $\{x, y\}$. Proof: Exercise 5.

The definition of greatest common divisor given in Section 3.9 "dualizes" to give the following definition:

Definition Let $a, b \in \mathbf{Z}$. An element $c \in \mathbf{N}$ is called a *least common multiple* of a and b provided that

   (1) $a|c$ and $b|c$

and (2) for all $d \in \mathbf{Z}$: if $a|d$ and $b|d$, then $c|d$.

Consider $\mathbf{N}$ under $|$. For any $x, y \in \mathbf{N}$, a least common multiple of x and y is a least upper bound for $\{x, y\}$. By Theorem 3.10.1, then, the least common multiple, if it exists, is unique. It is denoted by **lcm**(x, y). The existence of **lcm**(x, y) can be proved by showing $xy/\gcd(x, y)$ is an element of $\mathbf{N}$ that satisfies properties (1) and (2) of the definition above (Exercise 6).

Definition If P is a poset under $\leq$, we define the relation $<$ on P by:

For all $x, y \in P$: $x < y$ iff $x \leq y$ and $x \neq y$.

Theorem 3.10.3 Let x, y, z be elements of a poset P. If $x < y$ and $y < z$, then $x < z$.

Proof: Exercise 7.

3.10 Partially Ordered Sets

Theorem 3.10.4 Trichotomy: Let T be a totally ordered set. For any $x, y \in T$, precisely one of the following holds:

(1) $x < y$

(2) $x = y$

(3) $y < x$

Proof: Exercise 8.

Definition A poset in which each pair of elements has both a greatest lower bound and a least upper bound is called a **lattice**.

All the examples of posets we have considered so far have been lattices. It is not difficult to construct a poset that is not a lattice:

Example 7:
Let $P = \{a, b, c, d\}$ under the relation $\leq$ such that $a \leq c, b \leq c, a \leq d, b \leq d$. The pair a, b, for example, has neither lub nor glb.

Finite posets with a small number of elements can be easily pictured by a diagram called a Hasse diagram. The elements of the poset are drawn such that if $x < y$, then x is below y on the paper, with a path connecting x and y. The path consists of lines drawn from x to y if $x < y$ and there is no z such that $x < z < y$. Figure 3.10.1 is a Hasse diagram for the poset of Example 7. It is easy to see from the diagram that a and b have no common lower bound (and therefore no glb) and two incomparable common upper bounds (and therefore no lub).

Figure 3.10.2 (following page) is a Hasse diagram for the lattice of all subsets of $\{x, y, z\}$ under $\subseteq$.

Definition An element x is a **maximal element** of a poset P provided that for all $y \in P$: if $x \leq y$, then $x = y$. (Dually, x is a **minimal element** provided that $y \leq x$ implies $y = x$.)

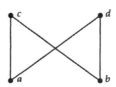

Figure 3.10.1

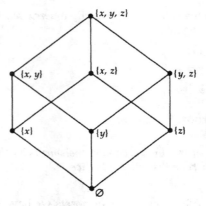

Figure 3.10.2

In Example 7, c and d are maximal elements. In the lattice of all subsets of $\{x, y, z\}$, the element $\{x, y, z\}$ is a maximal element and $\emptyset$ is a minimal element. In $\mathbf{Z}$ with the usual order, there are no maximal or minimal elements.

Additional Ideas

Definition   Let T and S be totally ordered sets with orders both denoted by $\leq$. A function $f: T \to S$ is called **nondecreasing**[†] provided that for all $x, y \in T$: if $x < y$, then $f(x) \leq f(y)$. The function f is called **increasing**[†] provided that for all $x, y \in T$: if $x < y$, then $f(x) < f(y)$.

Theorem 3.10.5   Let T and S be totally ordered sets with orders both denoted by $\leq$ and $f: T \to S$ one-to-one. Then the following conditions are equivalent for all $x, y \in T$:

(1) f is increasing.
(2) $x < y$ iff $f(x) < f(y)$.
(3) $x \leq y$ iff $f(x) \leq f(y)$.

Proof: Exercise 10.

Definition   A bijection $f: T \to S$ is called an **order isomorphism** if the conditions of Theorem 3.10.5 hold.

[†] See the footnote on page 255.

3.10 Partially Ordered Sets

Example 8:
Define an order on the set P_Z in Section 3.7 by the following rule: for all $[(a, b)]$, $[(c, d)] \in P_Z$: $[(a, b)] \leq [(c, d)]$ iff $a + d \leq b + c$. Then the definition of $\leq$ is independent of the representations of the equivalence classes; that is, $\leq$ is well defined (Exercise 12). Also, $\leq$ is a total order on P_Z (Exercise 12).

By Theorem 3.7.2 the function $f: P_Z \to \mathbb{Z}$, given by $f([a, b]) = a - b$, is a ring isomorphism; that is, it respects the operations of addition and multiplication. The following theorem states that f also respects $\leq$.

Theorem 3.10.6 *The function $f: P_Z \to \mathbb{Z}$, given by $f([(a, b)]) = a - b$, is an order isomorphism.*

Proof: Exercise 13.

Example 9:
Define an order on the set P_Q in Section 3.7 by the following rule:

For $[(a, b)], [(c, d)] \in P_Q$ with b and d of the same sign:

$$[(a, b)] \leq [(c, d)] \text{ iff } ad \leq bc$$

For $[(a, b)], [(c, d)] \in P_Q$ with b and d of opposite sign:

$$[(a, b)] \leq [(c, d)] \text{ iff } bc \leq ad$$

Then $\leq$ is well defined and is a total order on P_Q (Exercise 14).

The next theorem states that the ring isomorphism given by Theorem 3.7.3 is also an order isomorphism.

Theorem 3.10.7 *The function $f: P_Q \to \mathbb{Q}$, given by $f([(a, b)]) = \dfrac{a}{b}$, is an order isomorphism.*

Proof: Exercise 15.

Two sets A and B with operations and relations defined on them are called *isomorphic* if there exists an isomorphism $f: A \to B$. When we say that A and B are "isomorphic" without specifying the type of isomorphism involved, we mean isomorphic with respect to all operations and relations defined on both A and B.

The set P_Z has been defined in terms of $\mathbb{N}$. Also, the operations of addition and multiplication and the order $\leq$ on P_Z have been defined in terms of the operations and order on $\mathbb{N}$. In this section we have taken the set $\mathbb{Z}$ with its operations and order as already defined, and in this context we have proved that P_Z and $\mathbb{Z}$ are isomorphic. Since $\mathbb{Z}$ is just an isomorphic copy of P_Z, we can treat $\mathbb{Z}$ not as known and isomorphic to P_Z but as formally defined by P_Z.

Doing so reduces the number of undefined terms. Thus **Z** equals P_Z by definition of **Z**.

In following through with this approach, we need to rid the development of P_Z of properties of **Z** that have been assumed. (See Examples 3.2.6, 3.3.4, and 3.4.3.) "Property of **N**" will be allowable as justification for proof steps but not "property of **Z**". This causes some difficulty—in particular, in the proof that multiplication is well defined on P_Z (Example 3.4.3), where the existence of negatives is used: if a and b are any natural numbers, and therefore integers, then the integer $a - b$ is defined. We can stick to **N** and avoid negatives by using cases in our proofs: if a and b are natural numbers and $a < b$, then $b - a$ is defined; if $b < a$, then $a - b$ is defined. After defining **Z** this way, we can show that the properties of **Z** listed in Appendix 1 follow from the properties of **N**.

Having defined **Z** in terms of **N**, we can then define **Q** in terms of **Z** by defining **Q** to be P_Q. (See Examples 3.2.7, 3.3.5, and 3.4.4.) "Property of **Z**" will be a legitimate justification for proof steps in developing P_Q, but properties of **Q** or **R** will not. In Chapter 4 we will see how **R** can be defined in terms of **Q**.

Exercises 3.10

Problems

1. Show that in **P**(X) under $\subseteq$, $A \cup B$ is **lub**{A, B}.
2. Prove Theorem 3.10.1.
3. Show that the proof of Theorem 3.10.1 "dualizes" to give a proof of Theorem 3.10.2.
4. Show that in **P**(X) under $\subseteq$, $A \cap B$ is **glb**{A, B}.
5. Show that in **N** under $|$, **gcd**(x, y) is **glb**{x, y}.
6. Show that for $x, y \in$ **N**, $\dfrac{xy}{\gcd(x, y)} = \text{lcm}(x, y)$.
7. Prove Theorem 3.10.3. Compare your proof with the Additional Ideas discussion in Section 2.7, which covers the transitivity of the relation $<$ (smaller cardinality) between sets. Give an example that shows that the antisymmetric property (see the definition of poset) fails for the relation $\leq$ between sets.
8. Prove Theorem 3.10.4.
9. Sketch the Hasse diagrams for the following posets and identify maximal and minimal elements:
   (a) **P**({a, b}) under $\subseteq$
   (b) **P**({a, b, c, d}) under $\subseteq$
   (c) {1, 2, ..., 11, 12} under $|$

Supplementary Problems

10. Prove Theorem 3.10.5.
11. Show that the composition of two increasing functions is increasing.
12. Show that the relation $\leq$ defined on P_Z in Example 8 is a well-defined total order on P_Z.
13. Prove Theorem 3.10.6.
14. Show that the relation $\leq$ defined on P_Q in Example 9 is a well-defined total order on P_Q.
15. Prove Theorem 3.10.7.

Chapter 4

Proofs in Analysis

4.1 SEQUENCES

The ideas and approaches to discovering proofs introduced in the first chapters have application to a careful study of the functions seen in calculus. This study is known as either "advanced calculus" or an introduction to "real analysis". We will introduce this area by considering how some theorems from calculus can be proved using our proof style. For proofs in this chapter, "property of the real numbers" is an acceptable justification for proof steps. These properties include all the computational properties (use of which was previously limited to examples) but not properties of sets of real numbers.

Consider the sequence of terms:

$$\frac{1}{1}, \frac{1}{2}, \frac{1}{4}, \frac{1}{8}, \frac{1}{16}, \frac{1}{32}, \frac{1}{64}, \ldots$$

Each term in the sequence is obtained by multiplying the preceding term by $\frac{1}{2}$. Other examples of sequences are

$$1, 1, 1, 1, 1, 1, \ldots$$

$$1, 2, 3, 4, 5, 6, \ldots$$

$$1, 2, 1, 2, 1, 2, 1, \ldots$$

$$\frac{1}{1}, \frac{1}{2}, \frac{1}{3}, \frac{1}{4}, \frac{1}{5}, \ldots$$

The terms of a sequence are real numbers, and we distinguish the order in which the terms are listed. In order to make this idea of a sequence precise, the following definition is made.

Definition A *sequence* is a function $a: \mathbb{N} \to \mathbb{R}$. $a(n)$ is called the *n*th term of the sequence and is denoted a_n. The sequence itself is denoted by $\langle a_1, a_2, a_3, \ldots \rangle$ or merely by $\langle a_n \rangle$.

Thus the sequence $\langle \frac{1}{1}, \frac{1}{2}, \frac{1}{3}, \frac{1}{4}, \frac{1}{5}, \ldots \rangle$ is formally a function from $\mathbb{N}$ into $\mathbb{R}$ with range $\{\frac{1}{1}, \frac{1}{2}, \frac{1}{3}, \frac{1}{4}, \frac{1}{5}, \ldots\}$. We don't distinguish the order of listing in the set $\{\frac{1}{1}, \frac{1}{2}, \frac{1}{3}, \frac{1}{4}, \frac{1}{5}, \ldots\}$, but the notation $\langle \frac{1}{1}, \frac{1}{2}, \frac{1}{3}, \frac{1}{4}, \frac{1}{5}, \ldots \rangle$ implies that $\frac{1}{1}$ is the first term of the sequence, $\frac{1}{2}$ is the second, and so forth.

Definition A number $L \in \mathbb{R}$ is defined to be the **limit** of the sequence $\langle a_1, a_2, a_3, \ldots \rangle$ (written $\lim_{n \to \infty} \langle a_n \rangle = L$ or merely $a_n \to L$) provided that, given a real number $\epsilon > 0$, there exists a natural number N such that *for all* $n > N$: $|a_n - L| < \epsilon$. If L is the limit of $\langle a_n \rangle$, we say $\langle a_n \rangle$ **converges** to L. A sequence with no limit is said to **diverge**.

A drawn-out but completely formal language statement of the property defining "limit" would be **for all $\epsilon \in \mathbb{R}$ such that $\epsilon > 0$: there exists $N \in \mathbb{N}$ such that for all $n \in \mathbb{N}$ such that $n > N$:** $|a_n - L| < \epsilon$. We have used an informal abbreviation of the defining condition. You're supposed to know, when reading the informal version, that $n \in \mathbb{N}$—since a_n is used later.

If a and b are real numbers, then $|a - b|$ is just the distance from a to b along the real-number line. The definition states that the limit of $\langle a_n \rangle$ is L provided that, as we go out farther in the sequence, all terms become arbitrarily close to L. That is, no matter how close (within a distance ϵ) we first demand a_n be to L, it is possible to find a natural number N such that all terms a_n beyond the Nth term are within that distance ϵ of L.

Example 1:
Let $\langle a_n \rangle$ be defined by $a_n = n/n + 1$. Show that $a_n \to 1$. That is, show that the limit of the sequence $\langle \frac{1}{2}, \frac{2}{3}, \frac{3}{4}, \frac{4}{5}, \ldots \rangle$ is 1.

We need to show that, given $\epsilon > 0$, there exists $N \in \mathbb{N}$ such that for all $n > N$: $|(n/n + 1) - 1| < \epsilon$. Our proof outline is therefore:

Proof:
1. Let $\epsilon > 0$ be given.
2. (Define N here somehow, to prove $\exists$.)
3. Let n be arbitrary such that $n > N$.
    .
    .
    .
    k. $\left| \dfrac{n}{n+1} - 1 \right| < \epsilon$

4.1 Sequences

$k+1$. for all $n > N$: $\left|\dfrac{n}{n+1} - 1\right| < \epsilon$ (3—k; pr. $\forall$)

$k+2$. $\lim\limits_{n \to \infty} a_n = 1$ (1—$k+1$; def. lim)

Scrap Paper

The needed Step k, $\left|\dfrac{n}{n+1} - 1\right| < \epsilon$, is equivalent to $|n - (n+1)| < \epsilon(n+1)$, which is equivalent to $\dfrac{1}{\epsilon} < (n+1)$. By the Archimedean property of **R**, there is a natural number N greater than $\dfrac{1}{\epsilon}$. This N gives us Step 2.

Proof:

1. Let $\epsilon > 0$ be given.
2. $N > \dfrac{1}{\epsilon}$ for some $N \in \mathbf{N}$.   (Arch. prop. of **R**)
3. Let n be arbitrary such that $n > N$.
4. $\dfrac{1}{\epsilon} < N < n < (n+1)$ (2, 3; prop. **R**)
5. $1 < \epsilon(n+1)$ (4; prop. **R**)
6. $|n - (n+1)| < \epsilon(n+1)$ (5; prop. **R**)
7. $\left|\dfrac{n}{n+1} - 1\right| < \epsilon$ (6; prop. **R**)
8. for all $n > N$: $\left|\dfrac{n}{n+1} - 1\right| < \epsilon$ (3—7; pr. $\forall$)
9. $\lim\limits_{n \to \infty} a_n = 1$ (1—8; def. lim) ∎

Example 2:
Let $\langle a_n \rangle$ be defined by $a_n = \dfrac{1}{n}$. Show that $a_n \to 0$. That is, show that $\langle \dfrac{1}{1}, \dfrac{1}{2}, \dfrac{1}{3}, \dfrac{1}{4}, \ldots \rangle$ converges to 0.

Proof: Exercise 1.

Theorem 4.1.1 Let $\langle a_n \rangle$ be a sequence. If $a_n \to L$ and $a_n \to M$, then $L = M$.

This theorem, of course, asserts that if the limit exists, then it is unique. The proof is based on the fact that if $L \neq M$, then there is a definite distance, call it d, between L and M. If we pick $\epsilon = \frac{1}{3}d$, the definition of limit tells us that there exists a natural number N such that for all n greater than N, a_n is within $\frac{1}{3}d$ of L. Similarly, there is a natural number K such that for all n greater than K, a_n is within $\frac{1}{3}d$ of M. For any n greater than both N and K, then, a_n is both within $\frac{1}{3}d$ of L and within $\frac{1}{3}d$ of M—which is impossible. (Note: the use of $\frac{1}{3}d$ is arbitrary; $\frac{1}{2}d$ would do as well, but $\frac{2}{3}d$ would not. It is not necessary, for the proof, to take pains at finding the largest possible distance that does work.)

Proof:

We show $L = M$ by contradiction.

1. Assume (to get #) $L \neq M$.
2. Let $d = |L - M|$.
3. Let $\epsilon = \dfrac{1}{3} d$.
4. There exists $N \in \mathbb{N}$ such that **for all** $n > N$: $|a_n - L| < \epsilon$.   (hyp.; def. lim)
5. There exists $K \in \mathbb{N}$ such that **for all** $n > K$: $|a_n - M| < \epsilon$.   (hyp.; def. lim)
6. Define j to be 1 plus the larger of K and N.
7. $|a_j - L| < \epsilon$ (4, 6; us. $\forall$)
8. $|a_j - M| < \epsilon$ (5, 6; us. $\forall$)
9. $|L - a_j| + |a_j - M| < 2\epsilon$ (7, 8; prop. $\mathbb{R}$)
10. $|L - M| < 2\epsilon$ (9; prop. $\mathbb{R}$)
11. $d < 2\epsilon$ # Step 3.   (2, 10; sub.)
12. $L = M$ (1—11; #)

The property of $\mathbb{R}$ used in going from Step 9 to Step 10 is the triangle inequality: $|A - C| \leq |A - B| + |B - C|$ for all real A, B, C. This version follows from Theorem A.1.12 of Appendix 1.

4.1 Sequences

Example 3:
The following sequences diverge, that is, have no limit:

(a) $\langle 1, 2, 1, 2, \ldots \rangle$
(b) $\langle 1, 2, 3, 4, \ldots \rangle$
(c) $\langle -1, -2, -3, -4, \ldots \rangle$

In order to show that a sequence $\langle a_n \rangle$ is divergent, we need to show that, for all real numbers L, L is not the limit of $\langle a_n \rangle$. Showing this involves the negation of the defining condition for limit. For the moment let us denote the statement **for all $n > N$: $|a_n - L| < \epsilon$** by the identifier **condition(L)**. Then L is the limit of $\langle a_n \rangle$ if **for all $\epsilon > 0$: there exists N such that condition(L)**. The negation of this last statement is:

there exists $\epsilon > 0$ such that $\sim$(there exists N such that condition(L))

or **there exists $\epsilon > 0$ such that for all N: $\sim$condition(L).**

In order to show that L is not the limit of $\langle a_n \rangle$, we must therefore specify some $\epsilon > 0$ and show the negation of **condition(L)** for all N. Now examine the negation of **condition(L)**:

$\sim$(condition(L))

$\Leftrightarrow$ $\sim$(for all $n > N$: $|a_n - L| < \epsilon$)

$\Leftrightarrow$ there exists $n_0 > N$ such that $\epsilon \leq |a_{n_0} - L|$.

If the last statement is to hold for all N, then there must be an infinite number of a_{n_0} a distance at least ϵ away from L—and conversely.

Thus to show that $\langle a_n \rangle$ is divergent, we show that, for each real number L, there is an $\epsilon > 0$ such that an infinite number of a_n are at least a distance ϵ from L. To show that the sequence in Example 3a diverges, we can take $\epsilon = \frac{1}{2}$ for any real L. Any fixed ϵ will do to show (b) and (c) diverge.

Definition   Let $\langle a_n \rangle$ and $\langle b_n \rangle$ be sequences and $k \in \mathbb{R}$.

(a) *Define the sequence* $k \langle a_n \rangle = \langle k a_n \rangle$.
(b) *Define the sequence* $\langle a_n \rangle + \langle b_n \rangle = \langle a_n + b_n \rangle$.
(c) *Define the sequence* $\langle a_n \rangle - \langle b_n \rangle = \langle a_n \rangle + (-1)\langle b_n \rangle$.

The new sequences (a) and (b) are defined by specifying their nth terms using the terms of the given sequences. For example, $a_n + b_n$ is the nth term of $\langle a_n + b_n \rangle$.

Example 4:

$$5\left\langle \frac{1}{1}, \frac{1}{2}, \frac{1}{3}, \frac{1}{4}, \ldots \right\rangle = \left\langle \frac{5}{1}, \frac{5}{2}, \frac{5}{3}, \frac{5}{4}, \ldots \right\rangle$$

$$\langle 1, 2, 1, 2, \ldots \rangle + \langle 2, 1, 2, 1, \ldots \rangle = \langle 3, 3, 3, 3, \ldots \rangle$$

$$\left\langle \frac{1}{1}, \frac{1}{2}, \frac{1}{3}, \frac{1}{4}, \ldots \right\rangle + \langle 1, 2, 3, 4, \ldots \rangle = \left\langle 2, 2\frac{1}{2}, 3\frac{1}{3}, 4\frac{1}{4}, \ldots \right\rangle$$

Theorem 4.1.2 Let $\langle a_n \rangle$ and $\langle b_n \rangle$ be sequences and $k \in \mathbf{R}$.

(a) If $L = \lim_{n \to \infty} \langle a_n \rangle$, then $kL = \lim_{n \to \infty} k\langle a_n \rangle$.

(b) If $L = \lim_{n \to \infty} \langle a_n \rangle$ and $M = \lim_{n \to \infty} \langle b_n \rangle$, then $L + M = \lim_{n \to \infty} \langle a_n \rangle + \langle b_n \rangle$.

Proof: Exercise 2.

Example 5:
Let $a_1 = 0.5$, $a_2 = 0.55$, $a_3 = 0.555$, and, in general, $a_n = 0.55\ldots 5$, where the digit 5 appears n times. Then $\langle a_n \rangle$ has the limit $0.5555\ldots$ (that is, the real number given by the repeating decimal). Repeating decimals represent rational numbers, and a corresponding ratio of integers can be found by the following technique.

Let $A = 0.55555\ldots$. Then, multiplying both sides of this equality by **10**, we get $10A = 5.5555\ldots$. Subtracting, we get

$$10A = 5.5555\ldots$$
$$- A = 0.5555\ldots$$
$$9A = 5 \qquad \text{or} \quad A = \frac{5}{9}$$

In the computation above, we assumed the infinite decimal representation for real numbers. It is possible to *define* an **infinite decimal** as a sequence of finite decimals, each term agreeing in all its places with the places of the next term. With this definition, Theorem 4.1.2a can be taken as justification for multiplying both sides of $a = 0.55555\ldots$ by 10.

Alternatively, using customary computations with real numbers, you can show (Exercise 4) that $a_n \to 0.55555\ldots$ and $a_n \to \frac{5}{9}$, so that $\frac{5}{9} = 0.55555\ldots$ by Theorem 4.1.1. Note that the algorithm for performing long division does just the opposite of the technique above. It starts with the rational $\frac{5}{9}$ and produces the repeating decimal $0.55555\ldots$.

4.1 Sequences

Example 6:
Note that the ideas in Example 5 imply that $1.0 = 0.9999\ldots$ and, for example, $2.32 = 2.31999\ldots$. Thus the decimal representation of a real number is not unique unless we agree to some convention such as always replacing a repeating pattern of 9s with 0s and adding 1 to the place preceding the pattern of 9s.

Additional Ideas

It is possible, without bringing in the real numbers, to define a sequence as a function from $\mathbb{N}$ to $\mathbb{Q}$. A *decimal rational* is defined to be a rational number of the form $a/10^n$, where a and n are integers and $n \geq 0$. Note that numbers of the form $a/10^n$ can be written as terminating decimals with at most n decimal places. A real number can then be *defined* as a certain kind of sequence of decimal rationals. For example, the rational number $2\frac{1}{9}$ can be expressed as the repeating decimal $2.111\ldots$. In order to define the real numbers, we need to be able to say what a repeating decimal *is*—in terms of previously defined terms. We define $2.111\ldots$ to be the sequence $\langle 2, 2.1, 2.11, 2.111, \ldots \rangle$ of decimal rationals. The real number $2\frac{1}{2}$ is the sequence $\langle 2, 2.5, 2.5, 2.5, \ldots \rangle$.

In order to carry out a program of defining the real numbers $\mathbb{R}$ in this way, we need to specify which sequences of decimal rationals are to be real numbers, how to add and multiply these, and how to order these by the $\leq$ relation. We need to be able to find an isomorphic copy of $\mathbb{Q}$ as a subset S_Q of $\mathbb{R}$ so as to consider the reals as an extension of $\mathbb{Q}$. That is, the inclusion map $S_Q \to \mathbb{R}$ must respect addition, multiplication, and the order relation $\leq$, plus the isomorphism from $\mathbb{Q}$ to S_Q must also respect $\leq$. Although we did not mention it at the time, analogs of these considerations are needed to adequately define $\mathbb{Z}$ as P_Z and $\mathbb{Q}$ as P_Q as in the Additional Ideas portion of Section 3.7.

S_Q will be the set of sequences that correspond to repeating decimals, but this raises the question of just what a rational number *is*. Is it an element of the set $\mathbb{Q}$ used to define $\mathbb{R}$, or an element of the subset S_Q of $\mathbb{R}$? An advanced viewpoint would hold that it doesn't matter, since $\mathbb{Q}$ and S_Q are isomorphic. From our viewpoint of beginning logic, however, we want (1) the rationals to have a precise definition and (2) all properties of the rationals to follow from the definition. Thus we wish to have the rationals as a subset of the reals, the integers as the subset of the rationals, and the natural numbers as a subset of the integers. The sets used to "construct" the reals can be given such names as "prerationals" and subsequently forgotten. In fact, the formal definition of the reals is itself forgotten. It is only the properties of the reals with which we work that are relevant. For example, we do not think of a sequence of real numbers as a sequence of sequences of decimal prerationals. The situation is analogous to our having provided a formal definition of ordered pair, only to forget the formal definition after the theorem characterizing ordered pairs was proved.

Exercises 4.1

Problems

1. Prove that $\dfrac{1}{n} \to 0$.
2. Prove Theorem 4.1.2.
3. Determine whether each of the following sequences converges or diverges and prove your assertions:

   (a) $\left\langle \dfrac{2}{n} \right\rangle$ (b) $\left\langle \dfrac{n}{2^n} \right\rangle$ (c) $\left\langle \dfrac{2^n}{n!} \right\rangle$ (d) $\left\langle \dfrac{(-1)^n}{n} \right\rangle$ (e) $\langle \sqrt{n} \rangle$

   (f) $\left\langle \dfrac{n+1}{n} \right\rangle$

4. Let $a_n = 0.555\ldots 5$ (the digit 5 occurring n times). Use the definition of limit to prove that $\langle a_n \rangle \to 0.55555\ldots$ and $\langle a_n \rangle \to \dfrac{5}{9}$, so that $\dfrac{5}{9} = 0.55555\ldots$.
5. Find a ratio of integers equal to $0.272727\ldots$.

4.2 FUNCTIONS OF A REAL VARIABLE

In this section we turn our attention to functions of the kind studied in calculus—that is, functions whose domain and range are subsets of the real numbers. Since this text is only preparatory for an analysis course and since neither time nor our need to continue the flow of ideas permits a detailed technical investigation, we will limit our discussion to a simple context that will nevertheless reveal the nature of the proofs in the subject. In particular, we will avoid the need for considering several special cases (for instance, one-sided limits) in each of our proofs by restricting the domains of our functions to open intervals of the following forms:

$$\text{the set } \mathbb{R} \text{ itself}$$
$$(a, b) = \{x \in \mathbb{R}: a < x < b\}$$
$$(a, \infty) = \{x \in \mathbb{R}: a < x\}$$
$$(-\infty, b) = \{x \in \mathbb{R}: x < b\}$$

Thus, to say in this chapter that $f: D \to \mathbb{R}$ is a function will automatically mean that D is one of the sets above. The definition of limit for a function $f: D \to \mathbb{R}$ is similar to the definition of limit of a sequence. Proofs of theorems on limits of functions use techniques similar to those for sequences.

4.2 Functions of a Real Variable

Definition Let $f: D \to \mathbf{R}$ be a function and $a \in D$. A number $L \in \mathbf{R}$ is defined to be the **limit** of $f(x)$ as x approaches a (written $L = \lim\limits_{x \to a} f(x)$) provided that, given a real number $\epsilon > 0$, there exists a real number $\delta > 0$ such that **for all** $x \in D$: **if** $0 < |x - a| < \delta$, **then** $|f(x) - L| < \epsilon$.

A formal language statement of the property defining limit is **for all** $\epsilon \in \mathbf{R}$ such that $\epsilon > 0$: there exists $\delta \in \mathbf{R}$ such that for all $x \in D$: if $0 < |x - a| < \delta$, then $|f(x) - L| < \epsilon$. Our informal (and clearer) wording of the defining condition will be simplified even more in a moment.

The definition states that the limit of $f(x)$ as x approaches a is L, provided that as x gets close to a, we have $f(x)$ getting close to L. The word "as" in the previous sentence is made precise in the following way. No matter how close (within a distance ϵ) we demand $f(x)$ to be to L, it is possible to find a distance δ such that for all $x \neq a$ within a distance δ of a, we have that $f(x)$ is within ϵ of L.

Example 1:
Let $f: \mathbf{R} \to \mathbf{R}$ be defined by $f(x) = x^2$. Show $\lim\limits_{x \to 2} f(x) = 4$.

We need to show that, given $\epsilon > 0$, there exists $\delta > 0$ such that **for all** $x \in \mathbf{R}$: **if** $0 < |x - 2| < \delta$, **then** $|x^2 - 4| < \epsilon$. We have the following proof outline:

Proof:
1. Let $\epsilon > 0$ be given.
2. (Define δ here somehow.)
3. Let $0 < |x - 2| < \delta$ for $x \in \mathbf{R}$.
   .
   .
k. $|x^2 - 4| < \epsilon$
k+1. for all $x \in \mathbf{R}$: if $0 < |x - 2| < \delta$, then $|x^2 - 4| < \epsilon$. (3—k; pr. $\forall \Rightarrow$)

k+2. $\lim\limits_{x \to 2} x^2 = 4$ (1, 2, k+1; def. lim)

It is now up to us to define δ in the course of our proof and then show that the **for all: if-then** statement holds for the given ϵ and the defined δ. Defining δ can be done by working backward from the conclusion (so what's new?):

Scrap Paper
We want the absolute value (size of) $x^2 - 4$ to be less than ϵ. That is, $|(x-2)(x+2)| < \epsilon$. We are allowed to define δ any way we want, in terms of the given ϵ. $|x-2|$ gives the distance of x from 2. If we make sure our δ is ≤ 1, then $1 \leq x \leq 3$, and so $x + 2$ must be ≤ 5. Thus the equation we need to show becomes

$$|x-2||x+2| \leq |x-2| \cdot 5 < \epsilon$$

From $|x-2| \cdot 5 < \epsilon$, we get $|x-2| < \epsilon/5$. Now we need to insure that $|x-2| < \epsilon/5$, which we can do by making sure that $\delta \leq \epsilon/5$. With this information we can continue with the proof.

Proof:
1. Let $\epsilon > 0$ be given.
2. Let δ be the smaller of 1 and $\epsilon/5$.
3. Let $0 < |x-2| < \delta$ for $x \in \mathbb{R}$.
4. $|x-2| < 1$ (2, 3; def. $\delta \leq 1$)
5. $1 < x < 3$ (4; prop. **R**)
6. $|x+2| < 5$ (5; prop. **R**)
7. $|x-2| < \epsilon/5$ (2, 3; def. $\delta \leq \epsilon/5$)
8. $|x+2| \cdot |x-2| < 5 \cdot \epsilon/5$ (6, 7; prop. **R**)
9. $|x^2 - 4| < \epsilon$ (8; prop. **R**)
10. for all $x \in \mathbb{R}$: if $0 < |x-2| < \delta$, then $|x^2 - 4| < \epsilon$.   (3—9; pr. $\forall \Rightarrow$)
11. $\lim\limits_{x \to 2} x^2 = 4$ (1, 2, 10; def. lim)

The use of free variables in informal statements will make our statements much less cumbersome than their formal counterparts. The statement **for all** $x \in D$: **if** $0 < |x - a| < \delta$, **then** $|f(x) - L| < \epsilon$ found in the definition of limit can be written "if $0 < |x - a| < \delta$, then $|f(x) - L| < \epsilon$" using the idea of a free variable. (See the discussion on pages 137 and 138.) Recall that when an undefined symbol (free variable) crops up in a statement, it means that the symbol represents an element arbitrarily chosen out of a domain that makes sense—D in our statement here—in order that $f(x)$ make sense.

Thus we will reword our definition of limit:

4.2 Functions of a Real Variable

Definition Let $f: D \to \mathbb{R}$ be a function and let $a \in D$. A number $L \in \mathbb{R}$ is defined to be the **limit** of $f(x)$ as x approaches a (written $L = \lim_{x \to a} f(x)$) provided that, given a real number $\epsilon > 0$, there exists a real number $\delta > 0$ such that if $0 < |x - a| < \delta$, then $|f(x) - L| < \epsilon$.

The rules for proving or using the statement

if $0 < |x - a| < \delta$, then $|f(x) - L| < \epsilon$

are exactly the same as those for proving or using the statement

for all $x \in D$: if $0 < |x - a| < \delta$, then $|f(x) - L| < \epsilon$

These are abbreviated "pr. $\forall \Rightarrow$" and "us. $\forall \Rightarrow$". We will abbreviate these rules "pr. $\Rightarrow$" and "us. $\Rightarrow$" when dealing with the free-variable form of a statement, making the words "for all" implicit in both the statement and the justification.

Example 2:
Let $f: \mathbb{R} \to \mathbb{R}$ be defined by $f(x) = 3x + 1$. Show $\lim_{x \to 2} f(x) = 7$. Use the free-variable form of the definition.

Proof: Exercise 1.

Theorem 4.2.1 If $\lim_{x \to a} f(x) = L$ and $\lim_{x \to a} f(x) = M$, then $L = M$.

This theorem asserts that if the limit exists, then it is unique. As in the proof of Theorem 4.1.1, the proof of Theorem 4.2.1 depends on the fact that if $L \neq M$, then there is a definite distance, call it d, between L and M. If we pick $\epsilon = \frac{1}{3}d$, the definition of limit tells us that there exists a δ such that for all x within δ of a, $f(x)$ is both within $\frac{1}{3}d$ of L and within $\frac{1}{3}d$ of M, which is impossible. (Again, the use of $\frac{1}{3}d$ is arbitrary; $\frac{1}{2}d$ would do as well, but $\frac{2}{3}d$ would not.)

Proof:
We show $L = M$ by contradiction.
1. Assume (to get #) $L \neq M$.
2. Let $d = |L - M|$.
3. Let $\epsilon = \frac{1}{3}d$.
4. There exists $\delta_1 > 0$ such that if
   $0 < |x - a| < \delta_1$, then $|f(x) - L| < \epsilon$.   (hyp.; def. lim)
5. There exists $\delta_2 > 0$ such that if
   $0 < |x - a| < \delta_2$, then $|f(x) - M| < \epsilon$.   (hyp.; def. lim)

> 6. Define δ to be the smaller of δ_1 and δ_2.
> 7. $\delta > 0$ (4, 5; def. δ)
> 8. There exists $x_0 \in \mathbb{R}$ such that $x_0 \neq a$ and $|x_0 - a| < \delta$.   (prop. $\mathbb{R}$)
> 9. $|f(x_0) - L| < \epsilon$ (4, 8; us. ⇒)
> 10. $|f(x_0) - M| < \epsilon$ (5, 8; us. ⇒)
> 11. $|L - f(x_0)| + |f(x_0) - M| < 2\epsilon$ (9, 10; prop. $\mathbb{R}$)
> 12. $|L - M| < 2\epsilon$ (11; prop. $\mathbb{R}$)
> 13. $d < 2\epsilon$ # Step 3   (2, 12; sub.)
> 14. $L = M$ (1—13; #)

In going from Step 11 to Step 12, we again used the triangle inequality ($|A - C| \leq |A - B| + |B - C|$). Step 8 asserts that no matter how close we are to a real number a (the distance measured by δ), there is a real number x_0 (not just a) within that distance of a. This is true according to your understanding of the real numbers lying along a continuous, solid number line. It follows from Theorem A.1.10 in Appendix 1 or by defining $x_0 = a + \delta/2$. Steps 9 and 10 are obtained using the free-variable if–then statements in Steps 4 and 5.

Definition   Let $f: D \to \mathbb{R}$, $g: D \to \mathbb{R}$, and $k \in \mathbb{R}$.

(a) Define the function $kf: D \to \mathbb{R}$ by $(kf)(x) = k(f(x))$ for all $x \in D$.

(b) Define the function $f+g: D \to \mathbb{R}$ by $(f+g)(x) = f(x) + g(x)$ for all $x \in D$.

Theorem 4.2.2   Let $f, g: D \to \mathbb{R}$, $k \in \mathbb{R}$.

(a) If $L = \lim_{x \to a} f(x)$, then $kL = \lim_{x \to a} (kf)(x)$.

(b) If $L = \lim_{x \to a} f(x)$ and $M = \lim_{x \to a} g(x)$, then $L + M = \lim_{x \to a} (f+g)(x)$.

Proof of (a): Exercise 3.

Proof of (b):

> **Scrap Paper**
> We are to show that given $\epsilon > 0$, there exists $\delta > 0$ such that if $0 < |x - a| < \delta$, then $|(f+g)(x) - (L + M)| < \epsilon$. "Let $\epsilon > 0$ be given" will be the first line of the proof. Next, δ needs to be defined in the proof. Then show (for this δ):
>
> if $0 < |x - a| < \delta$, then $|(f+g)(x) - (L + M)| < \epsilon$

4.3 Continuity

To do this, assume (as a proof line):
$$0 < |x - a| < \delta$$
and show (as a proof line):
$$|(f+g)(x) - (L + M)| < \epsilon$$
Now, $|(f+g)(x) - (L + M)| \leq |f(x) - L| + |g(x) - M|$ (triangle inequality), and so we should choose δ to be small enough to get
$$|f(x) - L| < \frac{\epsilon}{2} \quad \text{and} \quad |g(x) - M| < \frac{\epsilon}{2}$$
using the definition of limit applied first to f and then to g.

Exercise 4 asks you for a proof of Theorem 4.2.2b based on this scrap-paper analysis.

The sort of proofs done in this section are called "delta–epsilon" proofs. They are somewhat more complex than the other proofs we have seen so far but are developed, as all the other proofs, from the routine method and the rules of inference.

Exercises 4.2

Problems

1. Let $f: \mathbb{R} \to \mathbb{R}$ be defined by $f(x) = 3x + 1$. Prove $\lim_{x \to 2} f(x) = 7$.
2. Let $f: \mathbb{R} \to \mathbb{R}$ be defined by $f(x) = 2$ (a constant function). Prove $\lim_{x \to a} f(x) = 2$ for any $a \in \mathbb{R}$.
3. Prove Theorem 4.2.2a.
4. Prove Theorem 4.2.2b. Use the scrap-paper analysis to set up the proof outline. Define δ similarly to the way it was defined in the proof of Theorem 4.2.1.

4.3 CONTINUITY

Definition *A function $f: D \to \mathbb{R}$ is **continuous** at a real number $c \in D$ provided that $\lim_{x \to c} f(x) = f(c)$.*

Example 1:
The following functions are continuous at any real number c:

(a) $f(x) = x^2$
(b) $f(x) = 6x + 2$

We show $f(x) = x^2$ is continuous at any c and leave (b) for Exercise 1.

Proof:

Let c be given and let $f(x) = x^2$. We need to show $\lim_{x \to c} x^2 = c^2$. Thus let $\epsilon > 0$ be given.

Scrap Paper

By the definition of limit, we need a $\delta > 0$ such that if $0 < |x - c| < \delta$, then $|x^2 - c^2| < \epsilon$. Working backward from $|x^2 - c^2| < \epsilon$, we get $|x - c||x + c| < \epsilon$. This we can obtain in the following way.

$|x - c|$ can be made as small as required by choosing δ small, since we know $|x - c| < \delta$. It is not necessary to make $|x + c|$ small, only to know that it doesn't become large enough to counteract the effect of $|x - c|$ becoming small. This can be done if we can replace $|x + c|$ in $|x - c||x + c| < \epsilon$ with a constant, say, M, that is always larger than $|x + c|$. For such an M, if we can show that $|x - c| \cdot M < \epsilon$, it will follow that $|x - c||x + c| < \epsilon$. We therefore seek a constant M such that $M > |x + c|$.

If x is close to c (since $x \to c$, we can choose δ to make x as close to c as we want), then $|x|$ may be a bit bigger than $|c|$ but not much. If we pick $\delta = 1$, then $0 < |x - c| < \delta$ implies $|x| < |c| + 1$. (This can be shown by rigorous computation using the order and algebraic properties of the reals given in Appendix 1. Alternately, we can visualize points on the number line, remembering that $|x - c|$ means the distance from x to c. Our level of rigor is maintained by using "property of **R**" for justification of proof steps.) Thus $|x + c| \leq |x| + |c| < 2|c| + 1$. Let $M = 2|c| + 1$. Then $M > |x + c|$ as desired.

We then need only pick δ such that $|x - c| < \dfrac{\epsilon}{M}$ (in other words, $|x - c| \cdot M < \epsilon$).

Resuming the proof:

Proof:

Let c be given and let $f(x) = x^2$. We need to show that $\lim_{x \to c} x^2 = c^2$. Thus let $\epsilon > 0$ be given. Define δ to be the smaller of 1 and $\dfrac{\epsilon}{2|c| + 1}$. Assume

4.3 Continuity

$0 < |x - c| < \delta$. Then, because $\delta \leq 1$, we have

$$|x + c| \leq |x| + |c| < 2|c| + 1$$

Because $\delta \leq \dfrac{\epsilon}{2|c| + 1}$, we have

$$|x - c| < \dfrac{\epsilon}{2|c| + 1}$$

By multiplying inequalities, we get

$$|x + c||x - c| < (2|c| + 1)\left(\dfrac{\epsilon}{2|c| + 1}\right) = \epsilon$$

Therefore $|x^2 - c^2| < \epsilon$. Hence if $0 < |x - c| < \delta$, then $|x^2 - c^2| < \epsilon$. Thus $\lim\limits_{x \to c} x^2 = c^2$. ∎

Theorem 4.3.1   If $f, g : D \to \mathbb{R}$ are continuous at c and if $k \in \mathbb{R}$, then

(a) $f + g : D \to \mathbb{R}$ is continuous at c

(b) $kf : D \to \mathbb{R}$ is continuous at c

Proof: Exercises 2 and 3.

Definition   If $f : D \to \mathbb{R}$ and $S \subseteq D$, then f is **continuous on S** if f is continuous at every $a \in S$. If f is continuous on D, we merely say that $f : D \to \mathbb{R}$ **is continuous**.

Corollary 4.3.2   If $f, g : D \to \mathbb{R}$ are continuous and if $k \in \mathbb{R}$, then

(a) $f + g : D \to \mathbb{R}$ is continuous

(b) $kf : D \to \mathbb{R}$ is continuous

Proof:

We need to show that the functions are continuous at c for all $c \in D$. Let $c \in D$ be arbitrary and apply Theorem 4.3.1. ∎

If $f : C \to \mathbb{R}$ and $g : D \to \mathbb{R}$ are functions and $f(C) \subseteq D$, then the composition $g \circ f : C \to \mathbb{R}$ is defined.

Theorem 4.3.3   Let $f : C \to \mathbb{R}$ be continuous at c and let $f(c) = d$. Suppose that $f(C) \subseteq D$. Let $g : D \to \mathbb{R}$ be continuous at d. Then $g \circ f : C \to \mathbb{R}$ is continuous at c.

> **Scrap Paper**
> By the definitions of continuity and limit, we need to show: given $\epsilon > 0$, there exists $\delta > 0$ such that if $0 < |x - c| < 0$, then $|g \circ f(x) - g \circ f(c)| < \epsilon$. After setting up routine steps, we will need to show (for a given ϵ):
>
> $$|g \circ f(x) - g \circ f(c)| < \epsilon \quad \text{that is,} \quad |g(f(x)) - g(f(c))| < \epsilon$$
>
> By the continuity of g, there exists some $0 < \delta_1$ such that if $0 < |f(x) - f(c)| < \delta_1$, then $|g(f(x)) - g(f(c))| < \epsilon$. (Of course, if $0 = |f(x) - f(c)|$, then also $|g(f(x)) - g(f(c))| < \epsilon$.) But then, by the continuity of f we can choose a $\delta > 0$ such that if $0 < |x - c| < \delta$, then $|f(x) - f(c)| < \delta_1$.

Exercise 4 asks you for a proof of Theorem 4.3.3 based on this scrap-paper analysis.

Corollary 4.3.4 *Let $f: C \to \mathbb{R}$ be continuous and $f(C) \subseteq D$. Let $g: D \to \mathbb{R}$ be continuous. Then $g \circ f: C \to \mathbb{R}$ is continuous.*

Proof: Exercise 5.

For given real numbers a and r, the set of all x satisfying $|x - a| < r$ is the set of all points on the number line whose distance from a is less than r. This set is called the neighborhood about a of radius r. It is denoted by "$N(a, r)$".

Definition *Given real numbers a and r, the **neighborhood about a of radius r** is the set $N(a, r) = \{x \in \mathbb{R} \mid |x - a| < r\}$.*

Theorem 4.3.5 *If $f: D \to \mathbb{R}$ is continuous, $a \in D$, and $f(a) > 0$, then there exists a neighborhood $N(a, r)$ about a of some radius $r > 0$ such that $f(x) > 0$ for all $x \in N(a, r)$.*

Proof: Exercise 6.

Corollary 4.3.6 *If $f: D \to \mathbb{R}$ is continuous, $a \in D$, and $f(a) < 0$, then there exist $r > 0$ and a neighborhood $N(a, r)$ about a such that $f(x) < 0$ for all $x \in N(a, r)$.*

> **Proof:**
> Define the function $g: D \to \mathbb{R}$ by $g = (-1)f$, according to the definition in Section 4.2. Then g is continuous and $g(a) > 0$. Applying Theorem 4.3.5 to g, we get a neighborhood $N(a, r)$ on which $g(x) > 0$. (This means that $g(x) > 0$ for all $x \in N(a, r)$.) It follows that $f(x) < 0$ for all $x \in N(a, r)$. ∎

Exercises 4.3

Problems

1. Prove that $f(x) = 6x + 2$ is continuous at any $c \in \mathbb{R}$.
2. Prove Theorem 4.3.1a. The techniques needed are similar to those useful in proving Theorem 4.2.2b.
3. Prove Theorem 4.3.1b.
4. Prove Theorem 4.3.3.
5. Prove Corollary 4.3.4.
6. Prove Theorem 4.3.5. In order to get $f(x)$ positive, get it close to the positive number $f(a)$.

4.4 AN AXIOM FOR SETS OF REALS

By graphing functions in calculus, we are able to visualize and thereby give meaning to our computations. Thus the idea of continuity can be pictured. For example, the function $f: \mathbb{R} \to \mathbb{R}$ defined by $f(x) = x^2$ has the graph given in Figure 4.4.1.

To say that the function $f: \mathbb{R} \to \mathbb{R}$ given by $f(x) = x^2$ is continuous at $x = 2$ is to say that the value of f at $x = 2$ (namely, 4) is the same as the limit of f as x gets close to 2.

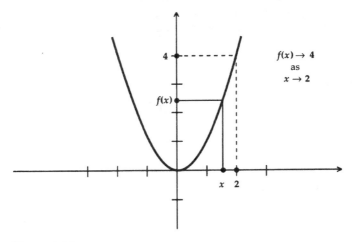

Figure 4.4.1

Continuity implies that there are no gaps in the graph of a function. Roughly, continuous functions are those whose graphs can be drawn without lifting the pencil. We can't prove this, of course, since—to be absurd—we have no formal definition of "lifting the pencil". Nevertheless, graphs contribute considerably to our understanding, and we want both our computations and our proofs to be meaningful in terms of this geometric representation. At the same time, we must realize that a picture can be considered as part of a rigorous proof only if the picture is a clear abbreviation for formal proof steps, that is, only if it is clear just which formal steps accomplish the argument suggested by the picture.

An important theorem in advanced calculus that illustrates this point is the Intermediate Value Theorem for Continuous Functions. A simple version of this theorem asserts that for $a, b \in D$, if $f: D \to \mathbf{R}$ is continuous and if $f(a) < 0$ and $f(b) > 0$, then there must a number c between a and b such that $f(c) = 0$. This is clearly suggested by the graph of f (Figure 4.4.2), for at a the graph is below the x-axis and at b it is above the x-axis. At some point, therefore, the lines representing the graph of f and the x-axis must cross and therefore presumably intersect.

The truth of the theorem depends on there being no points missing from either the graph or the x-axis—not even a single dimensionless point at which the lines might cross. The proof of the theorem (a **for some** statement) depends on establishing the existence of a point. This is not the sort of thing that is done by the computations you are familiar with from previous courses—the sort of thing justified by "property of $\mathbf{R}$".

The first "hole" in the number line was found by a member of the Greek society of Pythagoreans. The ancient Greeks at one time thought that any two line lengths were commensurate; in other words, that given a line of length l and a line of length m, there existed a smaller length d such that d went evenly into l and m—say, $l = pd$ and $m = qd$ for positive integers p and q. Someone applied this thinking to the sides of a $45°-45°-90°$ right triangle (Figure 4.4.3).

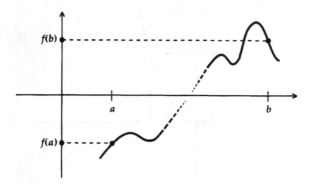

Figure 4.4.2

4.4 An Axiom for Sets of Reals

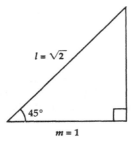

Figure 4.4.3

If a leg has length $m = 1$ and the hypotenuse has length $l = \sqrt{2}$, then

$$\frac{l}{m} = \frac{pd}{qd} = \frac{p}{q} = \sqrt{2}$$

We can assume that p/q is reduced (or else reduce it and call the result p/q). Thus p and q are relatively prime. Then

$$p^2 = 2q^2$$

Thus p^2 is even, so that p must be even, say, $p = 2r$. Then

$$4r^2 = 2q^2$$

so that $\quad 2r^2 = q^2$

so that q^2 and therefore q must be even. Thus p and q are both even, which contradicts the fact that p and q are relatively prime.

The idea that all line lengths are commensurate is equivalent to the idea that all line lengths can be expressed as the quotient of two integers—as fractions. The assumption that this is true has led to a contradiction.

Recall that a real number is called rational if it is the quotient of two integers and irrational if it is not rational. The argument above shows that the number line cannot be idealized as a solid line having any two of its points a rational distance apart. In terms of this idealization, we have found at least one "hole", $\sqrt{2}$, in the number line. A proof of some of the deeper results about functions defined on the real numbers, including the Intermediate Value Theorem, depends on a property of sets of real numbers that we will take as an axiom. This property ensures there there are no holes in the number line.

Recall from Section 3.10 that a real number M is called an upper bound for a set S of real numbers provided that **for all** $x \in S: x \leq M$. A real number L is called a least upper bound for S provided that (1) L is an upper bound for S and (2) for all M such that M is an upper bound for $S: L \leq M$. We now come to our axiom for sets of real numbers.

Chapter 4 Proofs in Analysis

LUB Axiom *If a nonempty set S of real numbers has an upper bound, then S has a least upper bound.*

> *Example 1:*
> The algorithm for taking square roots produces the decimal representation of $\sqrt{2}$ to any desired decimal place. Let $s = \langle 1, 1.4, 1.41, 1.414, 1.4142, \ldots \rangle$ be the sequence of n-place decimals determined by the algorithm and let $S = \{1, 1.4, 1.41, 1.414, 1.4142, \ldots\}$ be the set of terms of s. Then $\sqrt{2} = \text{lub}(S)$. Note that $\sqrt{2} \notin S$.

Theorem 4.4.1 *Suppose $f: D \to \mathbb{R}$ is continuous and $a, b \in D$ with $a < b$. If $f(a) < 0$ and $f(b) > 0$, then there exists some $c \in (a, b)$ such that $f(c) = 0$.*

> **Proof:**
> Define the set S by $S = \{r \in (a, b) \mid f(x) < 0 \text{ for all } x \in (a, r)\}$. We first claim that if $f(u) > 0$ for $u \in (a, b)$, then u is an upper bound for S. If u is not an upper bound, then there exists $t \in S$ such that $u < t$; that is, $u \in (a, t)$. Since $t \in S$, $f(x) < 0$ for all $x \in (a, t)$ (def. S), so that $f(u) < 0$, a contradiction that establishes our claim.
>
> By Corollary 4.3.6, S is nonempty. By hypothesis and the definition of S, b is an upper bound for S. By the LUB axiom, then, S has a least upper bound—call it L. By definition of L, $a \leq L \leq b$.
>
> If $f(L) > 0$, then by Theorem 4.3.5 there is a neighborhood $N(L, p)$ about L such that f is positive on $N(L, p)$. Thus $f(L - p/2) > 0$, so that $L - p/2$ is an upper bound for S, contradicting the fact that L is the least upper bound. If $f(L) < 0$, then by Corollary 4.3.6 there is a neighborhood $N(L, q)$ about L such that f is negative on $N(L, q)$. Note that if $a < x < L$, then $f(x) < 0$ or else x would be an upper bound for S. Thus f is negative on $(a, L + q)$, again contradicting the definition of L. Therefore $f(L) = 0$. Also, now $a < L < b$. ∎

The Intermediate Value Theorem is merely a generalization of Theorem 4.4.1, which contains all the mathematical "meat" in a special case that can easily be applied to get the more general result that follows.

Theorem 4.4.2 *Intermediate Value Theorem: Suppose $f: D \to \mathbb{R}$ is continuous and $a, b \in D$ with $a < b$. If C is a real number such that $f(a) < C < f(b)$ or $f(a) > C > f(b)$, then there exists $c \in (a, b)$ such that $f(c) = C$.*

Proof:

Define the function $g: D \to \mathbb{R}$ by $g(x) = f(x) - C$. Then g is continuous on D by Corollary 4.3.2.

Case 1. $f(a) < C < f(b)$. Here $g(a) < 0$ and $g(b) > 0$, so that we can apply Theorem 4.4.1 to g and obtain a number c such that $0 = g(c) = f(c) - C$. Therefore $f(c) = C$.

Case 2. $f(a) > C > f(b)$. Here $-g(a) < 0 < -g(b)$. The function $-g$ is continuous by Corollary 4.3.2, so that by Theorem 4.4.1 we obtain $c \in (a, b)$ such that $0 = -g(c) = -(f(c) - C)$. Therefore $f(c) = C$. ∎

Formal proofs of Theorems 4.4.1 and 4.4.2 have been written using the LUB axiom. After some experience is gained in writing formal proofs from representative pictures, one begins to see the connection—that is, from a picture one sees just how a formal proof could be framed. At this point (and not before), the picture and discussion can be construed as a proof since the two taken together abbreviate and suggest a formal proof.

Additional Ideas

Graphs and Definitions

If $f: D \to \mathbb{R}$ is a function defined on some domain D, then the graph of f is defined as the set $\{(x, y) \mid x \in D, y = f(x)\}$. Thus the formal, ordered-pair definition of function states that a function *is* its graph. (More completely, of course, the function is a triple consisting of the domain, the codomain, and the set of points that make up the graph.) Thus, in order to get a hold on the intuitive idea of function as a "rule" and provide a formal definition, we seize upon a representation (its graph) and formally declare the representation to be the thing itself. Note that in defining a real number to be an infinite decimal (that is, sequence of decimal rationals), we are again seizing upon a representation to make a formal definition.

Cuts

Theorem 4.4.3

Suppose A and B are subsets of $\mathbb{R}$ such that

(1) $x < y$ for all $x \in A$ and $y \in B$

and (2) Given $\epsilon > 0$, there exist $x \in A$ and $y \in B$ such that $|x - y| < \epsilon$.

Then there exists a unique number c such that $x \leq c \leq y$ for all $x \in A$ and $y \in B$.

Proof: Exercise 3.

Theorem 4.4.3 is the basis for an alternate way of *defining* the real numbers, assuming the rationals. We define a **cut** of the rationals to be a pair (A, B) of nonempty subsets of $\mathbf{Q}$ satisfying (1) and (2) of the theorem and also $A \cup B = \mathbf{Q}$ such that A contains no largest rational. Let C_Q be the set of all cuts of $\mathbf{Q}$. Define the function $f: C_Q \to \mathbf{R}$ by $f((A, B)) = c$, where c is the real number whose existence is determined from A and B by Theorem 4.4.3. Then f is a bijection (Exercise 4). An order relation $\leq$ is defined on C_Q by $(A_1, B_1) \leq (A_2, B_2)$ iff $A_1 \subseteq A_2$. Addition is defined on C_Q by $(A_1, B_1) + (A_2, B_2) = (A_3, B_3)$, where $A_3 = \{a_1 + a_2 \mid a_1 \in A_1, a_2 \in A_2\}$ and $B_3 = \mathbf{Q} - A_3$. With these definitions, f is an additive and order isomorphism between C_Q and $\mathbf{R}$ (Exercise 5). The isomorphism between C_Q and $\mathbf{R}$ prompts the formal *definition* of a real number as a cut of the rationals.

The LUB Axiom

If we define a real number to be a sequence of decimal rationals, or a cut of the rationals, then the LUB axiom must be proved from the definition. An outline of the proof for cuts is given in Exercise 6.

Exercises 4.4

Problems

1. Prove that $\sqrt{3}$ is irrational.
2. If n is a natural number that is not a perfect square, prove that $\sqrt{n}$ is irrational.

Supplementary Problems

3. Prove Theorem 4.4.3.
4. Let $f: C_Q \to \mathbf{R}$ be defined by $f((A, B)) = c$, where c is the real number whose existence is determined from A and B by Theorem 4.4.3. Prove that f is a bijection.
5. Prove that f from Exercise 4 is an additive and order isomorphism.
6. Let $S = \{(A_i, B_i) \mid i \in \mathcal{I}\}$ be a nonempty set of rational cuts that has some cut (A_0, B_0) as an upper bound. Define $A = \bigcup_{i \in \mathcal{I}} A_i$ and $B = \mathbf{Q} - A$. Prove that (A, B) is **lub**(S).

Challenging Problem

7. Recall from calculus that $e^x = 1 + x + x^2/2! + x^3/3! + \cdots$. Thus

$$e = 1 + 1 + \frac{1}{2!} + \frac{1}{3!} + \cdots$$

$$e^{-1} = 1 - 1 + \frac{1}{2!} - \frac{1}{3!} + \frac{1}{4!} - \cdots$$

The series for e^{-1} is an alternating series with terms of decreasing absolute value. For such a series, the error after n terms is greater than 0 and less than the absolute value of the $(n + 1)$st term. Thus (note $2k$ is even)

$$0 < e^{-1} - \sum_{n=0}^{2k-1} \frac{(-1)^n}{n!} < \frac{1}{(2k)!}$$

Use this expression to show that e^{-1} and therefore e are irrational.

4.5 SOME CONVERGENCE CONDITIONS FOR SEQUENCES

If the sequence $\langle x_n \rangle$ has the limit L, then the terms x_n become arbitrarily close to L and hence arbitrarily close to each other. This property of convergent sequences can be used to define a certain type of sequence.

Definition   A sequence $\langle x_n \rangle$ is a **Cauchy** sequence provided that, given $\epsilon > 0$, there exists $N \in \mathbb{N}$ such that for all $m, n > N$: $|x_m - x_n| < \epsilon$.

Theorem 4.5.1   If $\langle x_n \rangle$ converges, then $\langle x_n \rangle$ is a Cauchy sequence.

Proof:

Assume that $\langle x_n \rangle$ converges to, say, L and that $\epsilon > 0$ is given.

·
·
·

(Define N in here.)

·
·
·

Let $m, n \geq N$.

·
·
·

$|x_m - x_n| < \epsilon$

> **Scrap Paper**
> Since $x_n \to L$, there is a natural number N such that all terms beyond the Nth term are within a distance $\dfrac{\epsilon}{2}$ of L. Any two of these terms must therefore be within a distance ϵ of each other.

Proof:

Assume that $\langle x_n \rangle$ converges to, say, L and that $\epsilon > 0$ is given. Since $x_n \to L$, there exists N such that $|x_p - L| < \epsilon/2$ for all $p > N$. Let $m, n > N$. Then $|x_m - L| < \epsilon/2$ and $|x_n - L| < \epsilon/2$. Adding these inequalities gives $|x_m - x_n| \leq |x_m - L| + |L - x_n| < \epsilon/2 + \epsilon/2 = \epsilon$. ∎

Theorem 4.5.2 If $\langle a_n \rangle$ and $\langle b_n \rangle$ are Cauchy sequences and if $k \in \mathbb{R}$, then $\langle a_n \rangle + \langle b_n \rangle$ and $k \langle a_n \rangle$ are Cauchy sequences.

Proof: Exercise 2.

The converse to Theorem 4.5.1 states that if a sequence has the property that its terms become arbitrarily close to each other, then the sequence converges to some limit. In order to prove this, we first observe another property that characterizes convergent sequences: the sequence $\langle x_n \rangle$ has the limit L iff for any $\epsilon > 0$, there are only a finite number of terms greater than $L + \epsilon$ and a finite number of terms less than $L - \epsilon$.

Theorem 4.5.3 If $\langle x_n \rangle$ is a Cauchy sequence, then $\langle x_n \rangle$ converges.

Proof:

For each $r \in \mathbb{R}$, let $S(r) = \{n \in \mathbb{N} \mid x_n < r\}$. Let $S = \{r \mid S(r) \text{ is finite}\}$. Then S is nonempty and has an upper bound (Exercise 6) so that $\mathbf{lub}(S)$ exists. Let $L = \mathbf{lub}(S)$. We will show $x_n \to L$.

Let $\epsilon > 0$ be given.

.

.

(Get N here.)

Let $n > N$.

$|x_n - L| < \epsilon$

4.5 Some Convergence Conditions for Sequences

The next task in constructing the proof is to find N. For this we use the hypothesis.

Proof:

For each $r \in \mathbb{R}$, let $S(r) = \{n \in \mathbb{N} \mid x_n < r\}$. Let $S = \{r \mid S(r) \text{ is finite}\}$. Then S is nonempty and has an upper bound (Exercise 6) so that $\text{lub}(S)$ exists. Let $L = \text{lub}(S)$. We will show $x_n \to L$. Let $\epsilon > 0$ be given. By hypothesis, there exists N_1 such that for all $m, p > N_1$: $|x_m - x_p| < \epsilon/2$. Assuming $S(L + \epsilon/2)$ is finite contradicts the definition of L as an upper bound of such finite sets. Hence $S(L + \epsilon/2)$ is infinite. But $S(L - \epsilon/2)$ is finite: Otherwise $S(t)$ would be infinite for all $L - \epsilon/2 < t$, contradicting the definition of L as *least* upper bound. Let N_2 be larger than any element in $S(L - \epsilon/2)$ and let N be the larger of N_1 and N_2.

Let $n > N$. Since $S(L + \epsilon/2)$ is infinite, there exists $q > N$ such that $q \in S(L + \epsilon/2)$; that is, $x_q < L + \epsilon/2$, so that $x_q - L < \epsilon/2$. Since $q > N$, $L - \epsilon/2 < x_q$, so that $L - x_q < \epsilon/2$. Hence $|x_q - L| < \epsilon/2$. Since $N_1 \leq N$, $|x_n - x_q| < \epsilon/2$. Hence $|x_n - L| \leq |x_n - x_q| + |x_q - L| \leq \epsilon/2 + \epsilon/2 = \epsilon$. ∎

Suppose $\langle x_n \rangle$ converges to L. Then all but a finite number of terms are within any given distance, say, 1, of L. Let M be the maximum distance from L of each of the finite number of remaining terms. Then $|x_n - L| \leq M$ for all terms of the sequence, so that $|x_n| \leq |L| + M$. This shows that a convergent sequence is **bounded,** according to the following definition.

Definition A sequence $\langle x_n \rangle$ is **bounded** provided that there exists a number M such that $|x_n| \leq M$ for all n.

Definition A sequence $\langle x_n \rangle$ is called

(1) **nondecreasing** provided that if $m < n$, then $x_m \leq x_n$

(2) **increasing** provided that if $m < n$, then $x_m < x_n$

(3) **nonincreasing** provided that if $m < n$, then $x_m \geq x_n$

(4) **decreasing** provided that if $m < n$, then $x_m > x_n$

(5) **monotonic** if it is either nondecreasing or nonincreasing[†]

Theorem 4.5.4 A bounded monotonic sequence converges.

[†] There are two widely used conventions for the terms used to describe these sequences. One convention defines a sequence as *increasing* provided that if $m < n$, then $x_m \leq x_n$. This has the disadvantage of describing a constant sequence as increasing. Our term "nondecreasing" has the disadvantage that it does not describe the negation of "decreasing". Thus "nondecreasing" now has a technical meaning and must not be used for "~(decreasing)".

Proof: Exercise 3.

Definition A sequence $\langle x_n \rangle$ is called a *null sequence* provided that $x_n \to 0$.

Definition Define a relation $\simeq$ between convergent sequences by $\langle a_n \rangle \simeq \langle b_n \rangle$ iff $\langle a_n \rangle - \langle b_n \rangle$ is a null sequence.

Theorem 4.5.5 The relation $\simeq$ is an equivalence relation on the set of convergent sequences.

Proof: Exercise 4.

Theorem 4.5.6 Let $\mathscr{C}$ denote the set of $\simeq$ equivalence classes of convergent sequences. Let $f: \mathscr{C} \to \mathbf{R}$ be defined by $f([\langle x_n \rangle]) = \lim_{n \to \infty} x_n$. Then f is a bijection.

Proof: Exercise 5.

Additional Ideas

The definitions and theorems relating to sequences can be developed with the rational numbers $\mathbf{Q}$ replacing the real numbers $\mathbf{R}$. (In this context, not all Cauchy sequences converge—that is, converge to a rational number, the rational numbers being the only numbers that so far exist). The real numbers can be *defined* as the set $\mathscr{C}$ of equivalence classes of Cauchy sequences of rationals.

We have now indicated three different ways the real numbers can be formally defined in terms of the rationals. The fact that the reals have historically been developed in such disparate ways indicates that the essence of the reals is to be found not in their formal definition but in the properties that characterize them.

Exercises 4.5

Problems

1. The contrapositive of Theorem 4.5.1 can be used to show that a sequence diverges. Find the negation of the condition for a sequence to be Cauchy. (Recall the determination of the negation of the condition for convergence on page 235.) Use the negation of the Cauchy condition and the contrapositive of Theorem 4.5.1 to show that the sequences in Example 4.1.3 all diverge.
2. Prove Theorem 4.5.2.
3. Prove Theorem 4.5.4.

4. Prove Theorem 4.5.5.
5. Prove Theorem 4.5.6. First show that f is well defined.
6. Prove that the set S defined in the proof of Theorem 4.5.3 is nonempty and has an upper bound.

4.6 CONTINUOUS FUNCTIONS ON CLOSED INTERVALS

In this section we give some theorems and definitions that, except for their technical complexity, could have been given in an earlier section.

Definition Let $f: D \to \mathbb{R}$ and $g: D \to \mathbb{R}$. Define the function $f \cdot g: D \to \mathbb{R}$ by $(f \cdot g)(x) = f(x) \cdot g(x)$ for all $x \in D$.

Theorem 4.6.1 Let $f, g: D \to \mathbb{R}$ and $a \in D$. If $L = \lim_{x \to a} f(x)$ and $M = \lim_{x \to a} g(x)$, then $LM = \lim_{x \to a} f \cdot g(x)$.

Proof:

Assume: $\lim_{x \to a} f(x) = L$

$\lim_{x \to a} g(x) = M$

Show: $LM = \lim_{x \to a} f \cdot g(x)$

Scrap Paper
As usual, let's work back from the conclusion. We need to show that, given any $\epsilon > 0$, there is a $\delta > 0$ such that if $0 < |x - a| < \delta$, then $|f \cdot g(x) - LM| < \epsilon$. The quantity $|f \cdot g(x) - LM|$ is, by definition of $f \cdot g$, the same as $|f(x)g(x) - LM|$ and can be rewritten using the following technique (a technique is a trick used more than once):

$$|f(x)g(x) - LM| = |f(x)g(x) - f(x)M + f(x)M - LM|$$

Using the triangle inequality, we get

$$|f(x)g(x) - LM| \leq |f(x)||g(x) - M| + |M||f(x) - L|$$

We can get $|f(x)g(x) - LM| \leq \epsilon$ if we can get $|f(x)||g(x) - M| \leq \epsilon/2$ and $|M||f(x) - L| \leq \epsilon/2$. Since M is a fixed number and since $\lim_{x \to a} f(x) = L$, we can choose δ such that

$$|f(x) - L| \le \frac{\epsilon}{2|M|} \quad \text{(provided that } M \ne 0\text{)}$$

Since $|f(x)|$ is not fixed, we won't be able to choose δ to get

$$|g(x) - M| \le \frac{\epsilon}{2|f(x)|}$$

from the definition of the continuity of g. By using the continuity of f, however, we can insure that, for values of x close to a, $|f(x)| < R$ for some R. Thus

$$\frac{\epsilon}{2R} < \frac{\epsilon}{2|f(x)|}$$

We can take the cases where M or L is zero as special cases in the proof.

Returning to the proof:

Case 1. $M \ne 0$ and $L \ne 0$: Let $\epsilon > 0$ be given. By the continuity of f, there exists some $\delta_0 > 0$ such that if $0 < |x - a| < \delta_0$, then $|f(x) - L| < \frac{\epsilon}{2|M|}$. Also by the continuity of f, there exists some $\delta_1 > 0$ such that if $0 < |x - a| < \delta_1$, then $|f(x) - L| < |L|$. Let $R = 2|L|$. For $0 < |x - a| < \delta_1$ then, $|f(x)| < R$, whence $\frac{\epsilon}{2R} < \frac{\epsilon}{2|f(x)|}$. By the continuity of g, there exists some $\delta_2 > 0$ such that if $0 < |x - a| < \delta_2$, then $|g(x) - M| < \frac{\epsilon}{2R}$.

If we let δ be the minimum of δ_0, δ_1, and δ_2, then for all x such that $0 < |x - a| < \delta$, we have the following:

$$|M||f(x) - L| < \frac{\epsilon}{2}$$

$$|f(x)||g(x) - M| < \frac{\epsilon}{2}$$

Adding these two terms, we get

$$|f(x)g(x) - LM| = |Mf(x) - ML + f(x)g(x) - f(x)M|$$
$$\le |M||f(x) - L| + |f(x)||g(x) - M| < \epsilon$$

Thus $\lim_{x \to a} f \cdot g(x) = LM$.

Case 2. $M = 0$ or $L = 0$: Exercise 2. ∎

4.6 Continuous Functions on Closed Intervals

Theorem 4.6.2 Let $f, g: D \to \mathbb{R}$ be continuous. Then $f \cdot g$ is continuous.

Proof: Exercise 3.

Definition Let $f: D \to \mathbb{R}$ have the property that $f(x) \neq 0$ for all $x \in D$. Define the function $\frac{1}{f}: D \to \mathbb{R}$ by $\frac{1}{f}(x) = \frac{1}{f(x)}$ for all $x \in D$.

Theorem 4.6.3 Let $f: D \to \mathbb{R}$ be such that $f(x) \neq 0$ for all $x \in D$. Let $a \in D$. If $\lim_{x \to a} f(x) = L \neq 0$, then $\lim_{x \to a} \frac{1}{f}(x) = \frac{1}{L}$.

Scrap Paper

The usual proof analysis leads us to the need of showing $\left|\frac{1}{f(x)} - \frac{1}{L}\right| < \epsilon$ for a given $\epsilon > 0$. By adding fractions, we obtain

$$\left|\frac{1}{f(x)} - \frac{1}{L}\right| = \left|\frac{L}{Lf(x)} - \frac{f(x)}{Lf(x)}\right| = \left|\frac{1}{Lf(x)}\right| \cdot |L - f(x)|$$

The hypotheses enable us to make $|L - f(x)|$ as small as we wish. It suffices therefore to show that, for x close to a, $\left|\frac{1}{Lf(x)}\right|$ is bounded; in other words, that there is a constant M such that $\left|\frac{1}{Lf(x)}\right| = \left|\frac{1}{f(x)}\right| \cdot \left|\frac{1}{L}\right| \leq M$. In order to show that $\left|\frac{1}{f(x)}\right|$ does not become arbitrarily large, we need to show that $|f(x)|$ does not get arbitrarily close to zero. This we can do knowing that $f(x) \to L$ and $L \neq 0$; that is, we know there exists δ_0 such that if $0 < |x - a| < \delta_0$, then $|f(x) - L| < \frac{|L|}{2}$. From $|f(x) - L| < \frac{|L|}{2}$, we get $L - \frac{L}{2} \leq f(x) \leq L + \frac{L}{2}$, so that $|f(x)| \geq \frac{|L|}{2}$, whence $\frac{1}{|f(x)|} \leq \frac{2}{|L|}$, so that $\left|\frac{1}{f(x)}\right| \cdot \left|\frac{1}{L}\right| \leq \left|\frac{1}{L}\right| \cdot \frac{2}{|L|} = M$, as required.

In Exercise 4 you are asked to give a proof of Theorem 4.6.3 based on this analysis.

Theorem 4.6.4 Let $f: D \to \mathbb{R}$ be continuous and such that $f(x) \neq 0$ for all $x \in D$. Then $\frac{1}{f}: D \to \mathbb{R}$ is continuous.

Proof: Exercise 5.

Theorem 4.6.5 *If $f: D \to \mathbf{R}$ is continuous and if $[a, b] \subseteq D$, then $f([a, b])$ has an upper bound.*

Recall that $f([a, b]) = \{f(x) | x \in [a, b]\}$. If $f(S)$ has an upper bound for some set S, we usually say that f is bounded above on S. Since f is continuous at a, f is bounded above on some neighborhood of a (by definition of continuity, Exercise 1). The key to the proof is to extend this neighborhood as far as possible.

Proof:

Define the set $T = \{z \in [a, b] | f$ is bounded above on $[a, z]\}$. Since $a \in T$, T is not empty. Further, T (as a subset of $[a, b]$) is bounded above by b. By the LUB axiom, therefore, T has a least upper bound—call it c. By the definition of LUB, $c \leq b$. Suppose (in order to get a contradiction) that $c < b$. Since f is continuous at c, there is a neighborhood $N(c, r)$ on which f is bounded. Thus f is bounded above on $[c - r/2, c + r/2]$. By definition of T and c, f is bounded above on $[a, c - r/2]$. Thus f is bounded above on $[a, c + r/2]$, which contradicts the fact that c is the least upper bound. Therefore $c = b$ and f is bounded on $[a, b]$. ∎

Theorem 4.6.6 *If $f: D \to \mathbf{R}$ is continuous and if $[a, b] \subseteq D$, then f achieves a maximum value on $[a, b]$. That is, there is some $c \in [a, b]$ such that $f(x) \leq f(c)$ for all $x \in [a, b]$.*

Proof:

By Theorem 4.6.5, f is bounded above on $[a, b]$. Let L be $\text{lub}\{f(x) | x \in [a, b]\}$. Then $L - f(x)$ is a continuous function that is non-negative on $[a, b]$. In order to prove the theorem, we need only show that this function becomes zero on $[a, b]$. Suppose not. Then, since $L - f(a) > 0$ and $L - f(b) > 0$, there are neighborhoods of a and b on which $L - f$ is positive (Theorem 4.3.5). Thus $L - f$ is continuous and positive on a domain of which $[a, b]$ is a subset. By Theorem 4.6.4, $\dfrac{1}{L - f}$ is continuous on this domain and therefore bounded on $[a, b]$. Hence $\dfrac{1}{L - f(x)} \leq M$ for some $M > 0$ and for all $x \in [a, b]$. From this we get $f(x) \leq L - 1/M$ for all $x \in [a, b]$—which contradicts the definition of L as the least upper bound of $f([a, b])$. ∎

Additional Ideas

Definition A function $f: D \to \mathbb{R}$ is said to be **bounded on a subset** S of D provided that there exists a number M such that $|f(x)| \leq M$ for all $x \in S$.

Theorem 4.6.7 If $f: D \to \mathbb{R}$ is continuous and if $[a, b] \subseteq D$, then f is bounded on $[a, b]$.

Proof: Exercise 6.

Theorem 4.6.8 If $f: D \to \mathbb{R}$ is continuous and if $[a, b] \subseteq D$, then f achieves maximum and minimum values on $[a, b]$.

Proof: Exercise 7.

Exercises 4.6

Problems

1. Prove that if $f: D \to \mathbb{R}$ is continuous at a, then f is bounded on some neighborhood of a.
2. Prove Case 2 of Theorem 4.6.1. Note that Theorem 4.6.1 is symmetric with respect to L and M.
3. Prove Theorem 4.6.2.
4. Give a proof of Theorem 4.6.3 based on the scrap-paper analysis.
5. Prove Theorem 4.6.4.

Supplementary Problems

6. Prove Theorem 4.6.7.
7. Prove Theorem 4.6.8.

4.7 TOPOLOGY OF $\mathbb{R}$

Definition A real number a is called an **accumulation point** of a subset S of $\mathbb{R}$ if, for all neighborhoods $N(a, r)$ of a: $[N(a, r) - \{a\}] \cap S \neq \emptyset$.

Definition A subset S of $\mathbb{R}$ is **closed** if each accumulation point of S is in S.

Example 1:
That $\emptyset$ is closed follows from the fact that it has no accumulation points, and so the defining condition is vacuously satisfied.

Example 2:
That $\mathbb{R}$ is closed follows from the fact that every number is both an accumulation point and in $\mathbb{R}$.

Example 3:
For any $b \in \mathbb{R}$, the set $(-\infty, b] = \{x \in \mathbb{R} | x \leq b\}$ is closed. We can prove this contention by showing the contrapositive: any point not in $(-\infty, b]$ is not an accumulation point of $(-\infty, b]$. Thus let $c \notin (-\infty, b]$. Then $b < c$. Since $(c - |c - b|/2, c + |c - b|/2)$ is a neighborhood of c that misses $(-\infty, b]$, c is not an accumulation point of $(-\infty, b]$.

Example 4:
For any $a \in \mathbb{R}$, the set $[a, \infty) = \{x \in \mathbb{R} | a \leq x\}$ is closed. The proof of this is Exercise 1.

Theorem 4.7.1 If S and T are two closed subsets of $\mathbb{R}$, then $S \cap T$ is closed.

Proof: Exercise 2.

Example 5:
For any $a, b \in \mathbb{R}$, the set $[a, b]$ is closed. The proof of this follows from Examples 3 and 4 and Theorem 4.7.1. We finally see why $[a, b]$ is called the closed interval from a to b.

Definition A subset S of $\mathbb{R}$ is called *open* if $\mathbb{R} - S$ is closed.

Example 6:
By Example 1, $\mathbb{R}$ is open.

Example 7:
By Example 2, $\emptyset$ is open.

Example 8:
For any $b \in \mathbb{R}$, $(-\infty, b)$ is open. This follows from Example 4 since $(-\infty, b) = \mathbb{R} - [b, \infty)$.

Example 9:
For any $a \in \mathbb{R}$, (a, ∞) is open. This follows from Example 3 since $(a, \infty) = \mathbb{R} - (-\infty, a]$.

4.7 Topology of $\mathbb{R}$

Definition An element a of a subset S of $\mathbb{R}$ is called an **interior point** of S if there is some neighborhood $N(a, r) \subseteq S$.

Theorem 4.7.2 A subset S of $\mathbb{R}$ is open iff every point of S is an interior point of S.

Proof: Exercise 3.

Theorem 4.7.3 If S and T are two open subsets of $\mathbb{R}$, then $S \cap T$ is open.

Proof: Exercise 4.

Example 10:
For any $a, b \in \mathbb{R}$, (a, b) is open. This follows from Examples 8 and 9 and Theorem 4.7.3.

Theorem 4.7.4 Let S_i, $i = 1, \ldots, n$, be a finite family of open subsets of $\mathbb{R}$. Then $\bigcap_{i=1}^{n} S_i$ is open.

Proof: Exercise 5.

Corollary 4.7.5 Let S_i, $i = 1, \ldots, n$, be a finite family of closed subsets of $\mathbb{R}$. Then $\bigcup_{i=1}^{n} S_i$ is closed.

Proof:
By Theorem 4.7.4 and DeMorgan's Laws. ∎

Theorem 4.7.6 Let S_i, $i \in \mathscr{I}$, be an arbitrary family of open subsets of $\mathbb{R}$. Then $\bigcup_{i \in \mathscr{I}} S_i$ is open.

Proof:
Straightforward by Theorem 4.7.2. ∎

Corollary 4.7.7 Let S_i, $i \in \mathscr{I}$, be an arbitrary family of closed subsets of $\mathbb{R}$. Then $\bigcap_{i \in \mathscr{I}} S_i$ is closed.

> **Proof:**
> By Theorem 4.7.6 and DeMorgan's Laws. ∎

Theorem 4.7.8 Let $f: \mathbb{R} \to \mathbb{R}$ be continuous and let $T \subseteq \mathbb{R}$. Then
(a) If T is open, so is $f^{-1}(T)$.
(b) If T is closed, so is $f^{-1}(T)$.

Proof: Exercise 8.

Theorem 4.7.9 Let $f: \mathbb{R} \to \mathbb{R}$ be a function. Then f is continuous if either of the following conditions holds:
(1) For all $T \subseteq \mathbb{R}$: if T is open, then $f^{-1}(T)$ is open.
(2) For all $T \subseteq \mathbb{R}$: if T is closed, then $f^{-1}(T)$ is closed.

Proof: Exercise 9.

Exercises 4.7

Problems

1. Prove that for any $a \in \mathbb{R}$, The set $[a, \infty) = \{x \in \mathbb{R} \mid a \leq x\}$ is closed.
2. Prove Theorem 4.7.1.
3. Prove Theorem 4.7.2.
4. Prove Theorem 4.7.3.
5. Prove Theorem 4.7.4.

Challenging Problems

6. Show by example that if S_i, $i \in \mathscr{I}$, is an arbitrary family of closed subsets of $\mathbb{R}$, then $\bigcup_{i \in \mathscr{I}} S_i$ need not be closed.
7. Show by example that if S_i, $i \in \mathscr{I}$, is an arbitrary family of open subsets of $\mathbb{R}$, then $\bigcap_{i \in \mathscr{I}} S_i$ need not be open.
8. Prove Theorem 4.7.8.
9. Prove Theorem 4.7.9.

Chapter 5

Cardinality

5.1 CANTOR'S THEOREM

In Section 2.7, the ideas of *same cardinality* and *smaller cardinality* were introduced as applications of the function properties *one-to-one* and *onto*. In this chapter we wish to develop the idea of cardinality further. The material on doing proofs developed in earlier chapters is called upon in this chapter, but the focus changes to the mathematical content—that is, to the ideas about the mathematical objects studied. Proof is, as it always is in abstract mathematics, the foundation for the study.

Recall that sets A and B have the "same cardinality" if there is a one-to-one, onto function $f: A \to B$. In this case we write $A \sim B$. Set A has "smaller cardinality" than B if there is a one-to-one function $f: A \to B$ but no one-to-one function from A onto B. In this case we write $A < B$. The notation "$A \leq B$" means that either $A < B$ or $A \sim B$. Thus $A < B$ iff $A \leq B$ but $A \nsim B$. The notation "$A \leq B$" means simply that there is a one-to-one function from A to B.

We would like to define the idea of "cardinality" itself. Intuitively, the cardinality of a set is the "number of elements" in the set. Cardinality is the formally defined idea of "number of elements". We start with finite sets. By Corollary 3.1.5, a nonempty set S is finite iff $S \sim N_k$ for some k.

Definition   *The **cardinality** of the empty set is 0. The **cardinality** of a nonempty finite set S is k provided that $S \sim N_k$. We denote the cardinality of S by $|S|$.*

Our first task is to show that cardinality for finite sets is well defined; that is, if $S \sim N_k$ and $S \sim N_j$, then $k = j$. Thus suppose that $S \sim N_k$ and $S \sim N_j$ for $k, j \in N$. By the trichotomy property of Z (Appendix 1), either $k < j$ or $k = j$ or $k > j$. We seek the conclusion $k = j$. If this is not so, we may assume (in order to get a contradiction) that $k < j$ (without loss of generality—by symmetry). With

Chapter 5   Cardinality

this assumption, the function $g: \mathbb{N}_k \to \mathbb{N}_j$ defined by $g(i) = i$ is one-to-one but not onto. However, $\mathbb{N}_j \sim S$ and $S \sim \mathbb{N}_k$, so that $\mathbb{N}_j \sim \mathbb{N}_k$ (Theorem 2.7.4), and there exists a one-to-one, onto function $f: \mathbb{N}_j \to \mathbb{N}_k$. But then, $g \circ f: \mathbb{N}_j \to \mathbb{N}_j$ is one-to-one but not onto, contradicting the fact that $\mathbb{N}_j$ is finite. Thus for each finite set S, there is a unique, non-negative integer k such that $|S| = k$—that is, $|\ |$ is well defined.

So far we know that there are two kinds of sets, finite and infinite. The sets $\mathbb{N}_k$, for example, are finite. The set $\mathbb{N}$ is infinite, as are all sets that contain $\mathbb{N}$. The situation is more complicated than this, however—and more interesting. The following theorem necessitates the idea of different kinds of infinities.

Theorem 5.1.1   *Cantor's Theorem: For any set X, there is no onto function $f: X \to \mathbb{P}(X)$.*

This theorem was first proved by Georg Cantor (1845–1918), who developed the theory of set cardinality. It was given to you as a challenging problem in Section 2.4, but since it is central to the theory of set cardinality and since you now have had an opportunity to prove it for yourself, we will prove it here.

Scrap Paper
We need to show that an arbitrary function $f: X \to \mathbb{P}(X)$ is not onto. For each $x \in X$, $f(x)$ is a subset of X. It is natural to ask whether x is an element of $f(x)$. In fact, there are just two kinds of x in the domain of f: those x that map to sets $f(x)$ containing x and those that map to sets not containing x. In order to show that f is not onto, we need to find a set in $\mathbb{P}(X)$ that has nothing mapping to it. There are two candidates we might try: $S = \{x \mid x \in f(x)\}$ and $T = \{x \mid x \notin f(x)\}$. T looks promising. Suppose there is some $y \in X$ such that $f(y) = T$. Is $y \in T$? If so then $y \notin f(y)$ by definition of T—that is, $y \notin T$. The contradiction shows $y \notin T$. But then here $y \in T$ (by the definition of T), so we again get a contradiction.

Proof:
Let $f: X \to \mathbb{P}(X)$. Define $T = \{x \mid x \notin f(x)\}$. In order to get a contradiction, assume that there is some $y \in X$ such that $f(y) = T$. Then $y \in T$ or $y \notin T$. Assume $y \in T$. Then $y \notin f(y)$ by definition of T. Since $f(y) = T$, we have $y \notin T$ by substitution. This contradicts the assumption. Now assume $y \notin T$. But then $y \notin f(y)$ since $T = f(y)$. From $y \notin f(y)$ we get $y \in T$ by the definition of T—again contradicting the assumption. Therefore there is no $y \in X$ such that $f(y) = T$. Thus $f: X \to \mathbb{P}(X)$ is not onto. ∎

For a nonempty set X, there is an obvious one-to-one function $f: X \to \mathbb{P}(X)$ given by $f(x) = \{x\}$. Thus X has smaller cardinality than $\mathbb{P}(X)$. Applying Cantor's

5.1 Cantor's Theorem

Theorem to $\mathbb{N}$, we get $\mathbb{N} < \mathbb{P}(\mathbb{N})$. Applying the theorem to $\mathbb{P}(\mathbb{N})$, we have $\mathbb{P}(\mathbb{N}) < \mathbb{P}(\mathbb{P}(\mathbb{N}))$, and so on. This suggests an entire hierarchy of infinities, but this hierarchy depends on the transitivity of the relation $<$ on sets; that is, if $A < B$ and $B < C$, then $A < C$. The transitivity of the relation $\lesssim$ is asserted in Theorem 2.7.10. In the Additional Ideas discussion of Section 2.7, we showed, however, that the transitivity of $<$ is nontrivial and equivalent to the following condition: if $B \lesssim C$ and $C \lesssim B$, then $B \sim C$. This will be proved as Theorem 5.1.2 in a moment.

Cantor defined the set of *cardinal numbers*, in order to extend the idea of "number of elements" to infinite sets. Cardinal numbers form a totally ordered set that extends the non-negative integers. They can be added and multiplied in a way that extends the addition and multiplication of non-negative integers. The following theorem is key to the development of the cardinal numbers. The proof is not easy. In fact, Cantor was not able to find a proof. It was proved independently by Ernst Schroeder (1841–1902) and Felix Bernstein (1878–1956) and is known by their names.

Theorem 5.1.2

Schroeder–Bernstein Theorem: Let $f: A \to B$ and $g: B \to A$ be one-to-one functions. Then there exists a one-to-one, onto function $h: A \to B$.

Proof:

Let C be the set of all $x \in A$ that can be obtained by starting with an element y in $B - f(A)$ and repeatedly applying g and f. Formally, $C = \{x \in A \mid x = g \circ (f \circ g)^n(y) \text{ for some } y \in B - f(A) \text{ and some } n \geq 0\}$. (By definition, $(f \circ g)^n = (f \circ g)^{n-1} \circ (f \circ g)$, where $(f \circ g)^1 = f \circ g, (f \circ g)^0 = i_B$.) Then $C \subseteq g(B)$, so that the function g^{-1} is defined on C. Let $D = A - C$.

Define $h: A \to B$ by

$$h(x) = \begin{cases} f(x) & \text{for } x \in D \\ g^{-1}(x) & \text{for } x \in C \end{cases}$$

h is one-to-one: $h|_D$ and $h|_C$ are one-to-one, so the only way for h to fail to be one-to-one is for $h(x_1) = h(x_2)$ for some $x_1 \in D$ and $x_2 \in C$. Assume this is true in order to get a contradiction. Then

$$\begin{aligned} f(x_1) &= g^{-1}(x_2) \\ &= g^{-1}(g \circ (f \circ g)^n(y)) \text{ for some } y \in B - f(A), n \geq 0 \\ &= (f \circ g)^n(y) \quad (y \in B - f(A) \text{ implies } n \geq 1) \\ &= f \circ g \circ (f \circ g)^{n-1}(y) \end{aligned}$$

so $x_1 = g \circ (f \circ g)^{n-1}(y) \in C$, a contradiction

h is onto: Let $b \in B$.

Case 1. $g(b) \in C$: $h(g(b)) = g^{-1}(g(b)) = b$.

Case 2. $g(b) \notin C$: $b \in f(A)$ (def. C), so that $b = f(a)$ for some $a \in A$.

Case 2a. $a \in C$: $a = g \circ (f \circ g)^n(y)$ for some $y \in B - f(A)$, $n \geq 0$.

$g(b) = g(f(a)) = g \circ (f \circ g)^{n+1}(y) \in C$, a contradiction.

Case 2b. $a \in D$: $h(a) = f(a) = b$. ∎

Corollary 5.1.3 For sets A and B: if $A \lesssim B$ and $B \lesssim A$, then $A \sim B$.

Proof: By definition. ∎

Corollary 5.1.4 For sets A, B, and C: if $A < B$ and $B < C$, then $A < C$.

Proof: Exercise 1.

In terms of our new definition, the former ideas of "same cardinality" and "smaller cardinality" make sense; that is:

Theorem 5.1.5 For finite sets S and T:

(a) $|S| = |T|$ iff $S \sim T$
(b) $|S| < |T|$ iff $S < T$
(c) $|S| \leq |T|$ iff $S \lesssim T$

Proof: Exercise 2.

Additional Proof Ideas

Bertrand Russell (1872–1970) was skeptical of the application of Cantor's Theorem to infer a hierarchy of infinities. He asked the following question: What if we apply the theorem to the universal set **U**? If everything is a member of the universal set, how can **P(U)** in any sense be larger? Russell also questioned the idea of forming sets as collections of all things with a certain property. He applied the sort of reasoning Cantor used in proving Theorem 5.1.1 to the following situation.

Most sets consist of familiar objects—numbers, for example. It doesn't make sense to ask whether such a set can be a member of itself. Its members are numbers, not sets. But suppose we have a set of sets. This set could conceivably be a member of itself. At least we can ask whether it is or not. Define S to be the set that consists of all sets that are not members of themselves. Is S a member of itself? If so, then it must satisfy the defining condition and therefore not be a member of itself. Is S not a member of itself? If so, then it satisfies the defining condition and is therefore a member of itself. Either way, we get a contradiction.

This argument is called Russell's Paradox. It and similar arguments caused some agitation in the mathematical community. In order to preserve Cantor's valuable ideas and avoid the paradoxes, set theory was axiomatized. The cardinal numbers are defined in these axiomatizations, and Cantor's ideas form the basis for this secure branch of mathematics. Such axiomatic set theory is beyond our scope, however, and so we cannot present a satisfactory definition of cardinal number. Instead, in the next section, we define cardinality for two kinds of infinite sets—the integers and the real numbers.

One is tempted to use Theorem 2.7.4 to define cardinality. The relationship $\sim$ between sets is like an equivalence relation in that it is reflexive, symmetric, and transitive. It is not an equivalence relation between elements of a *set*, however, simply because there is no satisfactory set. We can't use the set of all sets because this leads us to Russell's Paradox. This is unfortunate because it would be very nice to define a cardinal number as an equivalence class of sets. This would tell us what a cardinal number *is*. For example, the cardinal number 2 would be the entire collection of all sets with two elements. Technical ways of avoiding the difficulties are worked out in the axiomatic treatments of set theory.

To avoid sets being members of sets, one of these axiomatic treatments uses the term "class" for a collection of sets. Rules for forming classes and sets are different. In everyday mathematics, however, it is useful to allow sets to be elements of other sets, such as the domains of functions. A real number, for example, can formally be defined as an equivalence class (set) of rationals, and a rational number can be defined as an equivalence class (set) of pairs (sets) of integers. It is customary to simultaneously consider the domains of functions as sets also—the set of real numbers, for example. Our definitions and terminology attempt to follow customary practice. Mathematicians know where the logical difficulties lie but don't allow their notation to be overly encumbered by the logicians' concerns.

Exercises 5.1

Problems

1. Prove Corollary 5.1.4.
2. Prove Theorem 5.1.5.
3. Prove this extension of Theorem 3.10.4.
   The following are equivalent for a poset P:

(a) For all $x, y \in P$: $x \leq y$ or $y \leq x$ (comparability).
(b) Exactly one of the following holds for all pairs $x, y \in P$: $x < y$ or $x = y$ or $y < x$ (trichotomy).

4. Let X be set on which is defined a relation $\leq$ such that for all $x, y, z \in X$:
   (1) $x \leq x$ (reflexivity).
   (2) If $x \leq y$ and $y \leq z$, then $x \leq z$ (transitivity of $\leq$).
   Show that the following conditions are equivalent for all $x, y, z \in X$:
   (a) If $x \leq y$ and $y \leq x$, then $x = y$ (antisymmetry).
   (b) If $x < y$ and $y < z$, then $x < z$ (transitivity of $<$).

5.2 CARDINALITIES OF SETS OF NUMBERS

Theorem 5.2.1 For sets A and B, if $A \sim B$, then $P(A) \sim P(B)$.

Proof:
Assume $f: A \to B$ is one-to-one and onto. The induced map $\bar{f}: P(A) \to P(B)$ is then one-to-one and onto (Exercises 2.3.11 and 2.4.10).

Theorem 5.2.2 For all $k \in \mathbb{N}$:
(a) $|P(\mathbb{N}_k)| = 2^k$
(b) $k < 2^k$

Proof: Exercise 1.

Corollary 5.2.3 If X is a finite set with cardinality k, then $P(X)$ has cardinality 2^k.

Proof:
If $|X| = 0$, then $X = \emptyset$ and $P(X) = \{\emptyset\}$, so that $|P(X)| = 1 = 2^0$. Otherwise $X \sim \mathbb{N}_k$ for some k and $P(X) \sim P(\mathbb{N}_k)$ by Theorem 5.2.1. By Theorem 5.1.5, $|P(X)| = |P(\mathbb{N}_k)|$, and the corollary follows by Theorem 5.2.2.

5.2 Cardinalities of Sets of Numbers

Definition The **cardinality** of a set X is defined to be $\aleph_0$ ("aleph null") provided that $X \sim \mathbb{N}$. In this case, we denote $|X| = \aleph_0$.

Example 1:
The set $\mathbb{N}$ has cardinality $\aleph_0$ by definition. The set $\mathbb{Z}$ (Theorem 2.7.6) and the set of even integers (Exercise 2.6.1) are equipotent with $\mathbb{N}$ and therefore have cardinality $\aleph_0$.

Using Theorem 5.2.2 as our guide for notation, we now make the following definition:

Definition The **cardinality** of a set X is defined to be $2^{\aleph_0}$ provided that $X \sim \mathbf{P}(\mathbb{N})$. In this case, we denote $|X| = 2^{\aleph_0}$.

As a final preparation for examining the cardinalities of sets of numbers, we obtain:

Theorem 5.2.4 Let A and B be sets.
(a) If $A \subseteq B$, then $A \preceq B$.
(b) If there exists an onto function $f: A \to B$, then $B \preceq A$.

> **Proof of (a):**
>
> Define the function $i: A \to B$ by $i(a) = a$ for all $a \in A$. i is called the *inclusion map*. It is clearly one-to-one, and the conclusion follows.
>
> **Proof of (b):**
>
> Let $f: A \to B$ be onto. By Theorem 2.5.2, f has a right inverse $g: B \to A$; that is, $f \circ g = i_B$. g is one-to-one since i_B is one-to-one. (This follows from Conjecture 2.3.3a—which is true and which you should now prove (Exercise 2) if you have not already done so.) ∎

Example 2:
Recall that $(0, 1) = \{x \in \mathbb{R} \mid 0 < x < 1\}$. We can get a one-to-one function $f: \mathbf{P}(\mathbb{N}) \to (0, 1)$ using the decimal representation for numbers in $(0, 1)$: An element $A \in \mathbf{P}(\mathbb{N})$ is a subset of $\mathbb{N}$. For any such A, define $f(A) = 0.a_1 a_2 a_3 \ldots$, where $a_j = 1$ if $j \in A$ and $a_j = 2$ if $j \notin A$. Then distinct subsets of $\mathbb{N}$ map to distinct real numbers in $(0, 1)$, so that f is one-to-one. To define a one-to-one function $g: (0, 1) \to \mathbf{P}(\mathbb{N})$, use a binary "decimal" representation (in base 2) for elements in $(0, 1)$: If $a \in (0, 1)$, then $a = 0.a_1 a_2 a_3 \ldots$, where each a_j is either 0 or 1. (Terminating "decimals" have all $a_j = 0$ beyond some point. There is no need to worry about uniqueness of representations here: if a has more than one rep-

Chapter 5 Cardinality

resentation, pick any one.) Define $g(a)$ to be the subset $A \subseteq \mathbb{N}$ defined by $j \in A$ iff $a_j = 1$. Then distinct representations give distinct subsets of $\mathbb{N}$, so that g is one-to-one. By the Schroeder–Bernstein Theorem, $(0, 1) \sim P(\mathbb{N})$. The cardinality of $(0, 1)$ is therefore $2^{\aleph_0}$ by definition. The interval $(0, 1)$ is called the *continuum*, and its cardinality is traditionally denoted by c. This example shows $c = 2^{\aleph_0}$.

Example 3:
Define the function $f: (0, 1) \to (-\pi/2, \pi/2)$ by $f(x) = \pi(x - 1/2)$. Then f is one-to-one and onto (Exercise 3), so that $|(0, 1)| = |(-\pi/2, \pi/2)|$.

Example 4:
The function $f: \mathbb{R} \to (-\pi/2, \pi/2)$ given by $f(x) = \arctan x$ is one-to-one and onto (Exercise 4), and so, by Example 3, the cardinality of $\mathbb{R}$ is c, that is, $2^{\aleph_0}$.

Example 5:
Denote the set of positive rationals by $\mathbb{Q}^+$. A function f from $\mathbb{N}$ onto $\mathbb{Q}^+$ can be described by the following diagrams. The left-hand array includes each positive rational—often many times, but always at least once. The right-hand array indicates which natural number maps to the rational number in the corresponding position.

1	2/1	3/1	4/1	5/1	...	$f(1)$	$f(2)$	$f(4)$	$f(7)$	$f(11)$	...
1/2	2/2	3/2	4/2	5/2	...	$f(3)$	$f(5)$	$f(8)$	$f(12)$		...
1/3	2/3	3/3	4/3		...	$f(6)$	$f(9)$	$f(13)$			...
1/4	2/4	3/4			...	$f(10)$	$f(14)$				...
1/5	2/5				...	$f(15)$					...
1/6					...						...

It is clear that $f: \mathbb{N} \to \mathbb{Q}^+$ is onto (but certainly not one-to-one), so that $\mathbb{Q}^+ \leq \mathbb{N}$ by Theorem 5.2.4b. Since $\mathbb{N} \subseteq \mathbb{Q}^+$, $\mathbb{N} \leq \mathbb{Q}^+$ by Theorem 5.2.4a. By Corollary 5.1.3 to the Schroeder–Bernstein Theorem, $\mathbb{N} \sim \mathbb{Q}^+$. $\mathbb{Q}^+$ therefore has cardinality $\aleph_0$.

Example 6:
Denote the negative rationals by $\mathbb{Q}^-$. The function $f: \mathbb{Q}^+ \to \mathbb{Q}^-$ given by $f(x) = -x$ is a bijection, so that $|\mathbb{Q}^-| = \aleph_0$. Therefore there exist bijections $h: \mathbb{N} \to \mathbb{Q}^-$ and $j: \mathbb{N} \to \mathbb{Q}^+$. Exercise 7 asks you to define a bijection $g: \mathbb{N} \to \mathbb{Q}$ in terms of h and j. From this it follows that $|\mathbb{Q}| = \aleph_0$.

Definition A set X is called *countably infinite* if it has the same cardinality as $\mathbb{N}$. It is called *countable* if it is either finite or countably infinite and *uncountable* if it is not countable.

From the examples above, we have that the integers and rationals are countable. The reals and the interval $(0, 1)$ are uncountable.

5.2 Cardinalities of Sets of Numbers

Theorem 5.2.5 *A subset of a countable set is countable.*

Proof:
Let B be countable and let $A \subseteq B$. If B is finite, then by Corollary 2.7.3, A is finite and therefore countable. Otherwise, there exists a bijection $f: B \to \mathbb{N}$. Consider two cases. If A is finite, then it is countable and we're done. If A is infinite, we can define a one-to-one function $g: \mathbb{N} \to A$ inductively as follows. Since A is not empty, choose an element $a_1 \in A$ and define $g_1(1) = a_1$. Assume that $g_n: \mathbb{N}_n \to A$ has been defined and is one-to-one. g_n is not onto since A is not finite. Thus there exists $a_{n+1} \in A - g_n(\mathbb{N}_n)$. Define $g_{n+1}: \mathbb{N}_{n+1} \to A$ by $g_{n+1}(n+1) = a_{n+1}$ and $g_{n+1}(i) = g_n(i)$ for all $i \in \mathbb{N}_n$. Finally, define $g: \mathbb{N} \to A$ by $g(i) = g_i(i)$ for all $i \in \mathbb{N}$. Then $g: \mathbb{N} \to A$ is one-to-one (Exercise 8). Also, the restriction of f to A, $f|_A : A \to \mathbb{N}$, is one-to-one. Thus $A \sim \mathbb{N}$ by the Schroeder–Bernstein Theorem, and A is therefore countable. ∎

Theorem 5.2.5 shows that $\aleph_0$ is the smallest kind of infinite cardinality. Any subset of a set with cardinality $\aleph_0$ either is finite or has cardinality $\aleph_0$.

Corollary 5.2.6 *Let X be a set.*

(a) *X is countable iff there exists a one-to-one function $f: X \to \mathbb{N}$.*

(b) *X is countable iff there exists an onto function $g: \mathbb{N} \to X$.*

(c) *X is countable iff $X \preceq \mathbb{N}$.*

Proof: Exercise 9. In doing proofs, you may freely use the obvious fact that if $X \sim Y$, then X is countable iff Y is countable.

Definition *If A_i, $i \in \mathscr{I}$, is a family of sets and if $\mathscr{I}$ is countable, then $\bigcup_{i \in \mathscr{I}} A_i$ is called a **countable union**.*

Theorem 5.2.7 *Suppose A_i, $i \in \mathscr{I}$, is a family of sets and that each A_i and $\mathscr{I}$ are countable. Then $\bigcup_{i \in \mathscr{I}} A_i$ is countable. That is, a countable union of countable sets is countable.*

Proof:
There is a one-to-one correspondence either between $\mathscr{I}$ and $\mathbb{N}$ or between $\mathscr{I}$ and $\mathbb{N}_k$ for some k. Since $\mathscr{I}$ serves only as an index set, it facilitates notation, without loss of generality, to take $\mathscr{I} = \mathbb{N}$ or $\mathscr{I} = \mathbb{N}_k$. By Corollary

5.2.6, for each $i \in \mathcal{I}$, there exist onto functions $g_i: \mathbb{N} \to A_i$. Thus the elements of each A_i are listed in the array

A_1: $g_1(1)$ $g_1(2)$ $g_1(3)$ $g_1(4)$ ...

A_2: $g_2(1)$ $g_2(2)$ $g_2(3)$ ...

A_3: $g_3(1)$ $g_3(2)$ ...

A_4: $g_4(1)$ ...

A_5: $g_5(1)$ ...

...

Define the onto function $f: \mathbb{N} \to \bigcup_{i \in \mathcal{I}} A_i$ as in Example 5: $f(1) = g_1(1)$, $f(2) = g_1(2)$, $f(3) = g_2(1)$, and so on. Then $\bigcup_{i \in \mathcal{I}} A_i$ is countable by Corollary 5.2.6b.

Example 7:
Each of $\mathbb{Q}^+$, $\mathbb{Q}^-$, and $\{0\}$ is countable (by Examples 5 and 6). Therefore $\mathbb{Q} = \mathbb{Q}^+ \cup \mathbb{Q}^- \cup \{0\}$ is countable.

A theorem analogous to Theorem 5.2.7 works for finite unions of finite sets. In order to prove this, we first need a lemma:

Lemma to Theorem 5.2.8   If A and B are finite, then $A \cup B$ is finite.

Proof:
Suppose $f: \mathbb{N}_k \to A$ and $g: \mathbb{N}_j \to B$ are bijections whose existence is given by Corollary 3.1.5. Define $h: \mathbb{N}_{k+j} \to A \cup B$ by $h(i) = f(i)$ for $1 \le i \le k$ and $h(i) = g(i - k)$ for $k + 1 \le i \le k + j$. Then h is onto, and so by Theorem 5.2.4 and Corollary 2.7.9, $A \cup B$ is finite.

Theorem 5.2.8   *Suppose A_i, $i \in \mathcal{I}$, is a family of sets and that each A_i and $\mathcal{I}$ are finite. Then $\bigcup_{i \in \mathcal{I}} A_i$ is finite. That is, a finite union of finite sets is finite.*

Proof: Exercise 10.

5.2 Cardinalities of Sets of Numbers

Example 8:
Another way to prove that (0, 1) is not countable is by "Cantor's diagonal argument", which goes as follows.

Suppose that (0, 1) is countable (to get a contradiction). Then there exists an onto function $f: \mathbb{N} \to (0, 1)$. For each $i \in \mathbb{N}$, list $f(i)$ as a decimal:

$$f(1) = 0.a_1 a_2 a_3 a_4 \ldots$$

$$f(2) = 0.b_1 b_2 b_3 b_4 \ldots$$

$$f(3) = 0.c_1 c_2 c_3 c_4 \ldots$$

Now define a decimal $d = 0.d_1 d_2 d_3 d_4 \ldots$ in (0, 1) by making $d_1 \neq a_1$, $d_2 \neq b_2$, $d_3 \neq c_3$, and so forth. In general, d_i can be set equal either to the ith digit of $f(i)$ plus 1 or to 0 if the ith digit is 9. Then, for all i: $d \neq f(i)$—contradicting the assumption that f is onto.

Example 9:
Another way to show $\mathbb{N} \sim \mathbb{Q}^+$ is by employing the function $g: \mathbb{Q}^+ \to \mathbb{N}$ defined as follows: Let $p/q \in \mathbb{Q}^+$, where p and q are in lowest terms. Define $g(p/q) = $ "ptq", where "ptq" is a natural number expressed in base 11 using the base-10 digits expressing p, then the symbol t expressing 10 in base 11, then the base-10 digits expressing q. (For example, $g(5/7) = 5t7$, which is the base-11 expression for 722 base 10.) The function $g: \mathbb{Q}^+ \to \mathbb{N}$ is one-to-one, as is the inclusion map $i: \mathbb{N} \to \mathbb{Q}^+$, so that $\mathbb{Q}^+ \sim \mathbb{N}$ by the Schroeder–Bernstein Theorem.

Additional Proof Ideas

We would like Theorem 5.1.5 to be true for all sets—at least the ones for which cardinality is defined—and not just for finite sets. This would give us:

Theorem 5.2.9 *Suppose S and T are sets for which the notion of cardinality has been defined. Then*

(a) $|S| = |T|$ iff $S \sim T$ *(S and T have the* same cardinality*)*

(b) $|S| < |T|$ iff $S < T$ *(S has smaller cardinality than T)*

(c) $|S| \leq |T|$ iff $S \leq T$

This theorem would tell us that the former ideas of same cardinality and smaller cardinality make sense in terms of the newly defined cardinality. As yet, however, we have no definition for what $|S| < |T|$ means for infinite sets. We therefore define the set C of cardinalities $\{0, 1, 2, 3, \ldots, \aleph_0, 2^{\aleph_0}\}$ to have an order given by $a < b$ for $a, b \in C$ provided that a precedes b in the list. With this definition and $a \leq b$ iff ($a < b$ or $a = b$), we can prove Theorem 5.2.9 (Exercise 14).

We could continue the list of cardinalities in C by adding $2^{2^{\aleph_0}}$, and so on, subject to avoiding logical paradoxes, and thus define the *cardinal numbers*. We know that any nonempty finite set is *equipotent* with $\mathbb{N}_k$ for some k. The question arises of whether every infinite set is equipotent with one of the sets N, P(N), P(P(N)), and so on. No one knows. Theorem 5.2.5 shows that there is no infinite set with cardinality less than $\aleph_0$. The assertion that there is no set with "cardinality" strictly between $\aleph_0$ and c (that is, no set S such that $N < S < (0, 1)$) is called the Continuum Hypothesis. Like the Axiom of Choice, it can be neither proved nor disproved from the usual axioms of set theory. Relative to the axioms, it is called *formally undecidable*. If you believe in the existence of infinite sets, quite apart from axiomatizations of set theory, then the Continuum Hypothesis is an unresolved conjecture. If you believe only in sets that exist in the framework of the usual axiomatization of set theory, then the Continuum Hypothesis has been resolved—it's undecidable.

Exercises 5.2

Problems

1. Prove Theorem 5.2.2 by induction.
2. Prove Conjecture 2.3.3a: if $f \circ g$ is one-to-one, then g is one-to-one.
3. Show that the function $f: (0, 1) \to (-\pi/2, \pi/2)$ defined by $f(x) = \pi(x - 1/2)$ is one-to-one and onto.
4. Show that the function $f: \mathbf{R} \to (-\pi/2, \pi/2)$ given by $f(x) = \arctan x$ is one-to-one and onto. Find an inverse function.
5. Prove that $[0, 1]$ has cardinality c. Use theorems such as the Schroeder-Bernstein.
6. Prove that if a and b are any real numbers with $a < b$, then
   (a) (a, b) has cardinality c.
   (b) $[a, b]$ has cardinality c.
7. From Examples 5 and 6 we know there exist bijections $j: \mathbf{N} \to \mathbf{Q}^+$ and $h: \mathbf{N} \to \mathbf{Q}^-$. Use these functions to define a one-to-one, onto function $f: \mathbf{N} \cup \{0\} \to \mathbf{Q}$. Let $f(0) = 0$ and define f differently on the even and odd numbers. Use f to define a bijection $g: \mathbf{N} \to \mathbf{Q}$.
8. Prove that the function g defined in the proof of Theorem 5.2.5 is one-to-one.
9. Prove Corollary 5.2.6.
10. Prove Theorem 5.2.8 by induction, using the lemma to Theorem 5.2.8.
11. Prove that $\mathbf{N} \times \mathbf{N}$ is countable.
12. Prove that if A and B are countable, then $A \times B$ is countable.
13. Arrange the following sets in order of increasing cardinality: $\{0\}$, $[-2, 2]$, $\mathbf{N}$, $\emptyset$, $[-4, 4]$, $\mathbf{Z}$, $\{2, 4, 6\}$, $\{\mathbf{Q}\}$, $\mathbf{P}(\mathbf{R})$, $\mathbf{P}(\mathbf{Z})$.

Supplementary Problem

14. Prove Theorem 5.2.9. The proof will use Theorem 5.1.5 and involve additional cases.

Chapter 6

Groups

6.1 SUBGROUPS

This chapter continues the study of groups begun in Section 3.5. Our goal is to specialize Exercises 3.7.5 and 3.7.6 to a group-theoretic context (Theorem 6.5.5). Thus the functions in these problems will be group homomorphisms, and the sets of equivalence classes will be groups.

For review purposes, we list the definition of a group and a few basic theorems from Section 3.5 before we continue the development.

Definition *A nonempty set G with a binary operation, represented by juxtaposition, is called a **group** provided that:*

    (1) *for all $x, y, z, \in G$: $x(yz) = (xy)z$*
    (2) *there exists $i \in G$ such that for all $x \in G$: $ix = x$ and $xi = x$*
and (3) *for each $x \in G$, there exists $y \in G$ such that $xy = i$ and $yx = i$.*

*A group G is called **Abelian** provided that*

    (4) $xy = yx$ *for all $x, y \in G$.*

Definition *Suppose G is a group. An element $i \in G$ that has the property $ix = x$ and $xi = x$ for all $x \in G$ is called an **identity element**.*

Theorem 3.5.1 *A group has a unique identity element.*

The symbol "i" will refer, without further mention, to the identity element of a group.

Theorem 3.5.2   Let G be a group and let $e \in G$. If $xe = x$ or $ex = x$ for some $x \in G$, then $e = i$.

Theorem 3.5.3   Let G be a group. For each element $x \in G$, there is a unique element $y \in G$ such that $xy = i$ and $yx = i$.

The element y given by Theorem 3.5.3 is called the **inverse** of x. Since we will use multiplicative notation in our development, the inverse of x will be denoted by x^{-1}.

Theorem 3.5.4   If x and y are elements of a group and if $xy = i$, then $y = x^{-1}$ and $x = y^{-1}$.

Theorem 3.5.5   Let x and y be elements of a group. Then

(a) $(x^{-1})^{-1} = x$

(b) $(xy)^{-1} = y^{-1}x^{-1}$

If H is a subset of a group G and if g and h are two elements in H, then by virtue of the fact that g and h are in G, the "product" gh is defined in G. This product may be in the set H. A subset H of a group G is said to be **closed** with respect to the group operation if for all $x, y \in H$: $xy \in H$. In this case, the operation can be considered a binary operation on H. We say that the operation on H has been **inherited** from G.

Definition   A nonempty subset H of a group G is a **subgroup** of G if H is a group under the operation inherited from G.

If H is a subgroup of a group G, H must have an identity (call it i_H). Denote the identity of G by i. Then in H we have $i_H i_H = i_H$. This is an equation between elements of G, however, so that by Theorem 3.5.2, $i_H = i$. Similarly, Theorem 3.5.4 shows that the inverse in H of an element in H is the same as that element's inverse in G.

Theorem 6.1.1   If G is a group and if H is a nonempty subset of G, then H is a subgroup of G iff

(a) for all $x, y \in H$: $xy \in H$

and   (b) for all $x \in H$: $x^{-1} \in H$.

> **Proof:**
>
> First suppose that conditions (a) and (b) hold. We wish to show that H is a group under the operation inherited from G. From (a) we see that this is indeed a binary operation on H. Group axiom (1) (from the definition of a group) holds for all elements in H because it holds for all elements in G,

6.1 Subgroups

including those in H. To show axiom (2), it suffices to show that the identity i of G is actually in H. Since H is nonempty, there is some element in H—call it x. x^{-1} is, then, the inverse of x in G. By (b) $x^{-1} \in H$, and so by (a) $xx^{-1} \in H$, that is, $i \in H$. Finally, axiom (3) follows from (b).

Next we need to show that if H is a subgroup of G, then (a) and (b) hold. Condition (a) holds since the assumption that H is a subgroup implies that the inherited operation is the operation on H. In order to show (b), we need to show that for each $x \in H$ the inverse, x^{-1}, of x in G is actually in H. Since H is itself a group, x has an inverse in H—call it y. Thus $xy = i$ in H and therefore in G. Therefore $y = x^{-1}$ by Theorem 3.5.4. Since $y \in H$, we have that $x^{-1} \in H$. ∎

Theorem 6.1.2 Let G be a group and let H be a nonempty subset of G. Then H is a subgroup of G iff for all $x, y \in H$: $xy^{-1} \in H$.

Proof: Exercise 2.

If G is a group and if $x \in G$, define $x^0 = i$, $x^1 = x$, $x^2 = xx$, $x^3 = xxx$, $x^{-2} = x^{-1}x^{-1}$, and so on. By Exercise 6 (a general associativity law), we can omit parentheses in group products. From the definition of our exponential notation follow the laws of exponents: $x^n \cdot x^k = x^{n+k}$ and $(x^n)^k = x^{nk}$. The set $\{x^k | k \in \mathbb{Z}\}$ is seen to be a subgroup of G by Theorem 6.1.1 and the laws of exponents.

Definition Let G be a group and let $x \in G$. The set $\{x^k | k \in \mathbb{Z}\}$ is called the **subgroup of G generated by x** and is denoted by $\langle x \rangle$. A group G is called **cyclic** if there is an element $x \in G$ such that every element in G is of the form x^k for some $k \in \mathbb{Z}$. Thus G is cyclic provided that $G = \langle x \rangle$ for some $x \in G$.

Example 1:
Consider the group $\mathbb{Z}$ with the operation of addition. If x is an element of $\mathbb{Z}$, then using additive notation $\langle x \rangle = \{kx | k \in \mathbb{Z}\}$. (Instead of writing x^2 for $x \cdot x$, we write $2x$ for $x + x$.) Then $\langle 5 \rangle = \{\ldots, -15, -10, -5, 0, 5, 10, 15, \ldots\}$ is the subgroup of $\mathbb{Z}$ generated by 5. Since $\mathbb{Z} = \langle 1 \rangle$ and $\mathbb{Z} = \langle -1 \rangle$, $\mathbb{Z}$ is cyclic and generated by 1 or -1.

Definition The **order** of an element x of G is the least positive integer n such that $x^n = i$ if such an integer exists. If no such positive integer exists, then x is of infinite order. If a group G is finite, then by the **order** of G we mean the number of elements in G. If G is not finite, then G is said to be of infinite order.

Theorem 6.1.3 Let G be a group and let x be an element of G. The following are equivalent:

(a) $\langle x \rangle$ is a finite set.

(b) x is of finite order.

(c) x^{-1} is equal to x^m for a non-negative integer m.

Proof:

We establish the chain of implications (a) ⇒ (b) ⇒ (c) ⇒ (b) ⇒ (a).

(a) ⇒ (b): Let $\langle x \rangle$ be finite. Consider the elements $x, x^2, x^3, x^4, \ldots$. These are all in $\langle x \rangle$. There are an arbitrarily large number of symbols $x, x^2, x^3, \ldots$. They cannot all be distinct elements of $\langle x \rangle$ since $\langle x \rangle$ is finite. Therefore at least two symbols must stand for the same element of $\langle x \rangle$, say $x^k = x^l$ for $k, l \in \mathbb{N}$ and $l < k$. Then $k - l$ is a non-negative integer, call it n. Multiplying $x^k = x^l$ by $(x^l)^{-1}$ (this is x^{-l} by the law of exponents), we get $x^k x^{-l} = x^l x^{-l}$ or $x^{k-l} = i$ (by the law of exponents). Thus $x^n = i$, and x is of finite order.

(b) ⇒ (c): Let n be the order of x. Then n is positive. Let $m = n - 1$, so that m is non-negative. Then $i = x^n = xx^m$, so that x^m is x^{-1}.

(c) ⇒ (b): Assume $x^{-1} = x^m$ for non-negative m. Let $n = m + 1$. Then $i = x^{-1}x = x^m x = x^n$, so that $\{n \in \mathbb{N} \mid x^n = i\}$ is nonempty. The order of x is the least element in this set.

(b) ⇒ (a): Let n be the order of x and let H be the set $\{i, x, x^2, \ldots, x^{n-1}\}$. $\langle x \rangle$ is the set $\{x^k \mid k \in \mathbb{Z}\}$. For an arbitrary $x^k \in \langle x \rangle$, use the Division Algorithm to divide k by n: $k = qn + r$, where $0 \leq r \leq n - 1$. Then $x^k = (x^n)^q \cdot x^r = i \cdot x^r = x^r \in H$. Hence $\langle x \rangle \subseteq H$, so that $\langle x \rangle$ is finite. Since $H \subseteq \langle x \rangle$ is clear, $\langle x \rangle = \{i, x, x^2, \ldots, x^{n-1}\}$ in this case. ∎

Corollary 6.1.4 *Let G be a group and let x be an element of G. Then the order of x is equal to the order of $\langle x \rangle$.*

Proof:

By Theorem 6.1.3a and 6.1.3b, the order of x is finite iff the order of $\langle x \rangle$ is finite. Suppose this is the case. Let n be the order of x. By the proof of Theorem 6.1.3, $\langle x \rangle = \{i, x, x^2, \ldots, x^{n-1}\}$. To show that the elements listed in this set are all distinct, assume $x^k = x^j$ for $0 \leq j \leq k \leq n - 1$. Then $x^{k-j} = i$. Since $k - j < n$, $k - j = 0$ by the definition of order of x. Therefore the order of $\langle x \rangle$ is n. ∎

Theorem 6.1.5 *Let G be a group and let H be a nonempty finite subset of G. Then H is a subgroup of G iff for all $x, y \in H$: $xy \in H$.*

Proof:

If H is a subgroup of G, then the condition holds by Theorem 6.1.1a. Conversely, if the condition holds, it suffices to show in addition that the

6.1 Subgroups

> condition given in Theorem 6.1.1b holds. Thus let x be given in H. In order to show that $x^{-1} \in H$, consider the elements $x, x^2, x^3, x^4, \ldots$. By the condition, these all are in H. There are an arbitrarily large number of symbols $x, x^2, x^3, \ldots$. They cannot all be distinct elements of H since H is finite. Therefore, as in the proof of Theorem 6.1.3, the order of x is finite and $x^{-1} = x^m$ for some $m \geq 0$. Hence $x^{-1} \in H$ by the condition on H. ∎

The argument in the proofs above concerning an infinite number of representations x^k for elements in a finite set, while certainly logical, cannot be said to follow from definitions and our inference rules. It ought not to be considered part of a valid proof unless we introduce a new rule that allows us to make this sort of argument. We are far enough removed from formal proofs at this stage to be beyond the need to call the following an inference rule—but we do list it, in the customary folksy terms, as a valid method of argument.

Pigeonhole Principle *If n objects are distributed among k boxes for $n > k$, then at least one box receives more than one object.*

Example 2:
This principle applies to the proof of Theorem 6.1.5 as follows. The order of H is k. Boxes correspond to elements in H. Objects correspond to symbols $x, x^2, x^3, \ldots, x^n$, where n can be taken to be $k+1$. By the Pigeonhole Principle, some element in H has two representations $x^i = x^j$ for $i \neq j$—from which the result follows, as in the proof of Theorem 6.1.3.

Theorem 6.1.6 *If H and J are subgroups of a group G, then $H \cap J$ is a subgroup of G.*

Proof: Exercise 3.

Every group has at least one element, the identity element. A group with just one element is called the *trivial* group. Every nontrivial group G has at least two subgroups, $\{i\}$ and G itself.

Exercises 6.1

Problems

1. Let x, y, z be elements of a group G. Prove the following cancellation properties (same problem as Exercise 3.5.5):
   (a) If $xy = xz$, then $y = z$.
   (b) If $yx = zx$, then $y = z$.

2. Prove Theorem 6.1.2.
3. Prove Theorem 6.1.6.
4. Prove or find a counterexample to the following conjecture: if H and J are subgroups of a group G, then $H \cup J$ is a subgroup of G.
5. Let $S = \mathbf{R} - \{1\}$ and define $*$ on S by $x*y = x + y - xy$. Show that $*$ is a binary operation on S and that $(S, *)$ is a group.

Supplementary Problem

6. There are just two ways the elements in the product $a_1 \cdot a_2 \cdot a_3$ can be grouped: $a_1 \cdot (a_2 \cdot a_3)$ and $(a_1 \cdot a_2) \cdot a_3$. The associative property asserts that these two expressions are equal, so that grouping is irrelevant. A generalized associative property, which can be proved by induction, states that grouping of n-fold products is irrelevant. How many distinct groupings of $a_1 \cdot a_2 \cdot a_3 \cdot \cdots \cdot a_n$ are there? (Same problem as Exercise 3.5.6.)

6.2 EXAMPLES

Example 1: S_3

In Example 2.6.1 we examined the set $\mathbf{B}(\{1, 2, 3\})$ of all bijections, or permutations, of the set $\{1, 2, 3\}$. Under the binary operation of composition, this set is a group called the symmetric group on three objects, or S_3. The identity of the group S_3 is the identity permutation $\binom{123}{123}$. The other elements of S_3 are $\binom{123}{132}$, $\binom{123}{213}$, $\binom{123}{231}$, $\binom{123}{312}$, and $\binom{123}{321}$. The following computation shows that S_3 is non-Abelian.[†]

$$\binom{123}{213} \circ \binom{123}{231} = \binom{123}{132}$$

$$\binom{123}{231} \circ \binom{123}{213} = \binom{123}{321}$$

Example 2: S_n

If n is any natural number, then S_n denotes the group of all permutations of the numbers 1 through n. The order of S_n is $n!$ (Example 2.6.2).

[†] If f and g are two elements in S_3, then, since the product $f \circ g$ represents composition of functions, g is applied first according to our convention. This is called "left-hand notation" since functions are written to the left of their arguments (elements in their domains). This means that when reading function compositions, we read from right to left in order to preserve the order in which the functions are applied. In abstract algebra, it is more usual to write functions to the right of their arguments so that function composition can be conveniently read from left to right. In texts that follow this convention, f is applied first in the product fg—which in this case is usually written by juxtaposition, without the symbol $\circ$ for composition.

6.2 Examples

Example 3: The Dihedral Groups

Consider a two-inch-square photographic slide in position to be correctly projected on a screen. Number the corners of the slide and adjacent reference positions on the paper **1, 2, 3, 4** clockwise starting from the upper left corner. In addition to the correct orientation, three other orientations can be obtained by turning the slide counterclockwise through 90°, 180°, and 270°, as shown in Figure 6.2.1.

In addition to these "motions", the slide can be flipped about four different axes. Starting from the correct position, each flip produces one of the orientations shown in Figure 6.2.2 (following page).

The motions that take the slide to a position that will fit the projector (the four shown in Figure 6.2.1 plus the four shown in Figure 6.2.2) correspond to the symmetries of the square, and there are just eight such motions. We can compose motions—rotate through 270° and then flip about a vertical axis, for instance—but any such composition of motions must leave us in one of the eight positions and has therefore the same effect as one of the originally described motions. Thus composition of motions is a binary operation on the set of all motions. The motions can be described by listing the new reference position to which a vertex goes. Thus the 90° counterclockwise rotation sends vertex **1** to reference position **4**—the place where vertex **4** used to be. It is given by the notation:

$$\begin{pmatrix} 1234 \\ 4123 \end{pmatrix} \quad \text{or} \quad \begin{pmatrix} 2341 \\ 1234 \end{pmatrix} \qquad \text{Thus } 1 \to 4,\, 2 \to 1,\, 3 \to 2,\, 4 \to 3$$

Note that **4123** (row 2 in the first notation) is the order of the new positions to which vertices **1234** have gone and that **2341** (row 1 in the second notation) is the order of the new vertices in reference positions **1234**. The eight motions of the slide are given in Figure 6.2.3 (following page).

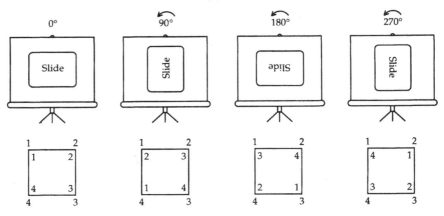

Figure 6.2.1

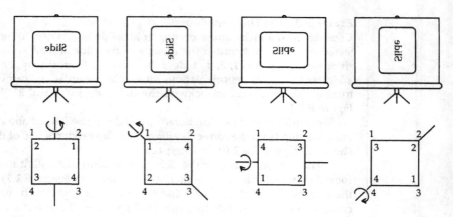

Figure 6.2.2

Let D_8 denote this group with elements *i, a, b, c, d, e, f, g*. It is called the *dihedral* group of order **8**. It is the group of all symmetries of the square. The group of all symmetries of a regular pentagon is of order **10**. The group of symmetries of a regular *n*-gon is called the *dihedral* group of order **2n**.

The notation we used for the elements of D_8 suggests that each element can be considered a permutation of the numbers **1, 2, 3, 4**. Consider the com-

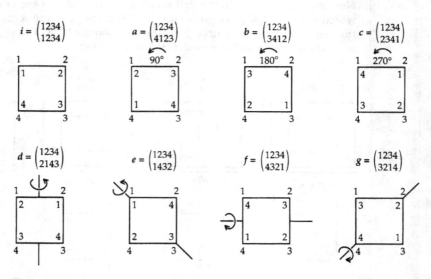

Figure 6.2.3

position of two elements, x and y, of D_8. Say y sends vertex j to reference position k and x sends vertex k to reference position l. Then the composition xy first sends (by the action of y) vertex j to reference position k, then sends j to reference position l—since the action of x sends the vertex (now j) in reference position k to reference position l. Hence composition of motions agrees with composition of the permutations of $\{1, 2, 3, 4\}$ indicated by the notation. That is to say, D_8 is isomorphic to a subgroup of S_4, the group of *all* permutations of $\{1, 2, 3, 4\}$. The group S_4 is of order $24 = 4!$. The *rotations* in D_8 (i, a, b, c) are those permutations in which the cyclic order of row 1 is the same as the cyclic order of row 2. The *reflections* (flips) have the cyclic order in row 2 exactly reversed (Figure 6.2.3).

Definition *Let G be a group. A subset S of G is called a **set of generators** for G if every element in G can be written as a product of elements of S and their inverses.*

A cyclic group is one that has at least one set of generators containing only one element. If a group G is cyclic with $S = \{x\}$ a set of generators, then x is called a *generator* of G and $G = \langle x \rangle$. To avoid confusion, we will not use the word "generator" for a single element in a set of generators unless the group is cyclic and generated by that one element.

The group D_8 is not cyclic, but the subgroup $H = \{i, a, b, c\}$ of rotations is cyclic. Indeed, $i = a^0$, $a = a^1$, $b = a^2$, $c = a^3$, so that a is a generator of H. The order of a is 4. The element c is also a generator of H, but i and b are not. The set $\{a, d\}$ is a set of generators for D_8 (Exercise 5).

Example 4: **Z**

$\mathbb{Z}$ is cyclic and generated by 1 or -1. Every element of $\mathbb{Z}$ except the identity 0 has infinite order.

Example 5: $\mathbb{Z}_n$

$\mathbb{Z}_n = \{[0], [1], [2], \ldots, [n-1]\}$. It is cyclic and generated by $[1]$. Certain other elements in $\mathbb{Z}_n$ are generators (Exercise 3).

Example 6:

Consider the problem of finding all the subgroups of $\mathbb{Z}$. By this we mean the following. A subgroup is a set. Knowing the set means having a specific rule for determining set membership—that is, knowing just which elements are in the set. Thus to find all subgroups, we need a representation for each subgroup that makes the rule for set membership clear.

Let H be an arbitrary subgroup of $\mathbb{Z}$. One possibility is $H = \{0\}$. If $H \neq \{0\}$ and if x is some nonzero element in H, then x or $-x$ (both of which are in H) is positive. Thus the set of natural numbers in H is nonempty. Let h be the least positive integer in H. Then H is cyclic and generated by h (Exercise 1):

$$H = \langle h \rangle = \langle -h \rangle = \{\ldots, -3h, -2h, -h, 0, h, 2h, 3h, \ldots\}$$

The subgroups of $\mathbb{Z}$ can be characterized as the sets $\langle x \rangle$ for x a non-negative integer. This characterization "finds" all subgroups of $\mathbb{Z}$.

Example 7:
For $n \in \mathbb{N}$, define $\varphi(n)$ to be the number of positive integers less than n and relatively prime to n. φ is called the Euler phi function. Let $\mathbf{E}_n = \{[1], [k], \ldots, [n-1]\}$ be the subset of $\mathbb{Z}_n$ of classes with representatives prime to n. (If $(k, n) = 1$ and $k \equiv p \bmod n$, then $(p, n) = 1$ (Exercise 6), so that this set is well defined.) Thus the order of $\mathbf{E}_n$ is $\varphi(n)$. If $k \equiv p \bmod n$ and $j \equiv q \bmod n$, then $kj \equiv pq \bmod n$ (Exercise 7), so that we have a well-defined operation $\cdot$ on $\mathbf{E}_n$ given by $[k] \cdot [j] = [kj]$ (multiplication modulo n). Associativity and commutativity in $\mathbf{E}_n$ are inherited from associativity and commutativity of multiplication in $\mathbb{Z}_n$. The class $[1]$ is the identity for $\mathbf{E}_n$ under $\cdot$. To find the inverse of some $[k] \in \mathbf{E}_n$, first write $ak + bn = 1$ by Theorem 3.9.2. Divide a by n: $a = qn + r$ for $0 \le r < n$. Substituting, $(qn + r)k + bn = 1$ or $rk + (qk + b)n = 1$. Hence $[rk] + [0] = [1]$, so that $[r] \cdot [k] = [1]$. Since $rk + (qk + b)n = 1$, any divisor of r and n divides 1, and hence $(r, n) = 1$. Thus $[r] \in \mathbf{E}_n$, and $[r]$ is the inverse of $[k]$.

Exercises 6.2

Problems

1. Show that if H is a nontrivial subgroup of $\mathbb{Z}$, then $H = \langle h \rangle$, where h is the least positive integer in H.
2. Find all the subgroups of D_8.
3. (a) Show that if x generates a group G and if $x = y^m$ for some m, then y generates G.
   (b) Show that if m and n are relatively prime, then $[m]$ is a generator for $\mathbb{Z}_n$. (Translate (a) into additive notation.)
4. Find all subgroups of $\mathbb{Z}_n$.
5. In the group D_8, pick x to be rotation through either 90° or 270°. Pick y to be some reflection. Show that $\{x, y\}$ is a set of generators for D_8 for your choice of x and y.
6. Prove that if $(k, n) = 1$ and $k \equiv p \bmod n$, then $(p, n) = 1$.
7. Prove that if $k \equiv p \bmod n$ and $j \equiv q \bmod n$, then $kj \equiv pq \bmod n$. (Same as Exercise 3.2.6b.)
8. In the set $\mathbf{E}_n$ of Example 7, let $n = 14$. Write out the set $\mathbf{E}_{14}$. Give the multiplication table for $\mathbf{E}_{14}$.
9. If A has cardinality greater than 2, find $f, g \in \mathbf{B}(A)$ such that $f \circ g \ne g \circ f$.

6.3 SUBGROUPS AND COSETS

Example 1:
Example 3.3.2 lists the equivalence classes for the relation of congruence modulo 5. These are the elements of Z_5:

$$[1] = \{\ldots, -19, -14, -9, -4, 1, 6, 11, 16, 21, \ldots\}$$
$$[2] = \{\ldots, -18, -13, -8, -3, 2, 7, 12, 17, 22, \ldots\}$$
$$[3] = \{\ldots, -17, -12, -7, -2, 3, 8, 13, 18, 23, \ldots\}$$
$$[4] = \{\ldots, -16, -11, -6, -1, 4, 9, 14, 19, 24, \ldots\}$$
$$[0] = \{\ldots, -15, -10, -5, 0, 5, 10, 15, 20, 25, \ldots\}$$

The class [0] is seen to be the subgroup $\langle 5 \rangle$ of Z. Thus [0] is an *element* of Z_5 but a *subgroup* of Z. $\langle 5 \rangle$ and [0] are equal as subsets of Z. None of the other classes is a subgroup of Z; none contains the identity 0, for example. The other classes have the property that they may be obtained by adding a single number to each of the elements in $\langle 5 \rangle$. If we add 2 to each element in

$$[0] = \{\ldots, -15, -10, -5, 0, 5, 10, 15, 20, 25, \ldots\}$$
we get $\qquad \{\ldots, -13, -8, -3, 2, 7, 12, 17, 22, 27, \ldots\}$

This is the class [2] above, and so [2] = [0] + 2, where by "[0] + 2" we mean "$\{x + 2 \mid x \in [0]\}$". Similarly, [1] = [0] + 1, [3] = [0] + 3, and so forth.

The classes [0], [1], [2], [3], and [4] are mutually disjoint, and their union is all of Z. That is, they give a partition of Z.

We can copy the ideas in this example in the setting of a general group—in multiplicative notation.

Definition   Let G be a group and let H be a subgroup of G. For any $x \in G$, define $Hx = \{hx \mid h \in H\}$. Then Hx is called the **right coset** of H determined by x.

For example, suppose H is finite and $H = \{h_0, h_1, h_2, \ldots, h_n\}$, where $h_0 = i$. Then $Hx = \{x, h_1 x, \ldots, h_n x\}$. If we suppose all the h_k in the list are distinct, then all the products $h_k x$ are distinct, since if $h_k x = h_j x$, then $h_k = h_j$ by the cancellation property. There are two cases: $x \in H$ and $x \notin H$.

If $x \in H$, then each product $h_k x$ is back in H, so that $Hx \subseteq H$. Since all the products are distinct, Hx and H have the same size, and so $Hx = H$. Thus the right coset of H determined by any element of H is just H itself.

In the other case ($x \notin H$), H and Hx have no elements in common. To see this, suppose $y \in H \cap Hx$. Then $y = h_k = h_j x$ for some k, j. Then $x = h_j^{-1} h_k$, and so $x \in H$, a contradiction.

Suppose we now have H and Hx for some $x \notin H$. Then $H \cup Hx \subseteq G$ and

$H \cap Hx = \emptyset$. Suppose there is some $y \in G - (H \cup Hx)$. Consider Hy. We have already seen that $H \cap Hy = \emptyset$. Is $Hy \cap Hx$ empty? Suppose not; then there is some $z \in Hx \cap Hy$; that is, $z = h_k x = h_j y$ for some k, j. However, $h_j^{-1} h_k$ is in H and $y = (h_j^{-1} h_k)x$, so that y is in Hx. This contradicts our choice of y. Therefore Hy does not intersect either H or Hx, and $H \cup Hx \cup Hy \subseteq G$. If G is finite, we can continue this process and thus exhaust G as a union of disjoint cosets of H. This gives a partition of G.

Recall that by the **order** of a finite group, we mean the number of elements in the group. If H is a subgroup of a finite group G, then our discussion above shows that each coset of H has the same number of elements as does H. The number of elements in G is then the product of the order of H by the number of cosets of H in G. We get the following theorem of Lagrange:

Theorem 6.3.1 *If G is a finite group and if H is a subgroup of G, then the order of H divides the order of G.*

Corollary 6.3.2 *Let G be a finite group of order m and let x in G have order n. Then*

(a) $n \mid m$

(b) $x^m = i$

(c) *If m is prime, then G is cyclic.*

Proof:

The subgroup $\langle x \rangle$ of G has order n by Corollary 6.1.4, proving (a).

By (a), $m = qn$ for some q. Hence $x^m = (x^n)^q = i^q = i$, proving (b). If m is prime and if $x \neq i$ in G, then by (a), $m = n$, so that $\langle x \rangle = G$. ∎

The demonstration above that a group is partitioned into cosets of a subgroup can be recast in a more refined way that also applies to infinite groups. We have gone through the more elementary computations first to make things more transparent. Thus for any fixed element x of a group G, we can define the function R_x from G to G by $R_x(g) = gx$ for all $g \in G$. For obvious reasons, R_x is called *right multiplication by x*.

Theorem 6.3.3 *Let $R_x : G \to G$ be right multiplication by x on a group G. Then*

(a) R_x *is one-to-one and onto.*

(b) *If $S \subseteq G$, then $R_x(S) = \{sx \mid s \in S\}$, which we call Sx.*

(c) *If H is a subgroup of G, then $H \sim Hx$ (H and Hx have the same cardinality).*

Proof: Exercise 1.

6.3 Subgroups and Cosets

Our discussion leading up to Theorem 6.3.1 shows that if G is a finite group and if H is a subgroup of G, then G is partitioned by the family of right cosets of H in G. That is, $G = \bigcup_{x \in G} Hx$ and $Hx \cap Hy = \emptyset$ for $Hx \ne Hy$. We can prove that this is also the case for arbitrary G.

Definition   If G is a group and if H is a subgroup of G, denote by $\dfrac{G}{H}$ (also written G/H) the family of right cosets of H in G.

Theorem 6.3.4   If H is a subgroup of a group G, then G/H partitions G.

> **Proof:**
>
> We need to show   (a) $G = \bigcup_{x \in G} Hx$
>
> and   (b) if $Hx \ne Hy$, then $Hx \cap Hy = \emptyset$.
>
> (a): Let $g \in G$. Then $g \in Hg$ since $i \in H$ and $g = ig$.
>
> (b): Assume $Hx \ne Hy$. Then by symmetry we can assume there is some hx in Hx but not in Hy. Assume $a \in Hx \cap Hy$ to get a contradiction. Then $a = h_1 x = h_2 y$ for $h_1, h_2 \in H$. Multiplying by hh_1^{-1} gives $hh_1^{-1}h_1 x = hh_1^{-1}h_2 y$, so that $hx \in Hy$, which is a contradiction. ∎

Definition   For a subgroup H of a group G, the number of right cosets of H in G is called the *index of H in G*.

The index of H in G may be finite or infinite. By Theorem 6.3.3 and 6.3.4, we have the following generalization of Lagrange's Theorem:

Theorem 6.3.5   Let H be a subgroup of a finite group G. The order of G is the product of the order of H and the index of H in G.

We now consider what must be true about x and y if they are to determine the same coset of H. Suppose that $Hx = Hy$. Note that $x \in Hx$ since $x = ix$ and $i \in H$. Therefore $x \in Hy$, so that $x = hy$ for some $h \in H$. Multiplying both sides of this equation on the right by y^{-1} gives $xy^{-1} = h$, so that $xy^{-1} \in H$. This shows that if $Hx = Hy$, then $xy^{-1} \in H$.

Conversely, suppose $xy^{-1} \in H$. To show that $Hx = Hy$, we first show that $Hx \subseteq Hy$. Let $hx \in Hx$ be arbitrary for some $h \in H$. Then, since H is a subgroup, $(h)(xy^{-1}) \in H$. Let $k = hxy^{-1}$. Since $k \in H$, $ky \in Hy$. However, $ky = hxy^{-1}y = hx$, so that $hx \in Hy$. Therefore $Hx \subseteq Hy$. The proof of $Hy \subseteq Hx$ is not exactly symmetric, since "xy^{-1}" is not symmetric in x and y, but it is similar and left for Exercise 2.

We have shown that $Hx = Hy$ if and only if $xy^{-1} \in H$. Once again things can be put more concisely if we use a more convenient notation. We first define $Sx = \{sx \mid s \in S\}$ and $xS = \{xs \mid s \in S\}$ for any subset S of a group G. If H is a subgroup, then xH is the **left coset** of H in G determined by x.

Lemma 6.3.6 For any subsets S and T and elements x and y of a group G:
  (a) $S(xy) = (Sx)y$ and $(xy)S = x(yS)$
  (b) $S \subseteq T$ iff $Sx \subseteq Tx$
  (c) $S \subseteq T$ iff $xS \subseteq xT$

Proof: Exercise 3.

Note that part (a) of this lemma allows us to multiply equations involving sets on the right or left exactly as we multiply equations between elements. Parts (b) and (c) assert that multiplication preserves set containment.

Theorem 6.3.7 Let G be a group and let H be a subgroup of G. For all $x, y \in G$:
  (a) $x \in H$ iff $H = Hx$
  (b) $Hx = Hy$ iff $Hxy^{-1} = H$ iff $H = Hyx^{-1}$
  (c) $Hx = Hy$ iff $xy^{-1} \in H$

Proof: Exercise 4.

Definition Let H be a subgroup of a group G. For $x, y \in G$, define x to be **congruent to y modulo H** (denoted $x \simeq y \bmod H$) if $xy^{-1} \in H$.

Theorem 6.3.8 Let H be a subgroup of a group G.
  (a) Congruence modulo H is an equivalence relation on G.
  (b) The $\simeq$ congruence classes are the right cosets of H in G.

Proof: Exercise 5.

From Theorem 3.3.4 we know that any equivalence relation on G partitions G into disjoint equivalence classes. Applying this to congruence modulo H, we see that G is the disjoint union of the right cosets of H. This applies whether or not G or H is finite. Our motivation for treating the finite case first and in detail was to shed light on the idea of "coset" and the partitioning of a group into the cosets of a subgroup.

Let n be a positive integer so that $\langle n \rangle$ is a subgroup of $\mathbf{Z}$. The equivalence relation $\simeq$ on $\mathbf{Z}$ defined by congruence modulo $\langle n \rangle$ is exactly the same as congruence modulo n defined in Chapter 3:

6.3 Subgroups and Cosets

Theorem 6.3.9 Let n be a positive integer. Then

(a) For $a, b \in \mathbb{Z}$: $a \equiv b \bmod n$ iff $a \simeq b \bmod \langle n \rangle$.

(b) For $c \in \mathbb{Z}$, the $\equiv$ congruence class $[c]$ is the same as the coset $\langle n \rangle + c$ of $\langle n \rangle$ in $\mathbb{Z}$ (with the coset given in additive notation).

Proof: Exercise 6.

The group theory we have considered so far has application in proving some theorems of number theory.

Theorem 6.3.10 If $m, n \in \mathbb{N}$ with $(m, n) = 1$, then $m^{\varphi(n)} \equiv 1 \bmod n$.

> **Proof:**
> Consider $[m]$ as an element in the group E_n of Example 6.2.7. The order of E_n is $\varphi(n)$. Hence $[m]^{\varphi(n)} = [1]$ by Corollary 6.3.2b. Thus $m^{\varphi(n)} \equiv 1 \bmod n$ by Theorem 6.3.9. ∎

Theorem 6.3.11 If p is prime and if $m \in \mathbb{Z}$, then $m^p \equiv m \bmod p$.

> **Proof:**
> Case 1. $(m, p) = 1$: Since p is prime, $\varphi(p) = p - 1$, so that $m^{p-1} \equiv 1 \bmod p$ by Theorem 6.3.10. Multiplying by m, we get $m^p \equiv m \bmod p$.
> Case 2. $(m, p) \neq 1$: Since p is prime, $p | m$, so that $p | m^p - m$. Whence $m^p \equiv m \bmod p$. ∎

Theorems 6.3.10 and 6.3.11 are nontrivial results in number theory due to Pierre de Fermat (1601–1665). We have obtained them here by applying the abstract axiomatic development of group theory to the specific example of the integers.

Exercises 6.3

Problems

1. Prove Theorem 6.3.3.
2. Let H be a subgroup of G. Finish the discussion on page 290 by showing that if $xy^{-1} \in H$, then $Hy \subseteq Hx$.

3. Prove Lemma 6.3.6.
4. Prove Theorem 6.3.7.
5. Prove Theorem 6.3.8.
6. Prove Theorem 6.3.9.
7. Prove that if H is a subgroup of G, then the number of right cosets of H in G is equal (or equipotent if this number is infinite) to the number of left cosets of H in G. Don't assume that G is finite.

6.4 NORMAL SUBGROUPS AND FACTOR GROUPS

Let n be a positive integer. Then Z_n is the set of equivalence classes $\{[0], [1], [2], \ldots, [n-1]\}$, where addition of classes is done by adding representatives. Thus $[a] + [b] = [a + b]$. According to Theorem 6.3.9, the equivalence class $[a]$ is the same as the coset $\langle n \rangle + a$ of $\langle n \rangle$. Thus to add cosets, we have the rule:

$$(\langle n \rangle + a) + (\langle n \rangle + b) = \langle n \rangle + (a + b)$$

With this addition, the set of all cosets of $\langle n \rangle$ in Z is itself a group by Theorem 3.4.2—the group we have denoted by $\dot{Z}_n$.

We now try to generalize this construction for a subgroup H of an arbitrary (perhaps non-Abelian) group G. That is, we want to make a group out of the set of right cosets of H in G. The group operation will be given by set multiplication, defined as follows:

Definition For subsets S and T of a group G, define $S \cdot T = \{st \mid s \in S, t \in T\}$.

Lemma 6.4.1 Let G be a group, $S, T \subseteq G$, and H a subgroup of G.

(a) $H \cdot H = H$
(b) For all $x \in G$: $(xS) \cdot T = x(S \cdot T)$
$(Sx) \cdot T = S \cdot (xT)$
$S \cdot (Tx) = (S \cdot T)x$

Proof: Trivial.

Let H be a subgroup of a group G. Analogous to the addition of cosets in Z_n, we would like to define the product of cosets of H in G as follows:

for all $a, b \in G$: $(Ha) \cdot (Hb) = Hab$

Let us suppose for the moment that such a multiplication is well defined and examine the consequences. Then we will see whether or not these consequences can, conversely, insure that the multiplication is well defined. First, from

6.4 Normal Subgroups and Factor Groups

$(Ha) \cdot (Hb) = Hab$, we get $(Ha \cdot H)b = (Ha)b$. Multiplying by b^{-1} gives $Ha \cdot H = H \cdot aH = Ha$. Since $i \in H$, the right-hand equation gives $i \cdot aH \subseteq Ha$ or just $aH \subseteq Ha$. Thus the left coset determined by a is contained in the right coset determined by a. If H is finite, the two cosets have the same size, and so $aH = Ha$. If H is infinite, then we can get the same result by starting with $Ha^{-1} \cdot Ha = Ha^{-1}a = H$—assuming multiplication is well defined. Hence $i \cdot a^{-1} \cdot H \cdot a \subseteq H$. Multiplying on the left by a gives $Ha \subseteq aH$. Hence $Ha = aH$.

We have shown that $Ha = aH$ is a *necessary* condition for multiplication to be well defined. That is, if multiplication is well defined, then $Ha = aH$ for all $a \in G$. The converse of this is that if $Ha = aH$ for all $a \in G$, then multiplication is well defined. In this case we say that $Ha = aH$ is a *sufficient* condition.

Theorem 6.4.2 Define multiplication on G/H, the set of right cosets of H in G, by $Hx \cdot Hy = Hxy$ for all $x, y \in G$. Multiplication is well defined iff $Ha = aH$ for all $a \in G$.

Proof:

Necessity: Already shown.
 Sufficiency: We need to show that if $Ha = aH$ for all $a \in G$, then $Hx \cdot Hy = Hxy$ for all $x, y \in G$. Assuming $Ha = aH$ for all $a \in G$, we get $Hx \cdot Hy = H \cdot xH \cdot y = H \cdot Hx \cdot y = Hxy$ by Lemma 6.4.1.

The last remaining step in our quest to define a multiplication on G/H will be to show the following:

Conjecture Let H be a subgroup of a group G. Then $Ha = aH$ for all $a \in G$.

Counterexample:

Let G be D_8 and H be $\langle d \rangle$, where $d = \begin{pmatrix} 1234 \\ 2143 \end{pmatrix}$ is reflection about a vertical axis. Then $\langle d \rangle = \{i, d\}$. Let $a = \begin{pmatrix} 1234 \\ 4123 \end{pmatrix}$ be 90-degree rotation. Then $Ha = \{a, da\}$ and $aH = \{a, ad\}$. Further,

$$ad = \begin{pmatrix} 1234 \\ 4123 \end{pmatrix} \circ \begin{pmatrix} 1234 \\ 2143 \end{pmatrix} = \begin{pmatrix} 1234 \\ 1432 \end{pmatrix}$$

but

$$da = \begin{pmatrix} 1234 \\ 2143 \end{pmatrix} \circ \begin{pmatrix} 1234 \\ 4123 \end{pmatrix} = \begin{pmatrix} 1234 \\ 3214 \end{pmatrix}$$

Thus $ad \in aH$ but $ad \notin Ha$.

We might think that even if the left coset aH is not the right coset Ha, it might be *some* right coset Hb. If $aH = Hb$, however, we can then show $Hb = Ha$ (Exercise 1).

Our attempt to define a multiplication on the set of right cosets of an arbitrary subgroup H of a group G has failed. Of course, we might make another definition of multiplication, but this is not our purpose. We want to generalize the operation on Z_n. That is, when we get done, we want the operation on Z_n to be a particular example of our general multiplication.

We have been involved in a process of discovery. This illustrates one way in which mathematics is developed. We attempt to generalize and fail. Then we back off. We are less ambitious. Even if we can't define a multiplication on G/H for an arbitrary H, we have found exactly those H for which a multiplication is possible. This motivates a new concept—a new definition.

Definition   *A subgroup H of a group G is called **normal** in G provided that $Ha = aH$ for all $a \in G$.*

Theorem 6.4.3   *The following conditions are equivalent to H being a normal subgroup of a group G:*

(a) $a^{-1}Ha = H$ for all $a \in G$.
(b) $a^{-1}Ha \subseteq H$ for all $a \in G$.
(c) For all $a \in G$, $h \in H$: $a^{-1}ha \in H$.
(d) For all $a \in G$, $h \in H$: $ha = ak$ for some $k \in H$.
(e) Each right coset of H is some left coset.
(f) Each left coset of H is some right coset.

Proof: Exercise 2.

The condition $Ha = aH$ asserts that a commutes with the set H. This is not the same as commuting with each element in H. In order that $Ha = aH$, all that is required is expressed in condition (d) of Theorem 6.4.3.

Theorem 6.4.4   *Let H and K be normal subgroups of G. Then*

(a) $H \cap K$ *is a normal subgroup of G.*
(b) $HK = KH$
(c) HK *is a normal subgroup of G.*
(d) $HK = \langle H \cup K \rangle$

Proof of (a): Exercise 3.

Proof of (b): Straightforward by Theorem 6.4.3d.

6.4 Normal Subgroups and Factor Groups

Proof of (c):

An arbitrary element of HK is of the form hk for $h \in H$, $k \in K$. Let $h_1 k_1$ and $h_2 k_2$ be arbitrary in HK. Then $h_1 k_1 h_2 k_2 = h_1 h_2 k' k_2$ for some $k' \in K$ by Theorem 6.4.3d (K being normal), and so $h_1 k_1 h_2 k_2$ is in HK. Also, $(h_1 k_1)^{-1} = k_1^{-1} h_1^{-1} = h_1^{-1} k'$ for some $k' \in K$, and so $(h_1 k_1)^{-1}$ is in HK. Therefore HK is a subgroup of G. To show that HK is normal in G, let $a \in G$ and $hk \in HK$. Then $a^{-1} hka = a^{-1} haa^{-1} ka \in HK$ by the normality of H and K.

Proof of (d):

The elements of $\langle H \cup K \rangle$ are finite products of elements in $H \cup K$. (An inverse of an element in $H \cup K$ is in $H \cup K$.) Using Theorem 6.4.3d, we can change such a product to a product with elements from H on the left and elements from K on the right. This product is therefore in HK. Hence $\langle H \cup K \rangle \subseteq HK$. The reverse inclusion is immediate. ∎

The condition of normality of H in G not only gives a good multiplication on G/H, but we even get more:

Theorem 6.4.5 *If H is a normal subgroup of a group G, then G/H is a group called the **quotient group** or **factor group** of G with respect to H, or simply G **modulo** H.*

Proof:

By Theorem 6.4.2, $Ha \cdot Hb = Hab$ is a binary operation on G/H.
Associativity: $(Ha \cdot Hb) \cdot Hc = Hab \cdot Hc = Habc = Ha \cdot Hbc = Ha \cdot (Hb \cdot Hc)$.
The identity of G/H is H: $Ha \cdot H = Ha \cdot Hi = Hai = Ha = H \cdot Ha$.
The inverse of Ha is Ha^{-1}: $Ha^{-1} \cdot Ha = Ha^{-1} a = H = Ha \cdot Ha^{-1}$. ∎

It is clear from Theorem 6.4.3 that every subgroup of an Abelian group is normal.

Example 1:
Any nontrivial subgroup of $\mathbb{Z}$ is of the form $\langle n \rangle$ for some non-negative integer n (Exercise 6.2.1). $\mathbb{Z}/\langle n \rangle$ is the group $\mathbb{Z}_n$.

There are normal subgroups of non-Abelian groups.

Theorem 6.4.6 *If H is of index 2 in a group G, then H is normal in G.*

Proof: Exercise 4.

Example 2:
The subgroup of rotations $H = \{i, a, b, c\}$ is a normal subgroup of D_8. The coset of H is the set of all rotations of a reflected square, which can be represented by Hd. The product of a rotation by a rotation is another rotation. Thus $H \cdot H = H$. The product of a rotation by a reflected rotation is a reflected rotation. Thus $Hd \cdot H = Hd = H \cdot Hd$. The product of two reflected rotations is a rotation. Thus $Hd \cdot Hd = H$.

The group table for D_8/H is given here, along with the table for $\mathbb{Z}_2$:

$\mathbb{Z}_2$

·	[0]	[1]
[0]	[0]	[1]
[1]	[1]	[0]

$\dfrac{D_8}{H}$

·	H	Hd
H	H	Hd
Hd	Hd	H

The isomorphism between $\mathbb{Z}_2$ and D_8/H is clear from the table. Recall the isomorphism between $\mathbb{Z}_2$ and $\mathbb{B}(\langle 1, 2 \rangle)$ expressed in Example 3.5.7.

Exercises 6.4

Problems

1. Let H be a subgroup of G. Prove that if $aH = Hb$, then $Hb = Ha$.
2. Prove Theorem 6.4.3, using a chain of implications. Note that the definition of normality is symmetric with respect to right and left.
3. Prove Theorem 6.4.4a.
4. Prove Theorem 6.4.6. Using an appropriate part of Theorem 6.4.3 can make this easy.
5. Let H_i for $i \in \mathscr{I}$ be a family of normal subgroups of a group G. Prove that $\bigcap_{i \in \mathscr{I}} H_i$ is a normal subgroup of G.
6. Let K be a normal subgroup of G and let H be a subgroup of G. Prove
    (a) $HK = KH$.
    (b) HK is a subgroup of G.
    (c) $H \cap K$ is normal in H.

6.5 FUNDAMENTAL THEOREMS OF GROUP THEORY

In this section we can state the Fundamental Homomorphism Theorem, which expresses Exercises 3.7.5 and 3.7.6 in the context of groups and group homomorphisms. The techniques we have developed throughout this book will guide you in your own proofs of major results in this section.

Recall that a function f from a group G to a group H is called a homomorphism provided that $f(xy) = f(x)f(y)$ for all $x, y \in G$. A one-to-one, onto homomorphism is called an isomorphism.

Theorem 6.5.1
Let $f: G \to H$ be a homomorphism from a group G into a group H.
(a) If i is the identity of G, then $f(i)$ is the identity of H.
(b) $f(x^{-1}) = (f(x))^{-1}$ for all $x \in G$.
(c) $f(G)$ is a subgroup of H.

Proof: Exercise 1.

Example 1:
Define the function $f: \mathbb{Z} \to \mathbb{Z}$ by $f(x) = 5x$ for all $x \in \mathbb{Z}$. Then f is a homomorphism that is one-to-one but not onto (Exercise 2).

Example 2:
Define the function $f: \mathbb{Z} \to \mathbb{Z}_5$ by $f(x) = [x]$. f is a homomorphism that is onto but not one-to-one (Exercise 3).

In view of the fact that $\mathbb{Z}_5$ is the name for $\mathbb{Z}/\langle 5 \rangle$, the homomorphism of Example 2 is typical of a general sort of homomorphism:

Theorem 6.5.2
Let H be a normal subgroup of a group G. Define the function $j: G \to G/H$ by $j(x) = Hx$ for all $x \in G$. Then j is a homomorphism of G onto G/H. It is called the natural *homomorphism onto G/H.*

Proof: Exercise 4.

To say that $f: G \to H$ is a homomorphism will also imply that G and H are groups. Similarly, to say that H is a subgroup of G will imply that G is a group.

Definition
Let $f: G \to H$ be a homomorphism and let i be the identity of H. The set $f^{-1}(i)$ is called the **kernel** of f and denoted by **ker(f)**.

Theorem 6.5.3 Let $f: G \to H$ be a homomorphism.
(a) **ker(f)** is a normal subgroup of G.
(b) **ker(f)** $= \{i\}$ iff f is one-to-one.

Proof: Exercise 5.

Theorem 6.5.4 Let $f: G \to H$ and $g: H \to K$ be homomorphisms. Then $g \circ f$ is a homomorphism.

Proof: Exercise 6.

Theorem 6.5.5 Fundamental Homomorphism Theorem: Let $f: G \to H$ be a homomorphism onto H. Let $j: G \to G/\text{ker}(f)$ be the natural homomorphism. There is a unique function $g: G/\text{ker}(f) \to H$ such that $f = g \circ j$. Further, g is an isomorphism.

Proof: Exercise 7. Diagram:

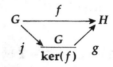

From Theorem 6.5.5 we have $H \cong G/\text{ker}(f)$. Hence any homomorphic image of a group G is essentially the same as G modulo a normal subgroup.

Theorem 6.5.6 Correspondence Theorem: Let $f: G \to H$ be a homomorphism onto H and let $K = \text{ker}(f)$.
(a) If N is a subgroup of H, then $f^{-1}(N)$ is a subgroup of G containing K. If N is normal in H, then $f^{-1}(N)$ is normal in G.
(b) If M is a subgroup of G, then $f(M)$ is a subgroup of H. If M is normal in G, then $f(M)$ is normal in H.
(c) There is a one-to-one correspondence F between subgroups of H and subgroups of G containing K. Normal subgroups correspond under F.

Proof of (a) and (b): Exercise 8.

Proof of (c):

Let $\mathcal{H}$ be the set of subgroups of H and let $\mathcal{G}$ be the set of subgroups of G containing K. Define $F: \mathcal{H} \to \mathcal{G}$ by $F(M) = f^{-1}(M)$ for all $M \in \mathcal{H}$. $F(M) \in \mathcal{G}$ by part (a).

To show that F is one-to-one, assume $F(M) = F(N)$. Let $x \in M$. Since f is onto, there exists $g \in G$ such that $f(g) = x$. Hence

$g \in f^{-1}(M) = F(M) = F(N) = f^{-1}(N)$, so that $f(g) \in N$—that is, $x \in N$. Hence $M \subseteq N$. By symmetry $N \subseteq M$, so that $M = N$. Therefore F is one-to-one.

To show that F is onto, let $M \in \mathscr{G}$. Then $f(M) \in \mathscr{H}$ by part (b). We claim that since $\ker(f) \subseteq M$, then $f^{-1}(f(M)) = M$. From this it follows that $F(f(M)) = M$, so that F is onto. To prove the claim, let $x \in f^{-1}(f(M))$. Then $f(x) \in f(M)$, so that $f(x) = f(m)$ for some $m \in M$. Whence $i = f(x)f(m)^{-1} = f(xm^{-1})$ by Theorem 6.5.1, and so $xm^{-1} \in K \subseteq M$. Therefore $xm^{-1} \cdot m \in M$; that is, $x \in M$. Thus $f^{-1}(f(M)) \subseteq M$. The reverse inclusion is clear from the definition of f^{-1}. Thus F is onto. ∎

Theorem 6.5.7 *First Isomorphism Theorem: If $N \subseteq H \subseteq G$ for subgroups N and H normal in G, then*

$$\frac{G/N}{H/N} \cong \frac{G}{H}$$

Proof: Exercise 9.

Theorem 6.5.8 *Second Isomorphism Theorem: If H and K are subgroups of a group G with K normal in G, then*

$$\frac{H}{H \cap K} \cong \frac{HK}{K}$$

Proof: Exercise 10.

In our quest to understand abstract groups, we need to consider many examples. The next theorem, due to Cayley, tells us that any abstract group G is isomorphic to a subgroup of a permutation group. Thus G can be isomorphically represented by a concrete example.

Theorem 6.5.9 *Every group is isomorphic to a subgroup of $\mathbf{B}(S)$ for some set S.*

Proof:

Let G be a group. For the set S, we take G itself as a set. By Theorem 6.3.3, multiplying elements in G by a fixed $x \in G$ gives a bijection of the set G. This bijection is called R_x—right multiplication by x. Thus $R_x \in \mathbf{B}(G)$. Define the function $f: G \to \mathbb{B}(G)$ by $f(x) = R_{x^{-1}}$—right multiplication by x^{-1}. Clearly f is onto $f(G)$.

To show that f is a homomorphism, let $x, y \in G$. Then $f(xy) = R_{(xy)^{-1}}$ and $f(x) \circ f(y) = R_{x^{-1}} \circ R_{y^{-1}}$. To show that these two bijections are equal, let $g \in G$ be arbitrary and consider

$$R_{(xy)^{-1}}(g) = g(xy)^{-1}$$
$$= gy^{-1}x^{-1}$$
$$= R_{x^{-1}}(R_{y^{-1}}(g))$$
$$= R_{x^{-1}} \circ R_{y^{-1}}(g)$$

Thus $R_{(xy)^{-1}} = R_{x^{-1}} \circ R_{y^{-1}}$, so that $f(xy) = f(x) \circ f(y)$; that is, f is a homomorphism.

Since f is a homomorphism, $f(G)$ is a subgroup of $\mathbb{B}(G)$.

To show that f is one-to-one, let $f(x) = f(y)$. Then $R_{x^{-1}} = R_{y^{-1}}$. Applying these bijections to $x \in G$ gives $i = xx^{-1} = xy^{-1}$, so that $x = y$. Therefore G is isomorphic to $f(G)$—a subgroup of $\mathbb{B}(G)$. ∎

Additional Ideas

Left multiplication can be defined from a group G to itself. Left multiplication by a fixed element $x \in G$ gives a bijection of G. In a manner analogous to our proof of Theorem 6.5.9, left multiplication can be used to define an isomorphism from G to a subgroup of $\mathbb{B}(G)$. Left multiplication by x is more natural than the right multiplication by x^{-1} used in the proof of Theorem 6.5.9. This is because we write functions on the left of their arguments. (In $f(x)$, x is the argument.) In the function composition $f \circ g$, g is applied first. This means that we read compositions from right to left if we follow their order of application. This convention is more widely used in applied mathematics than right-hand notation is.

Mathematicians who work in abstract algebra sometimes consider the composition of many functions, and it is convenient to write functions on the right because doing so enables us to read composition from left to right. Thus $f(x)$ is written xf. In group theory (part of abstract algebra), it is also customary to write coset representatives on the right of the subgroup they multiply. We have followed this custom in considering right cosets in defining factor groups instead of left cosets. The left cosets xH and yH are the same iff $y^{-1}x \in H$. That is, $x \simeq y \bmod H$ iff $y^{-1}x \in H$. This translates into $-y + x$ in additive notation.

The conventions of writing functions on the left and coset representatives on the right don't mix as well as doing both on the same side. In our approach, however, we have followed the most widely used conventions—even though they are not the most compatible.

6.5 Fundamental Theorems of Group Theory

Exercises 6.5

Problems

1. Prove Theorem 6.5.1.
2. Prove that the function $f: \mathbb{Z} \to \mathbb{Z}$ given by $f(x) = 5x$ is a one-to-one homomorphism.
3. Prove that the function $f: \mathbb{Z} \to \mathbb{Z}_5$ given by $f(x) = [x]$ is an onto homomorphism.
4. Prove Theorem 6.5.2.
5. Prove Theorem 6.5.3.
6. Prove Theorem 6.5.4.

Harder Problems

7. Prove Theorem 6.5.5 as follows: In order to simplify notation, let $K = \ker(f)$. An arbitrary element of G/K is of the form Kx for some $x \in G$. Define $g: G/K \to H$ by $g(Kx) = f(x)$.
   (a) Show g is well defined.
   (b) Show $f = g \circ j$.
   (c) Show g is the unique function such that $f = g \circ j$.
   (d) Show g is one-to-one.
   (e) Show g is onto.
   (f) Show g is a homomorphism.
8. Prove parts (a) and (b) of Theorem 6.5.6.
9. Prove Theorem 6.5.7. Define a function $G/N \to G/H$ by $Nx \to Hx$. Show that this function is a well-defined, onto homomorphism and use Theorem 6.5.5.
10. Prove Theorem 6.5.8. Use Exercise 6.4.6. Define a function
    $$KH \to \frac{H}{K \cap H}$$ by $kh \to (H \cap K)h$. Show that this function is a well-defined, onto homomorphism and use Theorem 6.5.5.

Appendix 1

Properties of Number Systems

The familiar computational properties of the real number system are used, without formal definition, in examples in Chapters 1 and 2. In Chapter 3, computational properties of the integers are used not only in examples but also in the development of definitions and theorems. In that chapter, ways of defining $\mathbb{Z}$ in terms of $\mathbb{N}$, and $\mathbb{Q}$ in terms of $\mathbb{Z}$ are given. In Chapter 4, three possible definitions of $\mathbb{R}$ in terms of $\mathbb{Q}$ are given. In turn, $\mathbb{N}$ can be defined in terms of set theory. Thus all our number systems can be defined in terms of sets. This leaves *set* as the only undefined mathematical object.

We have assumed the familiar properties of number systems in the development of the text for three reasons: (1) we wished to have number systems available as examples, (2) defining the systems at the outset presupposes precisely the mathematical sophistication we are trying to develop, and (3) it is not impossible that students have tired of a topic—the arithmetical properties of number systems—that has dominated their previous mathematical experience. In this appendix we therefore (1) indicate, very briefly, how the natural numbers can be defined in terms of sets and (2) collect those properties of the number systems we have assumed in our text and examples.

Definition A *Peano system* is a set X and a function $s: X \to X$ that together satisfy the following axioms:

(1) Axiom of Infinity: s is one-to-one but not onto.
(2) Axiom of Induction: If $Y \subseteq X$, $Y \nsubseteq s(X)$, and $s(X) \subseteq Y$, then $Y = X$. The function s is called a *successor* function.

Theorem A.1.1 Let (X, s) be a Peano system. Then $X - s(X)$ contains exactly one element.

> **Proof:**
>
> The set $X - s(X)$ is nonempty since s is not onto. Let $x \in X - s(X)$. Then $s(X) \cup \{x\}$ satisfies the hypotheses of the Axiom of Induction, so that $s(X) \cup \{x\} = X$. It follows that $\{x\} = X - s(X)$. ∎

The element x such that $\{x\} = X - s(X)$ in a Peano system (X, s) is called the *distinguished element* of (X, s).

Example 1:
The set $\{1, 2, 3, \ldots\} = \mathbb{N}$ we assumed for the natural numbers is a Peano system with distinguished element 1. The function $s: \mathbb{N} \to \mathbb{N}$ is given by $s(j) = j + 1$.

Example 2:
The set $\{12, 13, 14, \ldots\}$ is a Peano system with distinguished element 12 and successor function $s(j) = j + 1$.

Example 3:
The set $\{0, 2, 4, 6, 8, \ldots\}$ of even integers is a Peano system with distinguished element 0 and successor function $s(j) = j + 2$.

The preceding three examples all involve the set $\mathbb{N}$ of natural numbers and so are not useful for *defining* the natural numbers. Here are two examples of Peano systems constructed only from sets:

Example 4:
The set $X = \{\{\emptyset\}, \{\{\emptyset\}\}, \{\{\{\emptyset\}\}\}, \ldots\}$ is a Peano system with $\{\emptyset\}$ as distinguished element and $s: X \to X$ given by $s(j) = \{j\}$.

Example 5:
The set $X = \{\{\emptyset\}, \{\emptyset, \{\emptyset\}\}, \{\emptyset, \{\emptyset, \{\emptyset\}\}\}, \ldots\}$ is a Peano system with $\{\emptyset\}$ as distinguished element and successor function given by $s(j) = \{\emptyset, j\}$.

An *isomorphism* between two Peano systems (X, s) and (Y, t) is defined to be a bijection $f: X \to Y$, the action of which commutes with the successor functions; that is, $t(f(x)) = f(s(x))$ for all $x \in X$. Any two Peano systems can be shown to be isomorphic. We can define the natural numbers by picking an example of a Peano system involving only sets. For the sake of definiteness, we will use Example 4: Define the number 1 to be $\{\emptyset\}$. Define 2 to be $s(1) = \{1\}$, and, in general, define $n + 1$ to be $s(n) = \{n\}$. Addition and multiplication are defined, inductively, on $\mathbb{N}$ in terms of the successor function: $a + 1 = s(a)$, $a + s(n) = (a + n) + 1$; $b \cdot 1 = b$, $b \cdot s(n) = b \cdot n + b$.

Appendix 1   Properties of Number Systems

The following algebraic properties of $\mathbb{N}$ can be proved from the Peano axioms. The proofs of most theorems that follow involve a technical development that we will not consider.

Theorem A.1.2   For all $a, b, c \in \mathbb{N}$:

(a) $a + (b + c) = (a + b) + c$ (associativity of addition)
(b) $a \cdot (b \cdot c) = (a \cdot b) \cdot c$ (associativity of multiplication)
(c) $a + b = b + a$ (commutativity of addition)
(d) $a \cdot b = b \cdot a$ (commutativity of multiplication)
(e) If $a + b = a + c$, then $b = c$.   (cancellation for addition)
(f) If $a \cdot b = a \cdot c$, then $b = c$.   (cancellation for multiplication)
(g) $a \cdot (b + c) = a \cdot b + a \cdot c$ (distributivity)

Definition   Define an order $\leq$ on $\mathbb{N}$ by $a \leq b$ iff $a = b$ or $a = b + c$ for some $c \in \mathbb{N}$.

Theorem A.1.3   The set $\mathbb{N}$ is totally ordered under $\leq$, and $\leq$ is related to addition and multiplication by:

(a) $a \leq b$ iff $a + c \leq b + c$.
(b) $a \leq b$ iff $a \cdot c \leq b \cdot c$.

The algebraic and order properties of elements in $\mathbb{N}$ commonly used in computations are given in the theorems above. In addition, we have the following property of sets of natural numbers:

Theorem A.1.4   Every nonempty subset of $\mathbb{N}$ contains a *least* element *(glb)*.

We can now define $\mathbb{Z}$ to be P_Z as in Chapter 3. The subset of P_Z consisting of those classes $[(a, b)]$ with $b < a$ (in $\mathbb{N}$) is an isomorphic copy of $\mathbb{N}$. Thus we consider $\mathbb{Z}$ as containing $\mathbb{N}$ as a subset. From Theorems 3.4.4 and 3.4.5 we have:

Theorem A.1.5 $\mathbb{Z}$ is a commutative ring such that $1 \cdot a = a$ for all $a \in \mathbb{Z}$, where $1 \in \mathbb{N} \subseteq \mathbb{Z}$. Addition and multiplication, already defined on $\mathbb{N}$, agree with the operations inherited from $\mathbb{Z}$.

An order $\leq$ is defined on $P_Z = \mathbb{Z}$ in Section 3.10.

Theorem A.1.6   The relation $\leq$ already defined on $\mathbb{N}$ is the relation inherited from $\leq$ in $\mathbb{Z}$. $\mathbb{Z}$ is a totally ordered set under $\leq$ such that:

(a) For all $c \in \mathbb{Z}$: $a \leq b$ iff $a + c \leq b + c$.
(b) For all $c \in \mathbb{N}$: $a \leq b$ iff $a \cdot c \leq b \cdot c$.

(c) For all $c \in \mathbb{Z}$, one and only one of the following holds: $c = 0$; $c \in \mathbb{N}$; $-c \in \mathbb{N}$.

(d) If $-c \in \mathbb{N}$, then $a \leq b$ iff $bc \leq ac$.

(e) If $ab = ac$ and $c \neq 0$, then $b = c$.

(f) If $ab = 0$, then $a = 0$ or $b = 0$.

An element e in a ring R is said to be a **multiplicative identity,** or a **one,** provided that $e \cdot a = a$ for all $a \in R$. A multiplicative identity in a ring is unique and universally denoted by **1**. By Theorem A.1.5, this does not conflict with **1** denoting the unique nonsuccessor in $\mathbb{N}$. The identity of the additive group R is denoted by **0**. Given an element a in a ring R with a **1**, an element b such that $a \cdot b = b \cdot a = 1$ is called a **multiplicative inverse** of a. A **field** is a commutative ring with **1** such that each nonzero element has a multiplicative inverse. If we define the rational numbers $\mathbb{Q}$ to be P_Q and the real numbers $\mathbb{R}$ by any of the constructions in Chapter 4, then $\mathbb{Q}$ and $\mathbb{R}$ are fields.

Definition A field F is a ***totally ordered field*** if there is defined on F a total order $\leq$ such that:

(1) $0 < 1$

(2) If $a \leq b$, then $a + c \leq b + c$.

(3) If $a \leq b$ and $0 < c$, then $a \cdot c \leq b \cdot c$.

Both $\mathbb{R}$ and $\mathbb{Q}$ are totally ordered fields.

Definition A totally ordered field F is said to be ***complete*** if every nonempty subset of F that has an upper bound in F has a least upper bound in F.

$\mathbb{R}$ is a complete, totally ordered field. It can be shown that any two complete, totally ordered fields are isomorphic (by a continuous, additive, multiplicative, order isomorphism). It is shown in Chapter 4 that $\mathbb{Q}$ is not complete. From Chapter 5 we know, too, that $\mathbb{Q}$ and $\mathbb{R}$ have different cardinalities and so cannot be isomorphic. Therefore $\mathbb{Q}$ cannot be complete.

The properties shared by $\mathbb{R}$, $\mathbb{Q}$, and $\mathbb{Z}$, which follow from the axioms for a commutative ring, are listed in Theorem A.1.7.

Theorem A.1.7 Let R be $\mathbb{R}$, $\mathbb{Q}$, or $\mathbb{Z}$. Then

(a) There exists $0 \in R$ such that $0 + a = a$ for all $a \in R$.　　(additive identity)

(b) There exists $1 \in R$ such that $1 \cdot a = a$ for all $a \in R$.　　(multiplicative identity)

(c) For each $a \in R$, there exists a unique element $-a \in R$ such that $a + -a = 0$.　　(additive inverse)

For all $a, b, c \in \mathbb{R}$:

(d) $a + b = b + a$ *(commutativity of +)*
(e) $a \cdot b = b \cdot a$ *(commutativity of ·)*
(f) $a + (b + c) = (a + b) + c$ *(associativity of +)*
(g) $a \cdot (b \cdot c) = (a \cdot b) \cdot c$ *(associativity of ·)*
(h) $a \cdot (b + c) = (a \cdot b) + (a \cdot c)$ *(· distributes over +)*
(i) If $a + b = a + c$, then $b = c$.   *(cancellation for +)*
(j) For $c \neq 0$: if $a \cdot c = b \cdot c$, then $a = b$.   *(cancellation for ·)*
(k) $a \cdot 0 = 0$
(l) $-(-a) = a$
(m) $(-a) \cdot b = -(a \cdot b)$
(n) $(-a) \cdot (-b) = a \cdot b$

We will use juxtaposition to denote · and consider the operation · to take priority over +. Thus $(a \cdot b) + (a \cdot c)$ can be written $ab + ac$. The following theorem holds for $\mathbb{Q}$ and $\mathbb{R}$ since they are fields.

Theorem A.1.8   *Let F be $\mathbb{Q}$ or $\mathbb{R}$. Then for each element $a \neq 0$ in F, there exists a unique element $a^{-1} \in F$ such that:*

$$aa^{-1} = 1 \quad \text{(multiplicative inverse)}$$

The following theorem gives properties of elements of $\mathbb{R}$ that follow from its being a totally ordered field. The properties necessarily hold for elements in the subsets $\mathbb{Z}$ and $\mathbb{Q}$.

Theorem A.1.9   *For all $a, b, c \in \mathbb{R}$:*

(a) $0 < 1$
(b) *One and only one of the following holds:*

$a < b;\ a = b;\ a > b$ *(trichotomy)*

(c) If $a < b$ and $b < c$, then $a < c$.   *(transitivity of <)*
(d) If $a \leq b$ and $b \leq c$, then $a \leq c$.   *(transitivity of ≤)*
(e) If $a < b$, then $a + c < b + c$.
(f) If $a < b$ and $c > 0$, then $ac < bc$.
(g) If $a < b$ and $c < 0$, then $ac > bc$.
(h) If $ab = ac$ and $a \neq 0$, then $b = c$.
(i) If $ab = 0$, then $a = 0$ or $b = 0$.

Appendix 1 Properties of Number Systems

Definition *Define the binary operation of subtraction by $a - b = a + (-b)$ for all $a, b \in \mathbf{R}$.*

Definition *For $a, b \in \mathbf{R}$ and $b \neq 0$, define the binary operation of division by $a/b = a \cdot b^{-1}$.*

Our next theorem tells us how the set of rational numbers sits in $\mathbf{R}$.

Theorem A.1.10 *Density of $\mathbf{Q}$ in $\mathbf{R}$: For any real numbers $x < y$, there exists a rational number r such that $x < r < y$.*

The next theorem tells us how $\mathbf{N}$ sits in $\mathbf{R}$.

Theorem A.1.11 *Archimedean Property: For any $x \in \mathbf{R}$, there exists $n \in \mathbf{N}$ such that $x < n$.*

Definition *For any real number x, define the **absolute value of x** (denoted by $|x|$) by:*
$$|x| = x \text{ if } x \geq 0$$
$$|x| = -x \text{ if } x < 0$$

Theorem A.1.12 *Triangle Inequality: For real numbers x, y: $|x + y| \leq |x| + |y|$.*

Appendix 2

Truth Tables

A true statement is one that can be used as a proof step. If $\mathscr{P}$ is a true statement, then $\sim\mathscr{P}$ is false and conversely. There is an inference rule that allows us to write $\mathscr{P}$ **or** $\sim\mathscr{P}$ as a proof step. This means that $\mathscr{P}$ **or** $\sim\mathscr{P}$ is true regardless of the truth of $\mathscr{P}$. In general, however, the truth of most statements does depend on the truth of their constituent statements. For example, $\mathscr{P}$ **and** Q is true only if both $\mathscr{P}$ and Q are true. The rule for using **and** tells us that if $\mathscr{P}$ **and** Q is a proof step, then $\mathscr{P}$ can be written as a proof step and Q can be written as a proof step. This is symbolized by $\mathscr{P}$ **and** $Q \Rightarrow \mathscr{P}$ and $\mathscr{P}$ **and** $Q \Rightarrow Q$. The symbol "$\Rightarrow$" is part of our metalanguage. That is, "$\Rightarrow$" is used to make statements *about* proofs and can't be used in formal statements *in* proofs. Thus $\mathscr{P} \Rightarrow Q$ means that if $\mathscr{P}$ is a proof step, then Q can be written as a proof step. That is, if $\mathscr{P}$ is true, then Q is true.

The inference rules allow us to systematically determine the truth of a compound statement (like $\mathscr{P}$ **and** Q) from the truth of the constituent statements $\mathscr{P}$ and Q. First we list, in a table, all the combinations of truth and falsehood for $\mathscr{P}$ and Q:

$\mathscr{P}$	Q
T	T
T	F
F	T
F	F

The rule for proving **and** statements gives $\mathscr{P}$ and $Q \Rightarrow \mathscr{P}$ **and** Q. This means that the top row in the table can be filled out to be:

$\mathscr{P}$	Q	$\mathscr{P}$ **and** Q
T	T	T
T	F	
F	T	
F	F	

309

By the rule for using **and**, $\mathscr{P}$ **and** $Q \Rightarrow Q$. Thus if $\mathscr{P}$ **and** Q were true, then Q would be true. Since Q is false in rows 2 and 4 of the table, $\mathscr{P}$ **and** Q must be false:

$\mathscr{P}$	Q	$\mathscr{P}$ and Q
T	T	T
T	F	F
F	T	
F	F	F

Similarly, since $\mathscr{P}$ **and** $Q \Rightarrow \mathscr{P}$, we get row 3:

$\mathscr{P}$	Q	$\mathscr{P}$ and Q
T	T	T
T	F	F
F	T	F
F	F	F

This completed table is called the truth table for **and**. It completely determines the meaning of **and**, as do the inference rules. (The last paragraph on page 194 relates these two ideas.)

From the rules $\mathscr{P} \Rightarrow \mathscr{P}$ **or** Q and $Q \Rightarrow \mathscr{P}$ **or** Q, we get the first three rows of the truth table for **or** statements:

$\mathscr{P}$	Q	$\mathscr{P}$ or Q
T	T	T
T	F	T
F	T	T
F	F	F

The fourth row comes from assuming $\mathscr{P}$ **or** Q and the rule for using **or**:

> Assume $\mathscr{P}$ **or** Q.
> Case 1. Assume $\mathscr{P}$: Contradict F for $\mathscr{P}$ in row 4.
> Case 2. Assume Q: Contradict F for Q in row 4.
> $\sim(\mathscr{P}$ **or** $Q)$ (When all cases lead to a contradiction, we infer the negation of the most recently assumed statement.)

The table for **if** $\mathscr{P}$, **then** Q can be written:

$\mathscr{P}$	Q	if $\mathscr{P}$, then Q
T	T	T
T	F	F
F	T	T
F	F	T

Appendix 2 Truth Tables

The information in this table was used as the basis for justifying our inference rules for proving and using **if-then** statements (page 53).

Our definition of $\sim\mathcal{P}$ to mean that $\sim\mathcal{P}$ is true precisely when $\mathcal{P}$ is false gives the following table:

$\mathcal{P}$	$\sim\mathcal{P}$
T	F
F	T

Truth tables can be written in a nonredundant short form that allows us to write tables for complex statements (such as the example at the end of this appendix) on one page. For example, the short table for an **or** statement is formed by writing only the **or** statement on top, putting the possible values for $\mathcal{P}$ and $\mathcal{Q}$ under these two symbols, and putting the corresponding value for the **or** statement under the **or:**

$\mathcal{P}$	or	$\mathcal{Q}$
T	T	T
T	T	F
F	T	T
F	F	F

Short tables for **and** and **not** are:

$\mathcal{P}$	and	$\mathcal{Q}$
T	T	T
T	F	F
F	F	T
F	F	F

$\sim\mathcal{P}$
FT
TF

In order to write **if-then** statements in short form, we introduce a connective $\rightarrow$ between $\mathcal{P}$ and $\mathcal{Q}$. Thus $\mathcal{P} \rightarrow \mathcal{Q}$ means if $\mathcal{P}$, then $\mathcal{Q}$. Note that $\rightarrow$ is a language symbol, whereas $\Rightarrow$ is a metalanguage symbol. The short table for **if-then** is:

$\mathcal{P}$	$\rightarrow$	$\mathcal{Q}$
T	T	T
T	F	F
F	T	T
F	T	F

A statement that is always true regardless of the truth of its constituent statements is called a *tautology*. Thus $\mathcal{P}$ **or** $\sim\mathcal{P}$ is an example of a tautology. A table for $\mathcal{P}$ **or** $\sim\mathcal{P}$ is:

$\mathcal{P}$	$\sim\mathcal{P}$	$\mathcal{P}$ or $\sim\mathcal{P}$
T	F	T
F	T	T

Appendix 2   Truth Tables

In short:

𝒫	or	~𝒫
T	T	F T
F	T	T F

𝒫 **or** ~𝒫 is seen to be a tautology since below the final (top-level) connective (**or**) are all T's.

Even though the statement 𝒫 ⇒ 𝒬 is in our metalanguage and the statement 𝒫 → 𝒬 is in our language, the two statements are certainly related. In fact 𝒫 ⇒ 𝒬 is true if and only if 𝒫 → 𝒬 is true. To see this, assume 𝒫 → 𝒬; that is, **if** 𝒫, **then** 𝒬. In order to show 𝒫 ⇒ 𝒬, assume 𝒫 is true (is a proof step). By our rule for using **if-then**, 𝒬 is true (can be written as a proof step). Thus 𝒫 ⇒ 𝒬. Now assume 𝒫 ⇒ 𝒬. To show 𝒫 → 𝒬, by the rule for proving **if-then**, we need to assume 𝒫 and argue to 𝒬. But this is just what 𝒫 ⇒ 𝒬 says we can do.

It follows from the discussion above that 𝒫 **iff** 𝒬 is true if and only if 𝒫 ↔ 𝒬 is true. This forms the basis for using truth tables to show that two statements are equivalent. To show 𝒫 is equivalent to 𝒬, we show 𝒫 **iff** 𝒬 is a tautology.

The truth table for 𝒫 **iff** 𝒬 is:

𝒫	iff	𝒬
T	T	T
T	F	F
F	F	T
F	T	F

Example:
Exercises 4 and 5 of Section 3.3 involve showing the equivalence of:

(1) (𝒫 or 𝒬) and ~(𝒫 and 𝒬)

(2) (𝒫 → ~𝒬) and (~𝒫 → 𝒬)

and (3) 𝒫 iff ~𝒬

We can show (1) ↔ (3) by showing that

[(𝒫 or 𝒬) and ~(𝒫 and 𝒬)] iff [𝒫 iff ~𝒬]

is a tautology. By a truth table:

[(𝒫	or	𝒬)	and	~(𝒫	and	𝒬)]	iff	[𝒫	iff	~𝒬]
T	T	T	F	F T	T	T	T	T	F	F T
T	T	F	T	T T	F	F	T	T	T	T F
F	T	T	T	T F	F	T	T	F	T	F T
F	F	F	F	T F	F	F	T	F	F	T F

Appendix 2   Truth Tables

The following table shows (1) ⟺ (2):

[($\mathscr{P}$ or $\mathcal{Q}$)	and	~($\mathscr{P}$	and	$\mathcal{Q}$)]	iff	[($\mathscr{P}$	→	~$\mathcal{Q}$)	and	(~$\mathscr{P}$	→	$\mathcal{Q}$)]
T T T	F	F T	T	T	T	T F	FT	F	FT T T
T T F	T	T T	F	F	T	T T	TF	T	FT T F
F T T	T	T F	F	T	T	F T	FT	T	TF T T
F F F	F	T F	F	F	T	F T	TF	F	TF F F

The table showing (2) ⟺ (3) can be done as an exercise.

Appendix 3

Inference Rules

Section 1.1

def. **Set definition rule.** If an element is in a set, we may infer that it satisfies the defining property. Conversely, if it satisfies the defining property, we may infer that it is in the set.

prop. R **Computation rule.** Steps in a proof that follow from other steps by the familiar techniques of algebra applied to **R, Z,** and **N** are allowed. Justification for such steps is given, for example, as "property of the real numbers".

Section 1.2

def. **Definition rule.** Suppose some relationship has been defined. If the relationship holds (in some proof step or hypothesis), then the defining condition (only) may be inferred. Conversely, if the defining condition is true, then the relationship may be inferred.

pr. ∀ **Proving *for all* statements.** In order to prove a statement of the form **for all** x **such that** $\mathcal{P}(x)$**:** $\mathcal{Q}(x)$, assume that x is an arbitrarily chosen (general) element such that $\mathcal{P}(x)$ is true. Then establish that $\mathcal{Q}(x)$ is true.

Section 1.3

us. ∀ **Using *for all* statements.** If we know that the statement **for all** x **such that** $\mathcal{P}(x)$**:** $\mathcal{Q}(x)$ is true and if we have $\mathcal{P}(t)$ as a step in a proof for any variable t, then we may write $\mathcal{Q}(t)$ as a step in the proof.

us. or **Using *or* statements (preliminary version).** If we know that $\mathcal{P}$ **or** $\mathcal{Q}$ is true and if we can show that $\mathcal{R}$ is true assuming $\mathcal{P}$ and also that $\mathcal{R}$ is true assuming $\mathcal{Q}$, then we may infer $\mathcal{R}$ is true.

Section 1.4

def.²   **Extended definition rule.** When the statement $\mathcal{P}$ is the defining property for some definition, it is permissible to either use $\mathcal{P}$ or prove $\mathcal{P}$ (according to appropriate rules) without writing $\mathcal{P}$ itself as a step. For justification for the step inferred, give the definition and not the rule for using or proving $\mathcal{P}$.

pr. or   **Proving *or* statements.** If $\mathcal{P}$ has been established as a line in a proof, then $\mathcal{P}$ or Q may be written as a line. (Symmetrically, we may write $\mathcal{P}$ or Q if we have established Q.)

us. or   **Using *or* statements (preliminary version, second part).** If we know that $\mathcal{P}$ or Q is true and can establish a contradiction assuming $\mathcal{P}$ is true, then we know that Q is true.

us. or   **Using *or* statements.** If we know that $\mathcal{P}_1$ or $\mathcal{P}_2$ or $\cdots$ or $\mathcal{P}_k$ is true and if we prove that $\mathcal{R}$ is true in all cases that do not lead to a contradiction, then we infer that $\mathcal{R}$ is true. If all cases lead to a contradiction, then we infer the negation of the most recently assumed statement.

pr. or   **Proving *or* statements.** We may write $\mathcal{P}_1$ or $\mathcal{P}_2$ or $\cdots$ or $\mathcal{P}_k$ if we have established any one of $\mathcal{P}_1$ through $\mathcal{P}_k$.

Section 1.5

us. &   **Using *and* statements.** If $\mathcal{P}$ and Q is a step in a proof, then $\mathcal{P}$ can be written as a step and Q can be written as a step.

pr. &   **Proving *and* statements.** In order to show $\mathcal{P}$ and Q in a proof, show $\mathcal{P}$ and also show Q.

sym.   **Using symmetry.** If $\mathcal{P}(A_1, B_1, C_1, \ldots)$ is any statement that has been proved for arbitrary (that is, completely general) $A_1, B_1, C_1, \ldots$ in the hypotheses, and if $A_2, B_2, C_2, \ldots$ is any rearrangement of $A_1, B_1, C_1, \ldots$, then $\mathcal{P}(A_2, B_2, C_2, \ldots)$ is true. The foregoing also applies to variables inside a **for all** statement; that is, if **for all** $A_1, B_1, C_1, \ldots$: $\mathcal{P}(A_1, B_1, C_1, \ldots)$ is true, then **for all** $A_1, B_1, C_1, \ldots$: $\mathcal{P}(A_2, B_2, C_2, \ldots)$ is true.

Section 1.6

sub.   **Substitution.** Any name or representation of a mathematical object can be replaced by another name or representation for the same object. It is necessary to avoid using the same name for different objects.

thm. *   **Using theorems.** In order to apply a theorem to steps in a proof, find a language statement $\mathcal{P}$ equivalent to the statement of the theorem. Then $\mathcal{P}$ may be written as a new proof step or used, by substitution, to change a proof step.

Section 1.7

pr. ⇒   **Proving *if-then* statements.** In order to prove a statement of the form **if** $\mathscr{P}$, **then** $\mathscr{Q}$, assume that $\mathscr{P}$ is true and show that $\mathscr{Q}$ is true.

us. ⇒   **Using *if-then* statements.** If $\mathscr{P}$ is a step in a proof and if **if** $\mathscr{P}$, **then** $\mathscr{Q}$ is a step, then we may infer that $\mathscr{Q}$ is a step.

$\mathscr{P}$ **or** $\sim\mathscr{P}$ For any $\mathscr{P}$, $\mathscr{P}$ **or** $\sim\mathscr{P}$ is true.

pr. ⇔   **Proving equivalence.** In order to show that $\mathscr{P}$ is equivalent to $\mathscr{Q}$, first assume $\mathscr{P}$ and show $\mathscr{Q}$, and then assume $\mathscr{Q}$ and show $\mathscr{P}$.

us. ⇔   **Using equivalence.** Any statement may be substituted for an equivalent statement.

Section 1.8

def.   **Set definition rule (negation version).** If $S = \{x \mid \mathscr{P}(x)\}$, then from $x \notin S$ we may infer $\sim\mathscr{P}(x)$. Conversely, from $\sim\mathscr{P}(x)$ we may infer $x \notin S$.

def.   **Definition rule (negation version).** Suppose some relationship has been defined. If the negation of the relationship holds, then the negation of the defining condition holds. If the negation of the defining condition holds, then the negation of the relationship may be inferred.

Section 1.10

us. ∃   **Using *there exists* statements.** To use the statement $\mathscr{P}(j)$ **for some** $1 \le j \le n$ in a proof, immediately follow it with the step

     Pick $1 \le j_0 \le n$ such that $\mathscr{P}(j_0)$.

This defines the symbol j_0. The truth of both $1 \le j_0 \le n$ and $\mathscr{P}(j_0)$ may be used in the remainder of the proof (in appropriate steps).

pr. ∃   **Proving *there exists* statements (formal rule).** If $1 \le i \le n$ and $\mathscr{P}(i)$ are steps in a proof, then **for some** $1 \le j \le n$: $\mathscr{P}(j)$ can be written as a proof step.

pr. ∃   **Proving *there exists* statements (practical implementation).** To prove the statement $\mathscr{P}(j)$ **for some** $1 \le j \le n$, define j in your proof (in terms of previously defined symbols) and show that $\mathscr{P}(j)$ and $1 \le j \le n$ hold for your j.

Section 1.11

neg.   **Negation.** The negation of **for all** $1 \le i \le n$: $\mathscr{P}(i)$ is **for some** $1 \le i \le n$: $\sim\mathscr{P}(i)$.

neg.   **Negation.** The negation of **for some** $1 \le i \le n$: $\mathscr{P}(i)$ is **for all** $1 \le i \le n$: $\sim\mathscr{P}(i)$.

Section 1.12

us. ∃ **Using *for some* statements.** To use the statement $\mathcal{P}(j)$ for some $j \in \mathcal{I}$ in a proof, immediately follow it with the step

Pick $j_0 \in \mathcal{I}$ such that $\mathcal{P}(j_0)$.

This defines the symbol "j_0". The truth of $\mathcal{P}(j_0)$ and $j_0 \in \mathcal{I}$ may be used in the remainder of the proof.

pr. ∃ **Proving *for some* statements.** To prove the statement $\mathcal{P}(j)$ for some $j \in \mathcal{I}$, define j in your proof (in terms of previously defined symbols) and show that $\mathcal{P}(j)$ and $j \in \mathcal{I}$ hold for your j.

neg. **Negation.** The negation of $\mathcal{P}(i)$ for all $i \in \mathcal{I}$ is $(\sim\mathcal{P}(i))$ for some $i \in \mathcal{I}$.
The negation of $\mathcal{P}(i)$ for some $i \in \mathcal{I}$ is $(\sim\mathcal{P}(i))$ for all $i \in \mathcal{I}$.

Section 2.3

! **Uniqueness.** To show that x with property $\mathcal{P}(x)$ is **unique**, assume x_1 has property $\mathcal{P}$ and x_2 has property $\mathcal{P}$, then show $x_1 = x_2$. To use the fact that x with property $\mathcal{P}(x)$ is unique, we may infer $x = y$ from $\mathcal{P}(x)$ and $\mathcal{P}(y)$.

pr. ∀ ⇒ **Proving *for all: if-then* statements.** In order to prove a statement of the form for all $x \in A$: if $\mathcal{P}(x)$, then $\mathcal{Q}(x)$, choose an arbitrary x in A and assume $\mathcal{P}(x)$ is true for this x. (Either of $x \in A$ or $\mathcal{P}(x)$ may then be used in future steps.) Then prove that $\mathcal{Q}(x)$ is true. Analogous rules hold for more than one variable.

Meaning of free variables. If $\mathcal{P}(x, y)$ is an assertion in a proof involving previously undefined symbols "x" and "y", then x and y are taken to be arbitrarily chosen (general) elements subject only to the constraint of having $\mathcal{P}(x, y)$ make sense.

Section 2.5

Axiom of Choice. Let S be a nonempty set and let $\{S_i \mid i \in \mathcal{I}\}$ be a nonempty family of nonempty subsets of S. There exists a choice function $F: \mathcal{I} \to S$ such that $F(i) \in S_i$ for all $i \in \mathcal{I}$.

Section 3.1

Well-Ordering Axiom. Every nonempty subset of $\mathbb{N}$ has a smallest element.

Section 4.4

LUB Axiom. If a nonempty set S of real numbers has an upper bound, then S has a least upper bound.

Section 6.1

Pigeonhole Principle. If n objects are distributed among k boxes for $n > k$, then at least one box receives more than one object.

Appendix 4

Definitions

subset

Section 1.2

For sets A and B, A is a **subset** of B (written $A \subseteq B$) provided that **for all x such that $x \in A$: $x \in B$**.

union

Section 1.3

Given sets A and B, the **union** of A and B, written $A \cup B$, is defined by $A \cup B = \{x \mid x \in A \text{ or } x \in B\}$.

intersection

Section 1.5

For any sets A and B, the **intersection** of A and B is the set $A \cap B$ defined by $A \cap B = \{x \mid x \in A \text{ and } x \in B\}$.

set equality

A set A is **equal** to a set B (written $A = B$) provided that $A \subseteq B$ and $B \subseteq A$.

equivalent statements

Section 1.7

Statements $\mathcal{P}$ and $\mathcal{Q}$ in our language are called **equivalent** if it is possible to show that $\mathcal{P}$ follows from $\mathcal{Q}$ and that $\mathcal{Q}$ follows from $\mathcal{P}$ using our rules of inference and proof conventions.

relative complement

Section 1.8

For sets A and B, the **complement** of B in A (also called the **difference**) is the set $A - B$ (read "A minus B") defined by $A - B = \{x \mid x \in A \text{ and } x \notin B\}$.

complement

For any set B, $\overline{B} = \{x \in \mathsf{U} \mid x \notin B\}$.

319

Appendix 4   Definitions

Section 1.9

iff   The statement $\mathscr{P}$ **if and only if** $\mathcal{Q}$ is defined to mean the same as (**if** $\mathscr{P}$, **then** $\mathcal{Q}$) **and** (**if** $\mathcal{Q}$, **then** $\mathscr{P}$). The words "if and only if" are frequently contracted to "**iff**".

Section 1.10

Let $A_1, A_2, \ldots, A_n$ be n sets.

union $\bigcup_{i=1}^{n} A_i = \{x \mid x \in A_j \text{ for some } j \text{ such that } 1 \leq j \leq n\}$.

intersection $\bigcap_{i=1}^{n} A_i = \{x \mid x \in A_j \text{ for all } j \text{ such that } 1 \leq j \leq n\}$.

Section 1.12

Let $\mathscr{I}$ be any nonempty set whatever. For every $i \in \mathscr{I}$, suppose there is defined a set A_i. Define

union $\bigcup_{i \in \mathscr{I}} A_i = \{x \mid x \in A_i \text{ for some } i \in \mathscr{I}\}$.

intersection $\bigcap_{i \in \mathscr{I}} A_i = \{x \mid x \in A_i \text{ for all } i \in \mathscr{I}\}$.

indexed family $\{A_i \mid i \in \mathscr{I}\}$ is called an *indexed family* of sets.

Section 2.1

Let $f: A \to B$. Suppose $X \subseteq A$, $Y \subseteq B$, $y \in B$.

image $f(X) = \{b \in B \mid b = f(x) \text{ for some } x \in X\}$.

preimage $f^{-1}(Y) = \{a \in A \mid f(a) \in Y\}$.

preimage $f^{-1}(y) = \{a \in A \mid f(a) = y\}$.

power set   For any set X, the **power set** of X (denoted $\mathbb{P}(X)$) is the set of all subsets of X.

Section 2.2

composition of functions   Let $f: A \to B$, $g: B \to C$. Define $g \circ f: A \to C$ by $g \circ f(a) = g(f(a))$ for all $a \in A$.

equal functions   Two functions $f: A \to B$ and $g: A \to B$ are said to be **equal** (written $f = g$) provided that **for all** $x \in A$: $f(x) = g(x)$.

identity function   For any set A, define the function $i_A: A \to A$ by $i_A(a) = a$ for each $a \in A$. i_A is called the **identity function** on A.

Appendix 4 Definitions

Section 2.3

one-to-one A function $f: A \to B$ is called **one-to-one** provided that **for all** $a_1, a_2 \in A$: **if** $f(a_1) = f(a_2)$, **then** $a_1 = a_2$.

restriction Let $f: A \to B$ and $C \subseteq A$. The function $f|_C: C \to B$ defined by $f|_C(x) = f(x)$ for all $x \in C$ is called the **restriction of f to C**.

Section 2.4

onto A function $f: A \to B$ is called **onto** if for each $b \in B$ there exists some $a \in A$ such that $f(a) = b$.

Section 2.5

left, right inverse Let $f: A \to B$. A function $g: B \to A$ is called a **left inverse** of f if $g \circ f = i_A$ and a **right inverse** of f if $f \circ g = i_B$.

inverse Let $f: A \to B$. A function $g: B \to A$ is called an **inverse** of f if g is both a left and a right inverse of f, that is, if $g \circ f = i_A$ and $f \circ g = i_B$.

Section 2.6

bijection A function $f: A \to B$ is called a **bijection** if it is one-to-one and onto.

$\mathbb{B}(A)$ If A is any set, then the set of all bijections from A to A is denoted by $\mathbb{B}(A)$.

Sections 2.6 and 3.1

factorial For any $n \in \mathbb{N}$, the number n **factorial** is defined to be $1 \cdot 2 \cdot \cdots \cdot (n-1) \cdot n$ and is denoted by $n!$ In Section 3.1, n **factorial** is defined inductively by $0! = 1$ and $n! = n \cdot (n-1)!$.

Section 2.7

infinite A set A is **infinite** if there is a function $f: A \to A$ that is one-to-one but not onto.

finite A set that is not infinite is called **finite**.

same cardinality Sets A and B are said to be **equipotent**, or have the **same cardinality**, if there exists a bijection $f: A \to B$. We write $A \sim B$ to denote that A and B have the same cardinality.

proper subset For sets A and B, if $A \subseteq B$ but $A \neq B$, then A is called a **proper subset** of B.

smaller cardinality For sets A and B, we say that A has **smaller cardinality** than B if there exists a one-to-one function $f: A \to B$ but A and B do not have the same cardinality. We write $A < B$ to denote that A has smaller cardinality than B. We write $A \lesssim B$ to mean $A < B$ or $A \sim B$.

Section 2.8

ordered pair Let A and B be sets. For any $a \in A$, $b \in B$, the **ordered pair** (a, b) is the set $\{\{a\}, \{a, b\}\}$.

Cartesian product Let X and Y be sets. The **Cartesian product** of X and Y (denoted $X \times Y$) is the set $\{(x, y) \mid x \in X, y \in Y\}$.

function (formal definition) A function $f: A \to B$ is a triple (A, B, f), where f is a subset of $A \times B$ such that

(1) If $(x, y_1) \in f$ and $(x, y_2) \in f$, then $y_1 = y_2$

and (2) If $x \in A$, then $(x, z) \in f$ for some $z \in B$.

Here A is called the **domain** of $f: A \to B$, and B is called the **codomain**.

binary operation For any nonempty set A, a **binary operation** on A is a function from $A \times A$ to A.

Section 3.1

binomial coefficient $\binom{n}{i} = \dfrac{n \cdot (n-1) \cdot (n-2) \cdot \cdots \cdot (n-i+1)}{i!}$

Section 3.2

divides Let $a, b \in \mathbb{Z}$. We say b **divides** a iff $a = bc$ for some $c \in \mathbb{Z}$.

relation For any nonempty sets A and B, a **relation** from A to B is a triple (A, B, R) where $R \subseteq A \times B$.

equivalence relation A relation R on a set A is called an **equivalence relation** iff:

(1) for all $a \in A$: $a\,R\,a$ (reflexive property)

and (2) for all $a, b \in A$: if $a\,R\,b$, then $b\,R\,a$ (symmetric property)

and (3) for all $a, b, c \in R$: if $a\,R\,b$ and $b\,R\,c$, then $a\,R\,c$. (transitive property)

R_Z Define the relation R_Z on $\mathbb{N} \times \mathbb{N}$ by $(a, b)\,R_Z\,(c, d)$ iff $a + d = b + c$.

P_Z Define P_Z to be the set of R_Z equivalence classes (Section 3.4). Addition and multiplication on P_Z are defined by the following rules:

+ on P_Z $[(a, b)] + [(c, d)] = [(a + c, b + d)]$ and

· on P_Z $[(a, b)] \cdot [(c, d)] = [(ac + bd, ad + bc)]$ (Section 3.7).

$\leq$ on P_Z Define an order on the set P_Z by the following rule: for all $[(a, b)], [(c, d)] \in P_Z$: $[(a, b)] \leq [(c, d)]$ iff $a + d \leq b + c$ (Section 3.10).

R_Q Let $\mathbb{Z}^* = \mathbb{Z} - \{0\}$. Define the relation R_Q on $\mathbb{Z} \times \mathbb{Z}^*$ by $(a, b)\,R_Q\,(c, d)$ iff $ad = bc$.

Appendix 4 Definitions   323

P_Q — Define P_Q to be the set of R_Q equivalence classes (Section 3.4). Addition and multiplication on P_Q are defined by the following rules:

$+$ on P_Q — $[(a, b)] + [(c, d)] = [(ad + bc, bd)]$ and

$\cdot$ on P_Q — $[(a, b)] \cdot [(c, d)] = [(ac, bd)]$ (Section 3.7).

$\leq$ on P_Q — Define an order on the set P_Q by the following rule:
For $[(a, b)], [(c, d)] \in P_Q$ with b and d of the same sign:
$[(a, b)] \leq [(c, d)]$ iff $ad \leq bc$.
For $[(a, b)], [(c, d)] \in P_Q$ with b and d of opposite sign:
$[(a, b)] \leq [(c, d)]$ iff $bc \leq ad$ (Section 3.10).

rational number — A real number r is called **rational** if $r = a/b$ for some $a, b \in \mathbb{Z}$, $b \neq 0$. The set of rational numbers is denoted by $\mathbb{Q}$.

congruence modulo n — Let $n \in \mathbb{N}$. For $a, b \in \mathbb{Z}$, define a to be **congruent to b modulo n** (written $a \equiv b \bmod n$) iff $n | (a - b)$.

Section 3.3

equivalence class — Let A be a set and R an equivalence relation on A. For each $a \in A$, define $[a] = \{b \in A \mid b\, R\, a\}$. $[a]$ is called the **equivalence class of a** defined by R.

partition — Let A be a nonempty set and $\{A_i \mid i \in \mathscr{I}\}$ a family of nonempty subsets of A. The family is called a **partition** of A iff

(1) $A = \bigcup\limits_{i \in \mathscr{I}} A_i$

and (2) $A_i \cap A_j = \emptyset$ for $A_i \neq A_j$.

induced partition — If R is an equivalence relation on A, then the set of equivalence classes defined by R is a partition of A called the partition **induced** by R.

induced relation — Let $\{A_i \mid i \in \mathscr{I}\}$ be any partition of a set A. Define the relation R on A by: for $a, b \in A$, $a\, R\, b$ iff a and b are in the same set A_i. Then R is an equivalence relation on A. It is called the relation **induced** by the partition $\{A_i \mid i \in \mathscr{I}\}$.

Section 3.4

well-defined operation — Whenever an operation is defined in terms of representations for the elements being operated on, we must show that the rule specifying the operation gives the same value regardless of the different ways in which each element is represented. We say in this case that the operation is **well defined**.

$\mathbb{Z}_n$ — For any $n \in \mathbb{N}$, the set of congruence classes modulo n is denoted by $\mathbb{Z}_n$. Addition is defined on $\mathbb{Z}_n$ by $[a] + [b] = [a + b]$, multiplication by $[a] \cdot [b] = [ab]$.

Section 3.5

group — A nonempty set G with a binary operation $\cdot$ is called a **group** provided that:

(1) For all $x, y, z \in G$: $x \cdot (y \cdot z) = (x \cdot y) \cdot z$.

Appendix 4 Definitions

	and (2) There exists $i \in G$ such that for all $x \in G$: $i \cdot x = x$ and $x \cdot i = x$.
	(3) For each $x \in G$, there exists $y \in G$ such that $x \cdot y = i$ and $y \cdot x = i$.
Abelian	A group G is called **Abelian** provided that
	(4) $x \cdot y = y \cdot x$ for all $x, y \in G$.
identity element	Suppose $(G, \cdot)$ is a group. An element $i \in G$ that has the property $i \cdot x = x$ and $x \cdot i = x$ for all $x \in G$ is called an **identity element**.
group isomorphism	Let $(G, \oplus)$ and $(H, \otimes)$ be groups. A function $f: G \to H$ is a **group isomorphism** provided that:
	(1) f is one-to-one and onto.
	and (2) $f(x \oplus y) = f(x) \otimes f(y)$ for all $x, y \in G$.
isomorphic groups	Groups G and H are called **isomorphic** (denoted by $G \cong H$) if there is an isomorphism $f: G \to H$.
group homomorphism	A function f from a group $(G, \oplus)$ to a group $(H, \otimes)$ is called a **group homomorphism** provided that $f(x \oplus y) = f(x) \otimes f(y)$ for all $x, y \in G$.
ring	A set R with binary operations $+$ and $\cdot$ is called a **ring** provided that:
	(1) $(R, +)$ is an Abelian group.
	(2) For all $x, y, z \in R$: $x \cdot (y \cdot z) = (x \cdot y) \cdot z$.
	(3) For all $x, y, z \in R$: $x \cdot (y + z) = x \cdot y + x \cdot z$ and $(y + z) \cdot x = y \cdot x + z \cdot x$.
	R is called **commutative** if, in addition:
	(4) For all $x, y \in R$: $x \cdot y = y \cdot x$.
ring homomorphism	Let $(R, +, \cdot)$ and $(S, +, \cdot)$ be rings. A function $f: R \to S$ is a **ring homomorphism** provided that for all $x, y \in R$:
	(1) $f(x + y) = f(x) + f(y)$
	and (2) $f(x \cdot y) = f(x) \cdot f(y)$.
	A ring **isomorphism** is a one-to-one, onto ring homomorphism.

Section 3.6

additive, multiplicative homomorphism	A function $f: \mathbb{Z} \to \mathbb{Z}$ is called an **additive homomorphism** iff for all $a, b \in \mathbb{Z}$: $f(a + b) = f(a) + f(b)$. f is called a **multiplicative homomorphism** iff for all $a, b \in \mathbb{Z}$: $f(a \cdot b) = f(a) \cdot f(b)$.
zero-function	The function $\zeta: \mathbb{Z} \to \mathbb{Z}$ defined by $\zeta(x) = 0$ for all $x \in X$ is called the **zero-function**.

function respects multiplication	For any nonempty subset M of $\mathbb{Z}$, a function $f: \mathbb{Z} \to \mathbb{Z}$ **respects multiplication by M** provided that **for all $z \in \mathbb{Z}$, $m \in M$: $f(mz) = mf(z)$**.
closure under addition	A subset $X \subseteq \mathbb{Z}$ is called **closed under addition** if for all $x, y \in X$, we have $x + y \in X$.

Section 3.7

well-defined function	Whenever a function f is defined in terms of representations for the elements in its domain, we must show that the rule specifying the function gives the same value regardless of the different ways in which we represent each element. That is, if a and b are different representations for the same element, then we must show that $f(a) = f(b)$. We say in this case that the function f is **well defined**.

Section 3.8

ideal	A subset H of $\mathbb{Z}$ is called an **ideal** of $\mathbb{Z}$ provided that for all $a, b \in \mathbb{Z}$:

      (1) if $a, b \in H$, then $a + b \in H$.
and (2) if $z \in \mathbb{Z}$ and $a \in H$, then $za \in H$.

ideal $\mathbb{Z}a$	For $a \in \mathbb{Z}$, define $\mathbb{Z}a = \{za \mid z \in \mathbb{Z}\}$.
ideal sum	For $A, B \subseteq \mathbb{Z}$, define $A + B = \{a + b \mid a \in A, b \in B\}$.

Section 3.9

greatest common divisor	Let $a, b \in \mathbb{Z}$. An element $c \in \mathbb{N}$ is called a **greatest common divisor** of a and b provided that

      (1) $c \mid a$ and $c \mid b$.
and (2) for all $d \in \mathbb{Z}$: if $d \mid a$ and $d \mid b$, then $d \mid c$.

relatively prime	The numbers $a, b \in \mathbb{Z}$ are called **relatively prime** if $1 = \gcd(a, b)$.
prime	Let $p \in \mathbb{N}$, $p \neq 1$. Then p is called **prime** provided that for any $n \in \mathbb{Z}$: if $n \mid p$, then $n = \pm 1$ or $n = \pm p$.

Section 3.10

order	A relation $\leq$ on a set P is called a **partial order** on P provided that for all $x, y, z \in P$:

  (1) $x \leq x$ (reflexive property)
  (2) if $x \leq y$ and $y \leq x$, then $x = y$.   (antisymmetric property)
  (3) if $x \leq y$ and $y \leq z$, then $x \leq z$.   (transitive property)

poset	If $\leq$ is a partial order on P, then P is called a **partially ordered set** (or **poset**) under $\leq$.

Appendix 4 Definitions

comparable   If $x \leq y$ or $y \leq x$, then x and y are said to be **comparable**.

total order   A partial order $\leq$ on a set P is called a **total order** (or **linear order** or **chain**) if every two elements of P are comparable.

upper bound   Let P be a poset under $\leq$ and let $S \subseteq P$. An element $m \in P$ is called an **upper bound** for S provided that **for all $x \in S$: $x \leq m$**.

lub   An element $l \in P$ is called a **least upper bound** for S provided that

        (1) l is an upper bound for S

and   (2)   for all m such that m is an upper bound for S: $l \leq m$.

lower bound   Let P be a poset under $\leq$ and let $S \subseteq P$. An element $m \in P$ is called a **lower bound** for S provided that **for all $x \in S$: $m \leq x$**.

glb   An element $g \in P$ is called a **greatest lower bound** for S provided that

        (1) g is a lower bound for S

and   (2)   for all m such that m is a lower bound for S: $m \leq g$.

least common multiple   Let $a, b \in \mathbb{Z}$. An element $c \in \mathbb{N}$ is called a **least common multiple** of a and b provided that

        (1) $a|c$ and $b|c$

and   (2)   for all $d \in \mathbb{Z}$: if $a|d$ and $b|d$, then $c|d$.

<   If P is a poset under $\leq$, we define the relation $<$ on P by: for all $x, y \in P$: $x < y$ iff $x \leq y$ and $x \neq y$.

lattice   A poset in which each pair of elements has both a greatest lower bound and a least upper bound is called a **lattice**.

maximal element   An element x is a **maximal element** of a poset P provided that for all $y \in P$: if $x \leq y$, then $x = y$.

minimal   (Dually, x is a **minimal element** provided that $y \leq x$ implies $y = x$.)

nondecreasing   Let T and S be totally ordered sets with orders both denoted by $\leq$. A function $f: T \to S$ is called **nondecreasing** provided that for all $x, y \in T$: if $x < y$, then $f(x) \leq f(y)$.

increasing   The function f is called **increasing** provided that for all $x, y \in T$: if $x < y$, then $f(x) < f(y)$.

order isomorphism   A bijection $f: T \to S$ is called an **order isomorphism** if one of the following equivalent conditions holds:

(1) f is increasing.

(2)   For all $x, y \in T$: $x < y$ iff $f(x) < f(y)$.

(3)   For all $x, y \in T$: $x < y$ iff $f(x) \leq f(y)$.

Appendix 4 Definitions

Section 4.1

sequence A **sequence** is a function $a: \mathbb{N} \to \mathbb{R}$. $a(n)$ is called the nth term of the sequence and is denoted a_n. The sequence itself is denoted by $\langle a_1, a_2, a_3, \ldots \rangle$ or merely by $\langle a_n \rangle$.

limit of a sequence A number $L \in \mathbb{R}$ is defined to be the **limit** of the sequence $\langle a_1, a_2, a_3, \ldots \rangle$ (written $\lim\limits_{n \to \infty} \langle a_n \rangle = L$ or merely $a_n \to L$) provided that, given a real number $\epsilon > 0$, there exists a natural number N such that **for all** $n > N$: $|a_n - L| < \epsilon$. If L is the limit of $\langle a_n \rangle$, we say $\langle a_n \rangle$ **converges** to L. A sequence with no limit is said to **diverge**. Let $\langle a_n \rangle$ and $\langle b_n \rangle$ be sequences and $k \in \mathbb{R}$.

$k\langle a_n \rangle$ Define the sequence $k\langle a_n \rangle = \langle ka_n \rangle$.

$\langle a_n \rangle + \langle b_n \rangle$ Define the sequence $\langle a_n \rangle + \langle b_n \rangle = \langle a_n + b_n \rangle$.

$\langle a_n \rangle - \langle b_n \rangle$ Define the sequence $\langle a_n \rangle - \langle b_n \rangle = \langle a_n \rangle + (-1)\langle b_n \rangle$.

Section 4.2

limit of a function Let $f: D \to \mathbb{R}$ be a function and $a \in D$. A number $L \in \mathbb{R}$ is defined to be the **limit** of $f(x)$ as x approaches a (written $L = \lim\limits_{x \to a} f(x)$) provided that, given a real number $\epsilon > 0$, there exists a real number $\delta > 0$ such that if $0 < |x - a| < \delta$, then $|f(x) - L| < \epsilon$.

Let $f: D \to \mathbb{R}$, $g: D \to \mathbb{R}$, and $k \in \mathbb{R}$.

$kf: D \to \mathbb{R}$ Define the function $kf: D \to \mathbb{R}$ by $(kf)(x) = k(f(x))$ for all $x \in D$.

$f+g: D \to \mathbb{R}$ Define the function $f+g: D \to \mathbb{R}$ by $(f+g)(x) = f(x) + g(x)$ for all $x \in D$.

Section 4.3

continuous A function $f: D \to \mathbb{R}$ is **continuous** at a real number $c \in D$ provided that $\lim\limits_{x \to c} f(x) = f(c)$. If $f: D \to \mathbb{R}$ and $S \subseteq D$, then f is **continuous on** S if f is continuous at every $a \in S$. If f is continuous on D, we merely say that $f: D \to \mathbb{R}$ is **continuous**.

neighborhood Given a real number a and a real number r, the **neighborhood about** a **of radius** r is the set $N(a, r) = \{x \in \mathbb{R} \mid |x - a| < r\}$.

Section 4.5

Cauchy sequence A sequence $\langle x_n \rangle$ is a **Cauchy** sequence provided that, given $\epsilon > 0$, there exists $N \in \mathbb{N}$ such that for all $m, n > N$: $|x_m - x_n| < \epsilon$.

bounded sequence A sequence $\langle x_n \rangle$ is **bounded** provided that there exists a number M such that $|x_n| \leq M$ for all n.

A sequence $\langle x_n \rangle$ is called

nondecreasing	(1)	**nondecreasing** provided that if $m < n$, then $x_m \leq x_n$.
increasing	(2)	**increasing** provided that if $m < n$, then $x_m < x_n$.
nonincreasing	(3)	**nonincreasing** provided that if $m < n$, then $x_m \geq x_n$.
decreasing	(4)	**decreasing** provided that if $m < n$, then $x_m > x_n$.
monotonic	(5)	**monotonic** if it is either nondecreasing or nonincreasing.
null sequence		A sequence $\langle x_n \rangle$ is called a **null sequence** provided that $x_n \to 0$.
$\simeq$		Define a relation $\simeq$ between convergent sequences by $\langle a_n \rangle \simeq \langle b_n \rangle$ iff $\langle a_n \rangle - \langle b_n \rangle$ is a null sequence.

Section 4.6

$f \cdot g : D \to \mathbb{R}$ Let $f : D \to \mathbb{R}$ and $g : D \to \mathbb{R}$. Define the function $f \cdot g : D \to \mathbb{R}$ by $(f \cdot g)(x) = f(x) \cdot g(x)$ for all $x \in D$.

$\frac{1}{f} : D \to \mathbb{R}$ Let $f : D \to \mathbb{R}$ have the property that $f(x) \neq 0$ for all $x \in D$. Define the function $\frac{1}{f} : D \to \mathbb{R}$ by $\frac{1}{f}(x) = \frac{1}{f(x)}$ for all $x \in D$.

bounded function   A function $f : D \to \mathbb{R}$ is said to be **bounded on a subset** S of D provided that there exists a number M such that $|f(x)| \leq M$ for all $x \in S$.

Section 4.7

accumulation point   A real number a is called an **accumulation point** of a subset S of $\mathbb{R}$ if for all neighborhoods $N(a, r)$ of a: $[N(a, r) - \{a\}] \cap S \neq \emptyset$.

topologically closed   A subset S of $\mathbb{R}$ is **(topologically) closed** if each accumulation point of S is in S.

topologically open   A subset S of $\mathbb{R}$ is called **(topologically) open** if $\mathbb{R} - S$ is closed.

interior point   An element a of a subset S of $\mathbb{R}$ is called an **interior point** of S if there is some neighborhood $N(a, r) \subseteq S$.

Section 5.1

cardinality   The **cardinality** of the empty set is 0. The **cardinality** of a nonempty finite set S is k provided that $S \sim \mathbb{N}_k$. We denote the cardinality of S by $|S|$.

Section 5.2

$\aleph_0$ The **cardinality** of a set X is defined to be $\aleph_0$ ("aleph null") provided that $X \sim \mathbb{N}$. In this case, we denote $|X| = \aleph_0$.

Appendix 4   Definitions

$2^{\aleph_0}$ — The **cardinality** of a set X is defined to be $2^{\aleph_0}$ provided that $X \sim \mathbb{P}(\mathbb{N})$. In this case, we denote $|X| = 2^{\aleph_0}$.

countable — A set X is called **countably infinite** if has the same cardinality as $\mathbb{N}$. It is called **countable** if it is either finite or countably infinite and **uncountable** if it is not countable.

countable union — If A_i, $i \in \mathscr{I}$, is a family of sets and $\mathscr{I}$ is countable, then $\bigcup_{i \in \mathscr{I}} a_i$ is called a **countable union**.

Section 6.1

subgroup — A nonempty subset H of a group G is a **subgroup** of G if H is a group under the operation inherited from G.

group generator — Let G be a group and let $x \in G$. The set $\{x^k \mid k \in \mathbb{Z}\}$ is called the **subgroup of G generated by** x and is denoted by $\langle x \rangle$.

cyclic — A group G is called **cyclic** if there is an element $x \in G$ such that every element in G is of the form x^k for some $k \in \mathbb{Z}$. Thus G is cyclic provided that $G = \langle x \rangle$ for some $x \in G$.

order of an element — The **order** of an element x of G is the least positive integer n such that $x^n = i$ if such an integer exists. If no such positive integer exists, then x is of infinite order.

order of a group — If a group G is finite, then by the **order** of G we mean the number of elements in G. If G is not finite, then G is said to be of infinite order.

Section 6.2

set of generators — Let G be a group. A subset S of G is called a **set of generators** for G if every element in G can be written as a product of elements of S and their inverses.

Section 6.3

coset — Let G be a group and let H be a subgroup of G. For any $x \in G$, define $Hx = \{hx \mid h \in H\}$. Then Hx is called the **right coset** of H determined by x.

factor group — If G is a group and H is a subgroup of G, denote by G/H the group of right cosets of H in G. If H is normal in G, then G/H is a group called the **factor** or **quotient** group of G modulo H.

index — For a subgroup H of a group G, the number of right cosets of H in G is called the **index of H in G**.

congruence modulo a subgroup — Let H be a subgroup of a group G. For $x, y \in G$, define x to be **congruent to y modulo H** (denoted $x \simeq y \bmod H$) if $xy^{-1} \in H$.

Section 6.4

S · T   For subsets S and T of a group G, define $S \cdot T = \{st \mid s \in S, t \in T\}$.

normal subgroup   A subgroup H of a group G is called **normal** in G provided that $Ha = aH$ for all $a \in G$.

Section 6.5

kernel   Let $f: G \to H$ be a homomorphism and let i be the identity of H. The set $f^{-1}(i)$ is called the **kernel** of f and denoted by **ker(f)**.

Appendix 5

Theorems

1.3.1	Suppose A, B, and C are sets. If $A \subseteq B$ and $B \subseteq C$, then $A \subseteq C$.
1.3.2	For sets A, B, and C, if $A \subseteq C$ and $B \subseteq C$, then $A \cup B \subseteq C$.
1.4.1	For sets A, B, and C: if $A \subseteq B$ or $A \subseteq C$, then $A \subseteq B \cup C$.
1.5.1	For sets A, B, and C, if $A \subseteq B$ and $A \subseteq C$, then $A \subseteq B \cap C$.
1.5.2	For sets A and B, (a) $A \cap B = B \cap A$ (b) $A \cup B = B \cup A$
1.6.1	For sets A, B, and C, (a) $A \cap (B \cap C) = (A \cap B) \cap C$ (b) $A \cup (B \cup C) = (A \cup B) \cup C$
1.6.2	For sets A, B, and C, (a) $A \cap (B \cup C) = (A \cap B) \cup (A \cap C)$ (b) $A \cup (B \cap C) = (A \cup B) \cap (A \cup C)$
1.7.1	Let A, B, and C be sets. If $A \subseteq B$, then $A \cap C \subseteq B \cap C$.
1.7.2L	The statements $\mathcal{P}$ **or** $\mathcal{Q}$ and **if** $\sim\mathcal{P}$, **then** $\mathcal{Q}$ are equivalent.
1.7.3L	The statement $\mathcal{P}$ **or** $\mathcal{Q}$ is equivalent to the statement $\mathcal{Q}$ **or** $\mathcal{P}$.
1.7.4L	The statement $\mathcal{P}$ **and** $\mathcal{Q}$ is equivalent to the statement $\mathcal{Q}$ **and** $\mathcal{P}$.
1.7.5L	The statement **if** $\mathcal{P}$, **then** $\mathcal{Q}$ is true if and only if $\mathcal{P} \Rightarrow \mathcal{Q}$.
1.7.6L	The following are equivalent: (a) $(\mathcal{P}$ **or** $\mathcal{Q})$ **or** $\mathcal{R}$

(b) $\mathscr{P}$ or ($\mathscr{Q}$ or $\mathscr{R}$)
(c) $\mathscr{P}$ or $\mathscr{Q}$ or $\mathscr{R}$

1.7.7L The following are equivalent:
(a) ($\mathscr{P}$ and $\mathscr{Q}$) and $\mathscr{R}$
(b) $\mathscr{P}$ and ($\mathscr{Q}$ and $\mathscr{R}$)
(c) $\mathscr{P}$ and $\mathscr{Q}$ and $\mathscr{R}$

1.8.1 For sets A, B, and C:
(a) $A - (B \cup C) = (A - B) \cap (A - C)$
(b) $A - (B \cap C) = (A - B) \cup (A - C)$

1.8.2 For sets B and C:
(a) $\overline{B \cup C} = \overline{B} \cap \overline{C}$
(b) $\overline{B \cap C} = \overline{B} \cup \overline{C}$

1.8.3 For any sets A and B: $A - B = A \cap \overline{B}$

1.8.4L The statement $\sim(\mathscr{P}$ or $\mathscr{Q})$ is equivalent to $\sim\mathscr{P}$ and $\sim\mathscr{Q}$.

1.8.5L The statement $\sim(\mathscr{P}$ and $\mathscr{Q})$ is equivalent to $\sim\mathscr{P}$ or $\sim\mathscr{Q}$.

1.8.6L The statement $\sim$(if $\mathscr{P}$, then $\mathscr{Q}$) is equivalent to $\mathscr{P}$ and $\sim\mathscr{Q}$.

1.8.7L If $\mathscr{P}$ is equivalent to $\mathscr{Q}$, then $\sim\mathscr{P}$ is equivalent to $\sim\mathscr{Q}$.

1.8.8L The statement $\mathscr{P}$ or $\mathscr{Q}$ or $\mathscr{R}$ is equivalent to **if $\sim\mathscr{P}$ and $\sim\mathscr{Q}$, then $\mathscr{R}$.**

1.9.1 For sets A, B, and C:
(a) $A \subseteq B$ iff $A \cap B = A$
(b) $A \subseteq B$ iff $A \cup B = B$

1.9.2L The statement $\mathscr{P}$ iff $\mathscr{Q}$ is true if and only if $\mathscr{P}$ is equivalent to $\mathscr{Q}$.

1.9.3 For sets A and B, $A = B$ if and only if **for all x: $x \in A$ iff $x \in B$.**

1.10.1 Let $B, A_1, A_2, \ldots, A_n$ be sets.

(a) If $B \subseteq A_i$ for all $1 \leq i \leq n$, then $B \subseteq \bigcap_{i=1}^{n} A_i$.

(b) If $A_i \subseteq B$ for all $1 \leq i \leq n$, then $\bigcup_{i=1}^{n} A_i \subseteq B$.

1.10.2 Let $A_1, A_2, \ldots, A_n$ be sets.

(a) **for all $1 \leq j \leq n$:** $\bigcap_{i=1}^{n} A_i \subseteq A_j$

(b) **for all $1 \leq j \leq n$:** $A_j \subseteq \bigcup_{i=1}^{n} A_i$

Appendix 5 Theorems

1.10.3L If from $\mathscr{P}(i)$ and $\mathscr{Q}(i)$ we may infer $\mathscr{R}(i)$, then

from $\quad \mathscr{P}(i)$ for all $1 \le i \le n$
and $\quad \mathscr{Q}(i)$ for all $1 \le i \le n$
we may infer $\quad \mathscr{R}(i)$ for all $1 \le i \le n$

1.10.4L Let $\mathscr{P}(i)$ and $\mathscr{Q}(i)$ be language statements involving the free variable i. If $\mathscr{P}(i) \Rightarrow \mathscr{Q}(i)$, then $\mathscr{P}(i)$ for all $1 \le i \le n \Rightarrow \mathscr{Q}(i)$ for all $1 \le i \le n$.

1.11.1 Let $B, A_1, A_2, \ldots, A_n$ be sets.

(a) $B - \bigcup_{i=1}^{n} A_i = \bigcap_{i=1}^{n} (B - A_i)$

(b) $B - \bigcap_{i=1}^{n} A_i = \bigcup_{i=1}^{n} (B - A_i)$

1.12.1 Let A_i, $i \in \mathscr{I}$, be an indexed family of sets and let B be any set. Then

(a) $B - \bigcup_{i \in \mathscr{I}} A_i = \bigcap_{i \in \mathscr{I}} (B - A_i)$

(b) $B - \bigcap_{i \in \mathscr{I}} A = \bigcup_{i \in \mathscr{I}} (B - A_i)$

2.1.1 Let $f: A \to B$, $X, Y \subseteq A$ and $W, Z \subseteq B$.

(a) If $x \in X$, then $f(x) \in f(X)$.
(b) If $z \in Z$, then $f^{-1}(z) \subseteq f^{-1}(Z)$.
(c) If $X \subseteq Y$, then $f(X) \subseteq f(Y)$.
(d) If $W \subseteq Z$, then $f^{-1}(W) \subseteq f^{-1}(Z)$.
(e) If $c, d \in B$, $c \ne d$, then $f^{-1}(c) \cap f^{-1}(d) = \varnothing$.

2.1.2 Let $f: A \to B$, $C \subseteq B$, and $D \subseteq B$.

(a) $f^{-1}(C \cap D) = f^{-1}(C) \cap f^{-1}(D)$
(b) $f^{-1}(C \cup D) = f^{-1}(C) \cup f^{-1}(D)$

2.1.4 Let $f: A \to B$, $E \subseteq A$, and $F \subseteq A$.

(a) $f(E \cap F) \subseteq f(E) \cap (F)$
(b) $f(E \cup F) = f(E) \cup f(F)$

2.2.1 Let $f: A \to B$, $g: B \to C$, and $h: C \to D$. Then $(h \circ g) \circ f = h \circ (g \circ f)$.

2.2.2 For any $f: A \to B$:

(a) $f \circ i_A = f$
(b) $i_B \circ f = f$

2.3.1 Let $f: A \to B$. Then f is one-to-one iff for each b in the range of f there exists a unique $a \in A$ such that $f(a) = b$.

2.3.2	Let $f: A \to B$ and $g: B \to C$ be one-to-one functions. Then $g \circ f$ is one-to-one.
2.3.4	Let $f: A \to B$ and $g: B \to C$ be functions. If $g \circ f$ is one-to-one, then f is one-to-one.
2.3.5	If $f: A \to B$ is one-to-one and $C \subseteq A$, then $f|_C: C \to B$ is one-to-one.
2.3.6	A function $f: A \to B$ is one-to-one iff for all $a_1, a_2 \in A$: if $a_1 \neq a_2$, then $f(a_1) \neq f(a_2)$.
2.3.7L	A statement and its contrapositive are equivalent.
2.4.1	If $f: A \to B$ and $g: B \to C$ are onto, then $g \circ f$ is onto.
2.4.2	Let $f: A \to B$ and $g: B \to C$ be functions. If $g \circ f$ is one-to-one and f is onto, then g is one-to-one.
2.5.1	Let $f: A \to B$. Then f has a left inverse iff f is one-to-one.
2.5.1 (lemma)	For any set A, $i_A: A \to A$ is one-to-one.
2.5.2	Let $f: A \to B$. Then f has a right inverse iff f is onto.
2.5.3	If $f: A \to B$ has an inverse, then this inverse is unique.
2.5.4	A function $f: A \to B$ has an inverse iff f is one-to-one and onto.
2.5.5	If $f: A \to B$ is one-to-one and onto, then f^{-1} is one-to-one and onto.
2.5.6	If $f: A \to B$ has a left inverse g_1 and a right inverse g_2, then $g_1 = g_2$.
2.6.1	Let A be any set and $\mathbb{B}(A)$ the set of all bijections from A to A. Then (a) For all $f, g \in \mathbb{B}(A)$: $f \circ g \in \mathbb{B}(A)$. (b) For all $f, g, h \in \mathbb{B}(A)$: $f \circ (g \circ h) = (f \circ g) \circ h$. (c) There exists $i \in \mathbb{B}(A)$ such that for all $f \in \mathbb{B}(A)$: $i \circ f = f$ and $f \circ i = f$. (d) For each $f \in \mathbb{B}(A)$, there exists $g \in \mathbb{B}(A)$ such that $f \circ g = i$ and $g \circ f = i$.
2.7.1	There exists a one-to-one function $g: \mathbb{N} \to \mathbb{N}$ that is not onto.
2.7.1 (restated)	$\mathbb{N}$ is infinite.
2.7.2	If A is infinite and $A \subseteq B$, then B is infinite.
2.7.3	If B is finite and $A \subseteq B$, then A is finite.
2.7.4	For all sets A, B, and C: (a) $A \sim A$ (reflexive property) (b) If $A \sim B$, then $B \sim A$. (symmetric property) (c) If $A \sim B$ and $B \sim C$, then $A \sim C$. (transitive property)
2.7.5	If A is infinite and $A \sim B$, then B is infinite.
2.7.6	$\mathbb{Z} \sim \mathbb{N}$.

Appendix 5   Theorems

2.7.7   If A is infinite and $A \lesssim B$, then B is infinite.

2.7.8   If A is infinite and $A < B$, then B is infinite.

2.7.9   If B is finite and $A \lesssim B$, then A is finite.

2.7.10 For all sets A, B, and C: if $A \lesssim B$ and $B \lesssim C$, then $A \lesssim C$.

2.8.1   For $a_1, a_2 \in A$ and $b_1, b_2 \in B$, we have:
$(a_1, b_1) = (a_2, b_2)$ iff both $a_1 = a_2$ and $b_1 = b_2$.

2.8.2   For $a_1, a_2 \in A$, $b_1, b_2 \in B$, $c_1, c_2 \in C$, we have $(a_1, b_1, c_1) = (a_2, b_2, c_2)$ iff $a_1 = a_2$ and $b_1 = b_2$ and $c_1 = c_2$.

2.8.3   Two functions $f: A \to B$ and $g: A \to B$ are equal iff **for all** $x \in A$: $f(x) = g(x)$.

3.1.1   For each $k \in \mathbb{N}$, the set $\mathbb{N}_k$ is finite.

3.1.2   Let S be a subset of $\mathbb{N}$ such that

   (1) $1 \in S$

and   (2) **for all** $n \in \mathbb{N}$: **if** $n \in S$, **then** $n + 1 \in S$.

Then $S = \mathbb{N}$.

3.1.3   Suppose $k \in \mathbb{Z}$ and $S \subseteq \{k, k+1, k+2, \ldots\}$ is such that

   (1) $k \in S$

and   (2) **for all** $n \in \{k, k+1, k+2, \ldots\}$: **if** $n \in S$, **then** $n+1 \in S$.

Then $S = \{k, k+1, k+2, \ldots\}$.

3.1.4   Let S be a nonempty finite set. Then $S \sim \mathbb{N}_k$ for some $k \in \mathbb{N}$.

3.1.5   A set S is finite iff $S \sim \mathbb{N}_k$ for some natural number k.

3.1.6   Let $a, b \in \mathbb{R}$, $n \in \mathbb{N}$. Then

$$(a+b)^n = a^n + \binom{n}{1}a^{n-1}b + \binom{n}{2}a^{n-2}b^2 + \cdots + \binom{n}{1}a \cdot b^{n-1} + b^n.$$ That is,

$$(a+b)^n = \sum_{i=0}^{n} \binom{n}{i} a^{n-i} b^i.$$

Division Algorithm   Let $a, b \in \mathbb{Z}$ with $b > 0$. There exist unique integers q and r such that $a = bq + r$ and $0 \leq r < b$. q is called the **quotient** and r the **remainder** upon dividing a by b.

3.2.1   Let $a, b, c \in \mathbb{Z}$.

   (a) If $b \mid a$ and $b \mid c$, then $b \mid (a+c)$.

   (b) If $b \mid a$, then $b \mid ac$.

3.2.2   Congruence modulo n is an equivalence relation.

336 Appendix 5 Theorems

3.2.3   (a) If $a \equiv b \bmod n$ and $c \in \mathbb{Z}$, then $a + c \equiv b + c \bmod n$.
        (b) If $a \equiv b \bmod n$ and $c \in \mathbb{Z}$, then $ac \equiv bc \bmod n$.

3.3.1   Let R be an equivalence relation on A and $a, b \in A$. Then $[a] = [b]$ iff $a R b$.

3.3.2   Let $[a]$ and $[b]$ be two equivalence classes for the equivalence relation of congruence modulo n. Then $[a] = [b]$ iff $a \equiv b \bmod n$.

3.3.3   Let A be a nonempty set and R an equivalence relation on A. For $a, b \in A$, either $[a] \cap [b] = \emptyset$ or $[a] = [b]$ but not both.

3.3.4   If R is an equivalence relation on A, then the set of equivalence classes defined by R is a partition of A. It is called the partition *induced* by R.

3.3.5   For any positive integer n, the set of congruence classes modulo n forms a partition of $\mathbb{Z}$. There are n such classes, given by $[0], [1], [2], \ldots, [n-1]$.

3.3.6   Let $\{A_i | i \in \mathcal{I}\}$ be any partition of a set A. Define the relation R on A by: for $a, b \in A$, $a R b$ iff a and b are in the same set A_i. Then R is an equivalence relation on A. It is called the relation *induced* by the partition $\{A_i | i \in \mathcal{I}\}$.

3.4.1   (a) Addition on $\mathbb{Z}_n$ is well defined.
        (b) Multiplication on $\mathbb{Z}_n$ is well defined.

3.4.2   (a) For all $x, y \in \mathbb{Z}_n$: $x + y \in \mathbb{Z}_n$.
        (b) For all $x, y, z \in \mathbb{Z}_n$: $x + (y + z) = (x + y) + z$.
        (c) There exists $i \in \mathbb{Z}_n$ such that for all $x \in \mathbb{Z}_n$: $i + x = x$ and $x + i = x$.
        (d) For each $x \in \mathbb{Z}_n$, there exists $y \in \mathbb{Z}_n$ such that $x + y = i$ and $y + x = i$.
        (e) For all $x, y \in \mathbb{Z}_n$: $x + y = y + x$.

3.4.3   (a) For all $x, y \in \mathbb{Z}_n$: $x \cdot y \in \mathbb{Z}_n$.
        (b) For all $x, y, z \in \mathbb{Z}_n$: $x \cdot (y \cdot z) = (x \cdot y) \cdot z$.
        (c) For all $x, y, z \in \mathbb{Z}_n$: $x \cdot (y + z) = x \cdot y + x \cdot z$ and $(y + z) \cdot x = y \cdot x + z \cdot x$.
        (d) For all $x, y \in \mathbb{Z}_n$: $x \cdot y = y \cdot x$.

3.4.4   (a) For all $x, y \in P_Z$: $x + y \in P_Z$.
        (b) For all $x, y, z \in P_Z$: $x + (y + z) = (x + y) + z$.
        (c) There exists $i \in P_Z$ such that for all $x \in P_Z$: $i + x = x$ and $x + i = x$.
        (d) For each $x \in P_Z$, there exists $y \in P_Z$ such that $x + y = i$ and $y + x = i$.
        (e) For all $x, y \in P_Z$: $x + y = y + x$.

3.4.5   (a) For all $x, y \in P_Z$: $x \cdot y \in P_Z$.
        (b) For all $x, y, z \in P_Z$: $x \cdot (y \cdot z) = (x \cdot y) \cdot z$.
        (c) For all $x, y, z \in P_Z$: $x \cdot (y + z) = x \cdot y + x \cdot z$ and $(y + z) \cdot x = y \cdot x + z \cdot x$.
        (d) For all $x, y \in P_Z$: $x \cdot y = y \cdot x$.

Appendix 5 Theorems

3.4.6
(a) For all $x, y \in P_Q$: $x + y \in P_Q$.
(b) For all $x, y, z \in P_Q$: $x + (y + z) = (x + y) + z$.
(c) There exists $i \in P_Q$ such that for all $x \in P_Q$: $i + x = x$ and $x + i = x$.
(d) For each $x \in P_Q$, there exists $y \in P_Q$ such that $x + y = i$ and $y + x = i$.
(e) For all $x, y \in P_Q$: $x + y = y + x$.

3.4.7
(a) For all $x, y \in P_Q$: $x \cdot y \in P_Q$.
(b) For all $x, y, z \in P_Q$: $x \cdot (y \cdot z) = (x \cdot y) \cdot z$.
(c) For all $x, y, z \in P_Q$: $x \cdot (y + z) = x \cdot y + x \cdot z$ and $(y + z) \cdot x = y \cdot x + z \cdot x$.
(d) For all $x, y \in P_Q$: $x \cdot y = y \cdot x$.

3.5.1 A group has a unique identity element (denoted by i).

3.5.2 Let G be a group with identity i, and let $e \in G$. If $x \cdot e = x$ or $e \cdot x = x$ for some $x \in G$, then $e = i$.

3.5.3 Let G be a group with identity i. For each element $x \in G$, there is a unique element $y \in G$ such that $x \cdot y = i$ and $y \cdot x = i$.

3.5.4 If x and y are elements of a group and if $x \cdot y = i$, then $y = x^{-1}$ and $x = y^{-1}$.

3.5.5 Let x and y be elements of a group. Then
(a) $(x^{-1})^{-1} = x$
(b) $(x \cdot y)^{-1} = y^{-1} \cdot x^{-1}$

3.6.1 If $f: \mathbb{Z} \to \mathbb{Z}$ and $g: \mathbb{Z} \to \mathbb{Z}$ are additive homomorphisms, then $g \circ f: \mathbb{Z} \to \mathbb{Z}$ is an additive homomorphism.

3.6.2 Let $f: \mathbb{Z} \to \mathbb{Z}$ be an additive homomorphism.
(a) $f(0) = 0$
(b) $f(-a) = -f(a)$ for all $a \in \mathbb{Z}$
(c) $f(a - b) = f(a) - f(b)$ for all $a, b \in \mathbb{Z}$

3.6.3 The zero function $\zeta: \mathbb{Z} \to \mathbb{Z}$ is an additive and multiplicative homomorphism.

3.6.4 If $f: \mathbb{Z} \to \mathbb{Z}$ is an additive homomorphism, then f respects multiplication by $\mathbb{N}$.

3.6.5 If $f: \mathbb{Z} \to \mathbb{Z}$ is an additive homomorphism, then f respects multiplication by $\mathbb{Z}$.

3.6.6 The only functions $f: \mathbb{Z} \to \mathbb{Z}$ that are both additive and multiplicative homomorphisms are the identity $i_\mathbb{Z}$ and the zero-function.

3.6.7 If X and Y are closed under addition, then so is $X \cap Y$.

3.6.8 If $f: \mathbb{Z} \to \mathbb{Z}$ is an additive homomorphism and X is closed under addition, then
(a) $f(X)$ is closed under addition.
(b) $f^{-1}(X)$ is closed under addition.

338 Appendix 5 Theorems

3.7.1 Let $n \in \mathbb{N}$ be given and $f: \mathbb{Z} \to \mathbb{Z}$ be an (additive) homomorphism. Define the function $\bar{f}: \mathbb{Z}_n \to \mathbb{Z}_n$ by $\bar{f}([a]) = [f(a)]$. Then
(a) $\bar{f}$ is well defined.
(b) $\bar{f}$ is a homomorphism (that is, **for all** $a, b \in \mathbb{Z}$: $\bar{f}([a] + [b]) = \bar{f}([a]) + \bar{f}([b])$).

3.7.2 Let $f: P_\mathbb{Z} \to \mathbb{Z}$ be defined by $f([(a, b)]) = a - b$. Then f is a ring isomorphism.

3.7.3 Let $f: P_\mathbb{Q} \to \mathbb{Q}$ be defined by $f([(a, b)]) = a/b$. Then f is a ring isomorphism.

3.8.1
(a) If H and J are ideals of $\mathbb{Z}$, then so is $H \cap J$.
(b) For any $a \in \mathbb{Z}$, $\mathbb{Z}a$ is an ideal of $\mathbb{Z}$.
(c) For all $a, b \in \mathbb{Z}$: $\mathbb{Z}a \subseteq \mathbb{Z}b$ iff $b \mid a$.
(d) If H and J are ideals of $\mathbb{Z}$, then so is $H + J$.
(e) For all $a, b \in \mathbb{Z}$: $\mathbb{Z}a + \mathbb{Z}b$ is an ideal of $\mathbb{Z}$.

3.8.2 For any ideal H of $\mathbb{Z}$: $H = \mathbb{Z}a$ for some $a \geq 0$.

3.8.3 For any $a, b \in \mathbb{Z}$, there exists $c \geq 0$ such that $\mathbb{Z}a + \mathbb{Z}b = \mathbb{Z}c$.

3.9.1 If a and b have a greatest common divisor, then it is unique.

3.9.2 For any $a, b \in \mathbb{Z}$ (not both zero), $\gcd(a, b)$ exists and has the form $\gcd(a, b) = ma + nb$ for some $m, n \in \mathbb{Z}$.

3.9.3 Let p be prime and $n \in \mathbb{Z}$. If p does not divide n, then p and n are relatively prime.

3.9.4
(a) If $\gcd(a, b) = 1$ and $a \mid bc$, then $a \mid c$.
(b) If p is prime and $p \mid bc$, then $p \mid b$ or $p \mid c$.
(c) If p is prime and $p \mid b_1 b_2 \cdots b_k$, then $p \mid b_i$ for some $1 \leq i \leq k$.

3.9.5 If $ac \equiv bc \bmod n$ and if c and n are relatively prime, then $a \equiv b \bmod n$.

3.9.6 Let $S \subseteq \mathbb{N}$ with the following property: **for all** $n \in \mathbb{N}$: **if** $\{1, \ldots, n-1\} \subseteq S$, **then** $n \in S$. Then $S = \mathbb{N}$.

3.9.7 Any integer a greater than 1 has a unique factorization $a = p_1^{n_1} p_2^{n_2} \cdots p_k^{n_k}$ for primes $p_1 < p_2 < \cdots < p_k$ and $n_i \in \mathbb{N}$.

3.10.1 Let P be a poset under $\leq$ and let $S \subseteq P$. If x and y are least upper bounds for S, then $x = y$.

3.10.2 Let P be a poset under $\leq$ and let $S \subseteq P$. If x and y are greatest lower bounds for S, then $x = y$.

3.10.3 Let x, y, and z be elements of a poset P. If $x < y$ and $y < z$, then $x < z$.

3.10.4 Let T be a totally ordered set. For any $x, y \in T$, precisely one of the following holds:

Appendix 5 Theorems   339

(a) $x < y$
(b) $x = y$
(c) $y < x$

3.10.5 Let T and S be totally ordered sets with orders both denoted by $\leq$ and $f: T \to S$ one-to-one. Then the following conditions are equivalent for all $x, y \in T$:

(a) f is increasing.
(b) $x < y$ iff $f(x) < f(y)$.
(c) $x \leq y$ iff $f(x) \leq f(y)$.

3.10.6 The function $f: P_Z \to \mathbb{Z}$, given by $f([a, b]) = a - b$, is an order isomorphism.

3.10.7 The function $f: P_Q \to \mathbb{Q}$, given by $f([a, b]) = \dfrac{a}{b}$, is an order isomorphism.

4.1.1 Let $\langle a_n \rangle$ be a sequence. If $a_n \to L$ and $a_n \to M$, then $L = M$.

4.1.2 Let $\langle a_n \rangle$ and $\langle b_n \rangle$ be sequences and $k \in \mathbb{R}$.

(a) If $L = \lim_{n \to \infty} \langle a_n \rangle$, then $kL = \lim_{n \to \infty} k\langle a_n \rangle$.
(b) If $L = \lim_{n \to \infty} \langle a_n \rangle$ and $M = \lim_{n \to \infty} \langle b_n \rangle$, then $L + M = \lim_{n \to \infty} \langle a_n \rangle + \langle b_n \rangle$.

4.2.1 If $\lim_{x \to a} f(x) = L$ and $\lim_{x \to a} f(x) = M$, then $L = M$.

4.2.2 Let $f, g: D \to \mathbb{R}, k \in \mathbb{R}$.

(a) If $L = \lim_{x \to a} f(x)$, then $kL = \lim_{x \to a} (kf)(x)$.
(b) If $L = \lim_{x \to a} f(x)$ and $M = \lim_{x \to a} g(x)$, then $L + M = \lim_{x \to a} (f+g)(x)$.

4.3.1 If $f, g: D \to \mathbb{R}$ are continuous at c and if $k \in \mathbb{R}$, then

(a) $f+g: D \to \mathbb{R}$ is continuous at c
(b) $kf: D \to \mathbb{R}$ is continuous at c

4.3.2 If $f, g: D \to \mathbb{R}$ are continuous and $k \in \mathbb{R}$, then

(a) $f+g: D \to \mathbb{R}$ is continuous
(b) $kf: D \to \mathbb{R}$ is continuous

4.3.3 Let $f: C \to \mathbb{R}$ be continuous at c and let $f(c) = d$. Suppose that $f(C) \subseteq D$. Let $g: D \to \mathbb{R}$ be continuous at d. Then $g \circ f: C \to \mathbb{R}$ is continuous at c.

4.3.4 Let $f: C \to \mathbb{R}$ be continuous and $f(C) \subseteq D$. Let $g: D \to \mathbb{R}$ be continuous. Then $g \circ f: C \to \mathbb{R}$ is continuous.

4.3.5 If $f: D \to \mathbb{R}$ is continuous, $a \in D$, and $f(a) > 0$, then there exists a neighborhood $N(a, r)$ about a of some radius $r > 0$ such that $f(x) > 0$ for all $x \in N(a, r)$.

Appendix 5 Theorems

4.3.6 If $f: D \to \mathbf{R}$ is continuous, $a \in D$, and $f(a) < 0$, then there exist $r > 0$ and a neighborhood $N(a, r)$ about a such that $f(x) < 0$ for all $x \in N(a, r)$.

4.4.1 Suppose $f: D \to \mathbf{R}$ is continuous and $a, b \in D$ with $a < b$. If $f(a) < 0$ and $f(b) > 0$, then there exists some $c \in (a, b)$ such that $f(c) = 0$.

4.4.2 Suppose $f: D \to \mathbf{R}$ is continuous and $a, b \in D$ with $a < b$. If C is a real number such that $f(a) < C < f(b)$ or $f(a) > C > f(b)$, then there exists $c \in (a, b)$ such that $f(c) = C$.

4.4.3 Suppose A and B are subsets of $\mathbf{R}$ such that

    (1) $x < y$ for all $x \in A$ and $y \in B$

and (2) given $\epsilon > 0$, there exist $x \in A$ and $y \in B$ such that $|x - y| < \epsilon$.

Then there exists a unique number c such that $x \leq c \leq y$ for all $x \in A$ and $y \in B$.

4.5.1 If $\langle x_n \rangle$ converges, then $\langle x_n \rangle$ is a Cauchy sequence.

4.5.2 If $\langle a_n \rangle$ and $\langle b_n \rangle$ are Cauchy sequences and if $k \in \mathbf{R}$, then $\langle a_n \rangle + \langle b_n \rangle$ and $k \langle a_n \rangle$ are Cauchy sequences.

4.5.3 If $\langle x_n \rangle$ is a Cauchy sequence, then $\langle x_n \rangle$ converges.

4.5.4 A bounded monotonic sequence converges.

4.5.5 The relation $\simeq$ is an equivalence relation on the set of convergent sequences.

4.5.6 Let $\mathscr{C}$ denote the set of $\simeq$ equivalence classes of convergent sequences. Let $f: \mathscr{C} \to \mathbf{R}$ be defined by $f([\langle x_n \rangle]) = \lim_{n \to \infty} x_n$. Then f is a bijection.

4.6.1 Let $f, g: D \to \mathbf{R}$ and $a \in D$. If $L = \lim_{x \to a} f(x)$ and $M = \lim_{x \to a} g(x)$, then $LM = \lim_{x \to a} f \cdot g(x)$.

4.6.2 Let $f, g: D \to \mathbf{R}$ be continuous. Then $f \cdot g$ is continuous.

4.6.3 Let $f: D \to \mathbf{R}$ be such that $f(x) \neq 0$ for all $x \in D$. Let $a \in D$. If $\lim_{x \to a} f(x) = L \neq 0$, then $\lim_{x \to a} \frac{1}{f}(x) = \frac{1}{L}$.

4.6.4 Let $f: D \to \mathbf{R}$ be continuous and such that $f(x) \neq 0$ for all $x \in D$. Then $\frac{1}{f}: D \to \mathbf{R}$ is continuous.

4.6.5 If $f: D \to \mathbf{R}$ is continuous and if $[a, b] \subseteq D$, then $f([a, b])$ has an upper bound.

4.6.6 If $f: D \to \mathbf{R}$ is continuous and if $[a, b] \subseteq D$, then f achieves a maximum value on $[a, b]$. That is, there is some $c \in [a, b]$ such that $f(x) \leq f(c)$ for all $x \in [a, b]$.

4.6.7 If $f: D \to \mathbf{R}$ is continuous and if $[a, b] \subseteq D$, then f is bounded on $[a, b]$.

Appendix 5 Theorems   341

4.6.8 If $f: D \to \mathbf{R}$ is continuous and if $[a, b] \subseteq D$, then f achieves maximum and minimum values on $[a, b]$.

4.7.1 If S and T are two closed subsets of $\mathbf{R}$, then $S \cap T$ is closed.

4.7.2 A subset S of $\mathbf{R}$ is open iff every point of S is an interior point of S.

4.7.3 If S and T are two open subsets of $\mathbf{R}$, then $S \cap T$ is open.

4.7.4 Let S_i, $i = 1, \ldots, n$, be a finite family of open subsets of $\mathbf{R}$. Then $\bigcap_{i=1}^{n} S_i$ is open.

4.7.5 Let S_i, $i = 1, \ldots, n$, be a finite family of closed subsets of $\mathbf{R}$. Then $\bigcup_{i=1}^{n} S_i$ is closed.

4.7.6 Let S_i, $i \in \mathscr{I}$, be an arbitrary family of open subsets of $\mathbf{R}$. Then $\bigcup_{i \in \mathscr{I}} S_i$ is open.

4.7.7 Let S_i, $i \in \mathscr{I}$, be an arbitrary family of closed subsets of $\mathbf{R}$. Then $\bigcap_{i \in \mathscr{I}} S_i$ is closed.

4.7.8 Let $f: \mathbf{R} \to \mathbf{R}$ be continuous and let $T \subseteq \mathbf{R}$. Then
(a) If T is open, so is $f^{-1}(T)$.
(b) If T is closed, so is $f^{-1}(T)$.

4.7.9 Let $f: \mathbf{R} \to \mathbf{R}$ be a function. Then f is continuous if either of the following conditions holds:
(1) For all $T \subseteq \mathbf{R}$: if T is open, then $f^{-1}(T)$ is open.
(2) For all $T \subseteq \mathbf{R}$: if T is closed, then $f^{-1}(T)$ is closed.

5.1.1 For any set X, there is no onto function $f: X \to \mathbf{P}(X)$.

5.1.2 Let $f: A \to B$ and $g: B \to A$ be one-to-one functions. Then there exists a one-to-one, onto function $h: A \to B$.

5.1.3 For sets A and B: if $A \lesssim B$ and $B \lesssim A$, then $A \sim B$.

5.1.4 For sets A, B, and C: if $A < B$ and $B < C$, then $A < C$.

5.1.5 For finite sets S and T:
(a) $|S| = |T|$ iff $S \sim T$
(b) $|S| < |T|$ iff $S < T$
(c) $|S| \leq |T|$ iff $S \lesssim T$

5.2.1 For sets A and B, if $A \sim B$, then $\mathbf{P}(A) \sim \mathbf{P}(B)$.

5.2.2 For all $k \in \mathbf{N}$:
(a) $|\mathbf{P}(\mathbf{N}_k)| = 2^k$
(b) $k < 2^k$

5.2.3	If X is a finite set with cardinality k, then $\mathbb{P}(X)$ has cardinality 2^k.
5.2.4	Let A and B be sets. (a) If $A \subseteq B$, then $A \preceq B$. (b) If there exists an onto function $f: A \twoheadrightarrow B$, then $B \preceq A$.
5.2.5	A subset of a countable set is countable.
5.2.6	Let X be a set. (a) X is countable iff there exists a one-to-one function $f: X \to \mathbb{N}$. (b) X is countable iff there exists an onto function $g: \mathbb{N} \twoheadrightarrow X$. (c) X is countable iff $X \preceq \mathbb{N}$.
5.2.7	Suppose A_i, $i \in \mathscr{I}$, is a family of sets and that each A_i and $\mathscr{I}$ are countable. Then $\bigcup_{i \in \mathscr{I}} A_i$ is countable. That is, a countable union of countable sets is countable.
5.2.8 (lemma)	If A and B are finite, then $A \cup B$ is finite.
5.2.8	Suppose A_i, $i \in \mathscr{I}$, is a family of sets and that each A_i and $\mathscr{I}$ are finite. Then $\bigcup_{i \in \mathscr{I}} A_i$ is finite. That is, a finite union of finite sets is finite.
5.2.9	Suppose S and T are sets for which the notion of cardinality has been defined. Then (a) $\|S\| = \|T\|$ iff $S \sim T$ (S and T have the same cardinality) (b) $\|S\| < \|T\|$ iff $S < T$ (S has smaller cardinality than T) (c) $\|S\| \leq \|T\|$ iff $S \preceq T$
6.1.1	If G is a group and if H is a nonempty subset of G, then H is a subgroup of G iff (a) for all $x, y \in H$: $xy \in H$ and (b) for all $x \in H$: $x^{-1} \in H$.
6.1.2	Let G be a group and let H be a nonempty subset of G. Then H is a subgroup of G iff for all $x, y \in H$: $xy^{-1} \in H$.
6.1.3	Let G be a group and let x be an element of G. The following are equivalent: (a) $\langle x \rangle$ is a finite set. (b) x is of finite order. (c) x^{-1} is equal to x^m for a non-negative integer m.
6.1.4	Let G be a group and let x be an element of G. Then the order of x is equal to the order of $\langle x \rangle$.
6.1.5	Let G be a group and let H be a nonempty finite subset of G. Then H is a subgroup of G iff for all $x, y \in H$: $xy \in H$.

Appendix 5 Theorems

6.1.6	If H and J are subgroups of a group G, then $H \cap J$ is a subgroup of G.
6.3.1	If G is a finite group and if H is a subgroup of G, then the order of H divides the order of G.
6.3.2	Let G be a finite group of order m and let x in G have order n. Then (a) $n \mid m$ (b) $x^m = i$ (c) If m is prime, then G is cyclic.
6.3.3	Let $R_x: G \to G$ be right multiplication by x on a group G. Then (a) R_x is one-to-one and onto. (b) If $S \subseteq G$, then $R_x(S) = \{sx \mid s \in S\}$, which we call Sx. (c) If H is a subgroup of G, then $H \sim Hx$ (H and Hx have the same cardinality).
6.3.4	If H is a subgroup of a group G, then $\dfrac{G}{H}$ partitions G.
6.3.5	Let H be a subgroup of a finite group G. The order of G is the product of the order of H and the index of H in G.
6.3.6	For any subsets S and T and elements x and y of a group G: (a) $S(xy) = (Sx)y$ and $(xy)S = x(yS)$ (b) $S \subseteq T$ iff $Sx \subseteq Tx$ (c) $S \subseteq T$ iff $xS \subseteq xT$
6.3.7	Let G be a group and let H be a subgroup of G. For all $x, y \in G$: (a) $x \in H$ iff $H = Hx$ (b) $Hx = Hy$ iff $Hxy^{-1} = H$ iff $H = Hyx^{-1}$ (c) $Hx = Hy$ iff $xy^{-1} \in H$
6.3.8	Let H be a subgroup of a group G. (a) Congruence modulo H is an equivalence relation on G. (b) The $\simeq$ congruence classes are the right cosets of H in G.
6.3.9	Let n be a positive integer. Then (a) For $a, b \in \mathbb{Z}: a \equiv b \bmod n$ iff $a \simeq b \bmod \langle n \rangle$. (b) For $c \in \mathbb{Z}$, the $\equiv$ congruence class $[c]$ is the same as the coset $\langle n \rangle + c$ of $\langle n \rangle$ in $\mathbb{Z}$ (with the coset given in additive notation).
6.3.10	If $m, n \in \mathbb{N}$ with $(m, n) = 1$, then $m^{\varphi(n)} \equiv 1 \bmod n$.
6.3.11	If p is prime and $m \in \mathbb{Z}$, then $m^p \equiv m \bmod p$.

6.4.1 Let G be a group, $S, T \subseteq G$, and H a subgroup of G.
  (a) $H \cdot H = H$
  (b) For all $x \in G$: $(xS) \cdot T = x(S \cdot T)$
  $(Sx) \cdot T = S \cdot (xT)$
  $S \cdot (Tx) = (S \cdot T)x$

6.4.2 Define multiplication on G/H, the set of right cosets of H in G, by $Hx \cdot Hy = Hxy$ for all $x, y \in G$. Multiplication is well defined iff $Ha = aH$ for all $a \in G$.

6.4.3 The following conditions are equivalent to H being a normal subgroup of a group G:
  (a) $a^{-1}Ha = H$ for all $a \in G$.
  (b) $a^{-1}Ha \subseteq H$ for all $a \in G$.
  (c) For all $a \in G, h \in H$: $a^{-1}ha \in H$.
  (d) For all $a \in G, h \in H$: $ha = ak$ for some $k \in H$.
  (e) Each right coset of H is some left coset.
  (f) Each left coset of H is some right coset.

6.4.4 Let H and K be normal subgroups of G. Then
  (a) $H \cap K$ is a normal subgroup of G.
  (b) $HK = KH$
  (c) HK is a normal subgroup of G.
  (d) $HK = \langle H \cup K \rangle$

6.4.5 If H is a normal subgroup of a group G, then G/H is a group called the **quotient group** or **factor group** of G with respect to H, or simply G **modulo** H.

6.4.6 If H is of index 2 in a group G, then H is normal in G.

6.5.1 Let $f: G \to H$ be a homomorphism from a group G into a group H.
  (a) If i is the identity of G, then $f(i)$ is the identity of H.
  (b) $f(x^{-1}) = (f(x))^{-1}$ for all $x \in G$.
  (c) $f(G)$ is a subgroup of H.

6.5.2 Let H be a normal subgroup of a group G. Define the function $j: G \to G/H$ by $j(x) = Hx$ for all $x \in G$. Then j is a homomorphism of G onto G/H. It is called the *natural* homomorphism onto G/H.

6.5.3 Let $f: G \to H$ be a homomorphism.
  (a) $\ker(f)$ is a normal subgroup of G.
  (b) $\ker(f) = \{i\}$ iff f is one-to-one.

6.5.4 Let $f: G \to H$ and $g: H \to K$ be homomorphisms. Then $g \circ f$ is a homomorphism.

Appendix 5 Theorems

6.5.5 Let $f: G \to H$ be a homomorphism onto H. Let $j: G \to \dfrac{G}{\ker(f)}$ be the natural homomorphism. There is a unique function $g: \dfrac{G}{\ker(f)} \to H$ such that $f = g \circ j$. Further, g is an isomorphism.

6.5.6 Let $f: G \to H$ be a homomorphism onto H and let $K = \ker(f)$.

(a) If N is a subgroup of H, then $f^{-1}(N)$ is a subgroup of G containing K. If N is normal in H, then $f^{-1}(N)$ is normal in G.

(b) If M is a subgroup of G, then $f(M)$ is a subgroup of H. If M is normal in G, then $f(M)$ is normal in H.

(c) There is a one-to-one correspondence F between subgroups of H and subgroups of G containing K. Normal subgroups correspond under F.

6.5.7 If $N \subseteq H \subseteq G$ for subgroups N and H normal in G, then

$$\frac{G/N}{H/N} \cong \frac{G}{H}$$

6.5.8 If H and K are subgroups of a group G with K normal in G, then

$$\frac{H}{H \cap K} \cong \frac{HK}{K}$$

6.5.9 Every group is isomorphic to a subgroup of $\mathbf{B}(S)$ for some set S.

Appendix 6

A Sample Syllabus

Class	Problems for Assignment (* = extra credit)		
1.	1.1: all	21.	2.4: 1, 4, 5(a), 11*
2.	1.2: all	22.	2.5: 1, 2, 3
3.	1.3: 1, 2	23.	2.5: 4, 5, 6
4.	1.3: 3	24.	2.6: 1, 2; 2.7: 1
5.	1.4: 1–4	25.	2.7: 2, 6*
6.	1.5: 1–7	26.	2.8: 1, 3, 4, 6(a) (7 in class)
7.	1.6: 1–4	27.	3.1: 1(b) and (c)
8.	1.7: 1, 2, 3, 5	28.	3.1: 3, 4
9.	1.7: 4; 1.8: 1, 3	29.	3.2: 2, 4, 5, 8, 9
10.	1.8: 2, 6, 8	30.	3.3: 1, 3
11.	1.9: 1	31.	3.4: 1, 3, 4
12.	1.10: 1, 3	32.	3.4: 5, 6, 7
13.	1.11: 1, 2	33.	3.5: 2, 3, 4
14.	1.12: 1, 7(b) and (e)	34.	3.6: 1, 2, 4
15.	2.1: 1, 2, 6, 7	35.	3.6: 7, 8, 10
16.	2.1: 8, 9(f)	36.	3.7: 1, 2, 3
17.	2.1: 9(g) and (h) (9(e) in class)	37.	3.8: 1, 3, 4
18.	2.2: 1, 2, 3, 9	38.	3.9: 1–5
19.	2.3: 1, 3, 4	39.	3.10: 1–6
20.	2.3: 5, 6, 7		

Omit "Additional Proof Ideas" in all sections.

Answers to Practice Exercises

Chapter 1

Section 1.1

1. Define $A = \{z \in R \mid z \neq 1\}$.
   (a) Suppose we are given Step 4 and the reason for Step 5 in some proof:

   4. $b \in A$
   5. $b \neq 1$ (4; def. A)

   What *must* Step 5 be?

   (b) Suppose we are given the following:

   4. $b \neq 1$
   5. $b \in A$ (4; def. A)

   What *must* Step 4 be?

2. Suppose we are given the following:

   5. $x \in B$
   6. $x \leq 7$ (5; def. B)
   7. $x \leq 8$ (6; prop. $\mathbb{R}$)
   8. $x \in C$ (7; def. C)

   What *must* the definitions of sets B and C be?
   $B = \{t \in \mathbb{R} \mid t \leq 7\}$. $C = \{t \in \mathbb{R} \mid t \leq 8\}$.

3. Suppose we are given the following:

   5. $1 \leq a \leq 3$
   6. $a \in X$ (5; def. X)
   7. $b \in X$
   8. $1 \leq b \leq 3$ (7; def. X)

   What must the definition of X be? What must Step 8 be?
   $X = \{t \mid 1 \leq t \leq 3\}$.

4. Each of the following metalanguage statements contains pieces that are language statements. For each metalanguage statement, decide which language statements we may assume in a proof and which we need to show. Label them appropriately.

(a) For all real numbers x, y, z, if $x < y$ and $y < z$, then $x < z$.

Assume: x, y, z real
$x < y$
$y < z$
Show: $x < z$

(b) Let A and B be sets. Suppose $x \in A$. Prove $x \in B$.

Assume: A, B sets
$x \in A$
Show: $x \in B$

(c) Let $C = \{y \in Z \mid y \neq 0\}$. If $a = 1$, then $a = C$.

Assume: $C = \{y \in Z \mid y \neq 0\}$
$a = 1$
Show: $a \in C$

(d) If $x \leq y \leq z$, then $y \leq 2$.

Assume: $x \leq y \leq z$
Show: $y \leq 2$

(e) For sets X and Y: if $x \in X$, then $x \in Y$.

Assume: X, Y sets
$x \in X$
Show: $x \in Y$

5. What metalanguage statements might be interpreted by the following hypotheses and conclusions?

(a)

Assume: $x, y \in \mathbb{R}$
$x > 0$
$y > 0$
Show: $xy > 0$

Example: If $x > 0$ and $y > 0$ for $x, y \in \mathbb{R}$, then $xy > 0$.

(b)

Assume: A, B sets
$x \notin A$
Show: $x \in B$

Answers to Practice Exercises   351

Example: Let A and B be sets. If $x \notin A$, then $x \in B$.

(c)
  Assume: $x, y \in \mathbb{R}$
      $0 < x < y$
  Show: $x^2 < y^2$

Example: Let $x, y \in \mathbb{R}$. If $0 < x < y$, then $x^2 > y^2$.

Section 1.2

1. In the proof steps below, what must Step 5 be?

 (a) 1. Let x be arbitrary such that $x < 1$.
   2. ...
   3. ...
   4. $x \notin C$
   5. **for all $x < 1$: $x \notin C$**   (1—4; pr. $\forall$)

 (b) 1. $X = \{x \in \mathbb{Z} | x < 0\}$
   2. ...
   3. ...
   4. $y \in X$
   5. **$y < 0$**   (1, 4; def. X)

Note: From Steps 1 and 4 we also know that $y \in \mathbb{Z}$. This is understood but not written down as a proof step. Only the defining *property* of X, applied to y, is written down—not the universal set from whose elements X is constructed.

 (c) 1. ...
   2. ...
   3. ...
   4. $S \subseteq T$
   5. **for all $x \in S$: $x \in T$**   (4; def. $\subseteq$)

 (d) 1. ...
   2. ...
   3. ...
   4. for all $t \in X$: $t \in Y$
   5. **$X \subseteq Y$**   (4; def. $\subseteq$)

 (e) 1. Let $t \in B$ for an arbitrary t.
    ·
    ·
   5. $t \in X$
   6. **for all $t \in B$: $t \in X$**   (1—5; pr. $\forall$)

(f) 1. $t \in N$
2. ...
3. ...
4. $t \in A$
5. Let p be arbitrary such that $p \notin T$.
6. ...
7. $p \in S$.
8. for all p such that $p \notin T$: $p \in S$ (5—7; pr. ∀)

2. In each case below, give the justification for Step 5.

(a) 1. Let $x \in A$ be arbitrary.
2. ...
3. ...
4. $x \in B$
5. for all $x \in A$: $x \in B$ (1—4; pr. ∀)

(b) 1. ...
2. for all $t \in A$: $t \in B$
3. $x \in A$
4. $x \in B$
5. $A \subseteq B$ (2; def. ⊆)

(c) 1. ...
2. ...
3. ...
4. $A \subseteq B$
5. for all $t \in A$: $t \in B$ (4; def. ⊆)

3. Give all the proof steps you can (with appropriate gaps) that lead to showing $S \subseteq T$.

1. Let $x \in S$ be arbitrary.
   ·
   ·
   ·
k. $x \in T$
k+1. for all $x \in S$: $x \in T$ (1—k; pr. ∀)
k+2. $S \subseteq T$ (k+1; def. ⊆)

Section 1.3

1. In the proof fragment (a) below, Step 3 is to follow from Steps 1 and 2 and the rule for using **for all**. There is only one statement that can replace the

Answers to Practice Exercises

asterisk to make logical proof steps that follow our rules of inference. The same is true for Steps 4 and 7. Fill in Steps 3, 4, and 7 in such a way as to make logical proof steps. In the other proof fragments, similarly fill in the proof steps marked by an asterisk. In cases where more than one logical replacement is possible, any one will do.

(a)
1. for all $t \in A$: $t \in B$
2. $x \in A$
3. $x \in B$ (1, 2; us. ∀)
4. $y \in A$
5. $y \in B$ (1, 4; us. ∀)
6. $t \in A$
7. $t \in B$ (1, 6; us. ∀)

(b)
1. Let $x \in Y$ be arbitrary.
2. ...
3. $x \in Z$
4. for all $x \in Y$: $x \in Z$ (1—3; pr. ∀)
5. $a \in Y$
6. $a \in Z$ (4, 5; us. ∀)

(c)
1. $x \in A$ or $x \in B$
  Case 1   2. Assume $x \in A$.
3. ...
4. $a \in B$
  Case 2   5. Assume $x \in B$.
6. ...
7. $a \in B$
8. $a \in B$ (1—7; us. or)

(d)
1. $a \in A$ or $b \in A$
  Case 1   2. Assume $a \in A$.
3. ...
4. $b \in C$
  Case 2   5. Assume $b \in A$.
6. ...
7. $b \in C$
8. $b \in C$ (1—7; us. or)

2. (a) 1. Let $x \in A$ and $y \in B$ be arbitrary.
2. ...
3. $x^2 + 1 < y$
4. for all $x \in A, y \in B$: $x^2 + 1 < y$ (1—3; pr. ∀)

(b) 1. $x \in S$
2. for all $t \in S$: $t \in T$
3. $x \in T$ (1, 2; us. ∀)

(c) 1. $x \in S \cup T$
2. $x \in S$ or $x \in T$ (1; def. ∪)

(d) 1. $s \in M \cup N$
2. $s \in M$ or $s \in N$ (1; def. ∪)

(e) 1. $x \in S$ or $x \in T$
2. $x \in S \cup T$ (1; def. ∪)

(f) 1. Let A and B be arbitrary sets.
2. ...
3. $A \subseteq A \cup B$
4. for all sets A, B: $A \subseteq A \cup B$ (1—3; pr. ∀)

(g) 1. Let A and B be arbitrary sets.
2. ...
3. $Q(A, B)$
4. for all sets A, B: $Q(A, B)$ (1—3; pr. ∀)

(h) 1. $a \in B \cup C$
2. $a \in B$ or $a \in C$ (1; def. ∪)
Case 1   3. Assume $a \in B$.
4. ...
5. $a \in X$
Case 2   6. Assume $a \in C$.
7. ...
8. $a \in X$
9. $a \in X$ (2—8; us. **or**)

(i) 1. $z < 1$ or $x > 5$
Case 1   2. Assume $z < 1$.
3. ...
4. $x \in A$

Answers to Practice Exercises

Case 2	5. Assume $x > 5$.	
	6. $\cdots$	
	7. $x \in A$	
	8. $x \in A$	(1—7; us. **or**)

(j)   1. Let $x \in A$ be arbitrary.

   $\cdot$

   $\cdot$

   k. $x \in B \cup C$

   k+1.   **for all** $x \in A$: $x \in B \cup C$ (1—k; pr. $\forall$)

Section 1.4

1. Use the extended definition rule to fill in the proof steps marked by an asterisk. Then fill in Step "$2\frac{1}{2}$", the step that has not been written but has been applied to get Step 3. There may be more than one correct response; if so, any one will do.

(a)   1. $A \subseteq B$
   2. $x \in A$
   3. $x \in B$ (1, 2; def.^{2} $\subseteq$)
   ($2\frac{1}{2}$). **for all** $t \in A$: $t \in B$

(b)   1. $X \subseteq Y$
   2. $t \in X$
   3. $t \in Y$ (1, 2; def.^{2} $\subseteq$)
   ($2\frac{1}{2}$). **for all** $a \in X$: $a \in Y$

(c)   1. Let $x \in A$ be arbitrary.
   2. $x \in B$
   3. $A \subseteq B$ (1, 2; def.^{2} $\subseteq$)
   ($2\frac{1}{2}$). **for all** $t \in A$: $t \in B$

(d)   1. $\cdots$
   2. $x \in B$
   3. $x \in B \cup C$ (for example)   (2; def.^{2} $\cup$)
   ($2\frac{1}{2}$). $x \in B$ **or** $x \in C$

(e)   1. $\cdots$
   2. $t \in X$ (or 2. $t \in Y$)
   3. $t \in X \cup Y$ (2; def.^{2} $\cup$)
   ($2\frac{1}{2}$). $t \in X$ **or** $t \in Y$

2. In the proof fragments below, fill in the proof steps marked with an asterisk. There may be more than one correct response; if so, any one will do.

(a) 1. $x \in A \cup B$
     2. $x \in A$ or $x \in B$ (1; def.[1] $\cup$)

(b) 1. $x \in A$ or $x \in B$
     2. $x \in A \cup B$ (1; def.[1] $\cup$)

(c) 1. for all $x \in A$: $x \in B$
     2. $A \subseteq B$ (1; def.[1] $\subseteq$)

(d) 1. $x < 10$
     2. $x \in A \cup B$
     3. $x \in A$ or $x \in B$ (2; def. $\cup$)
Case 1   4. Assume $x \in A$.
         .
         .
         j. $x \geq 10$ # Step 1.
Case 2 j+1. Assume $x \in B$.
         .
         .
         k. $x = 5$
k+1. $x = 5$ (3—k; us. **or**)

Section 1.5

1. 1. for all $x, y \in \mathbb{R}$: $x + y = z$
   2. for all $x, y \in \mathbb{R}$: $y + x = z$ (1; symmetry)

2. Assume, in the following proof fragment, that A and B are hypothesized as arbitrary sets.

   k. $A \cap B \subseteq B$
   k+1. $B \cap A \subseteq A$ (k; symmetry)

3. 1. for all $a, b, c \in \mathbb{R}$: $(a + b) + c = a + (b + c)$
   2. for all $a, b, c \in \mathbb{R}$: $(b + a) + c = b + (a + c)$ (1; symmetry)
or 2. for all $a, b, c \in \mathbb{R}$: $(c + b) + a = c + (b + a)$
or 2. for all $a, b, c \in \mathbb{R}$: $(a + c) + b = a + (c + b)$, etc.

4. 1. for all $x, y \in \mathbb{R}$: $x^4 + y^2 \geq 0$
   2. for all $x, y \in \mathbb{R}$: $y^4 + x^2 \geq 0$ (1; symmetry)

5. Give all the proof steps you can leading up to a proof of $S = T$, where S and T are sets.

    1. Let $x \in S$ be arbitrary.
        ·
        ·
    k. $x \in T$
k+1. $S \subseteq T$ $(1 \text{---} k; \text{def.}^2 \subseteq)$
    k+2. Let $x \in T$ be arbitrary.
        ·
        ·
    j. $x \in S$
j+1. $T \subseteq S$ $(k+2 \text{---} j; \text{def.}^2 \subseteq)$
j+2. $T = S$ $(k+1, j+1; \text{def.}^2 =)$

Section 1.6

1. 1. $X \subseteq Y$
    2. $Y \subseteq Z$
    3. $X \subseteq Z$ (1, 2; Thm. 1.3.1)

Section 1.7

1. (a)   1. Assume $q \in R$.
           2. $\ldots$
           3. $q \in S$
        4. if $q \in R$, then $q \in S$ $(1 \text{---} 3; \text{pr.} \Rightarrow)$

   (b)   1. $q \in R$
           2. if $q \in R$, then $q \in S$
           3. $q \in S$ $(1, 2; \text{us.} \Rightarrow)$

   (c)   1. $A \subseteq B$
           2. if $A \subseteq B$, then $B \subseteq C$
           3. $B \subseteq C$ $(1, 2; \text{us.} \Rightarrow)$

   (d)   1. Assume $x \in B$.
           2. $A \subseteq B$
        3. if $x \in B$, then $A \subseteq B$ $(1, 2; \text{pr.} \Rightarrow)$

   (e)   1. $x \notin B$ (or 1. $A \subseteq B$)

		2.	$x \notin B$ or $A \subseteq B$	(1; pr. **or**)
		3.	if $x \in B$, then $A \subseteq B$	(2; Thm. 1.7.2L)
(f)		1.	if $x \geq 5$, then $A \subseteq B$	
		2.	$x < 5$ or $A \subseteq B$	(1; Thm. 1.7.2L)
(g)		1.	$\mathcal{P}$ and $\mathcal{Q}$	
		2.	if $\mathcal{P}$, then $\mathcal{R}$	
		3.	$\mathcal{P}$	(1; us. **&**)
		4.	$\mathcal{R}$	(2, 3; us. $\Rightarrow$)

2. Suppose **if $x \in A$, then $x < 10$** is a step you need to show in some proof. Give all the "inevitable" proof steps that lead up to this statement.

    1. Assume $x \in A$.

       .
       .

    k. $x < 10$

k+1. if $x \in A$, then $x < 10$ (1—k; pr. $\Rightarrow$)

Section 1.8

1. (a) 1. $x \notin A$
       2. $x \in B$
       3. $x \in B - A$ (1, 2; def. $-$)

  (b) 1. $C = \{x \mid x \leq 2\}$
       2. $y \notin C$
       3. $y > 2$ (1, 2; def. C, prop. $\mathbb{R}$)
                               (The negation version of the definition rule was used.)

  (c) 1. $D = \{x \mid \mathcal{P}(x)\}$
       2. $y \notin D$
       3. $\sim \mathcal{P}(y)$ (1, 2; def. D)
       4. $z \in D$
       5. $\mathcal{P}(z)$ (1, 4; def. D)

  (d) 1. $A \subseteq B$
       2. for all $x \in A$: $x \in B$ (1; def. $\subseteq$)

  (e) 1. $A \nsubseteq B$
       2. $\sim$(for all $x \in A$: $x \in B$)   (1; def. $\subseteq$)

Answers to Practice Exercises

(f) 1. $x \notin A \cap B$
2. $\sim(x \in A$ and $x \in B)$ (1; def. $\cap$)

Section 1.10

1. (a) 1. there exists j such that $1 \leq j \leq 9$ such that $x \in A_j$
2. Pick i such that $1 \leq i \leq 9$ and such that $x \in A_i$.   (1; us. $\exists$)

(b) 1. there exists $y \in A$ such that $1 \leq y \leq n$
2. Pick $x \in A$ such that $1 \leq x \leq n$.   (1; us. $\exists$)

(c) 1. Let $1 \leq j \leq n$ be arbitrary.
.
.
.
k. $x \in A_j$
k+1. $x \in \bigcap_{i=1}^{n} A_i$ (1—k; def.^{2} $\cap$)

(d) 1. $x \in A_7$ (for example)
2. there exists $1 \leq i \leq 9$ such that $x \in A_j$ (1; pr. $\exists$)

Section 1.11

1. (a) 1. $x \in A_2$ (for example)
2. $x \in A_j$ for some $1 \leq j \leq 3$ (1; pr. $\exists$)

(b) 1. $x \in A_j$ for some $1 \leq j \leq 3$
2. Pick $1 \leq i \leq 3$ such that $x \in A_i$.   (1; us. $\exists$)

(c) 1. $\sim(x \in A_j$ for some $1 \leq j \leq 3)$
2. $x \notin A_j$ for all $1 \leq j \leq 3$ (1; negation)

(d) 1. $\sim(x \in A_j$ for all $1 \leq j \leq 3)$
2. $x \notin A_j$ for some $1 \leq j \leq 3$ (1; negation)

(e) 1. $\sim(x \notin A_j$ for all $1 \leq i \leq n)$
2. $x \in A_j$ for some $1 \leq i \leq n$ (1; negation)

(f) 1. $\sim(x \notin A_i$ for some $1 \leq i \leq n)$
2. $x \in A_i$ for all $1 \leq i \leq n$ (1; negation)

Section 1.12

1. (a) 1. $x \in A_i$ for some $i \in \mathcal{I}$
2. $x \in \bigcup_{i \in \mathcal{I}} A_i$ (1; def. $\cup$)

(b) 1. $x \notin \bigcup_{i \in \mathscr{I}} A_i$
   2. $\sim(x \in A_i \text{ for some } i \in \mathscr{I})$ (1; def. $\cup$)
   3. $x \notin A_i \text{ for all } i \in \mathscr{I}$ (2; negation)

(c) 1. $\sim(x \in A \text{ for all } A \subseteq B)$
   2. $x \notin A \text{ for some } A \subseteq B$ (1; negation)

(d) 1. $x \in A_i \text{ for all } i \in \mathscr{I}$
   2. $\sim(x \notin A_i \text{ for some } i \in \mathscr{I})$ (1; negation)

Chapter 2

Section 2.1

1. Define a function $f: A \to B$ where
   (a) A has two elements and B has four. Example:
       Let $f: \{a, b\} \to \{1, 2, 3, 4\}$ be defined by $f(a) = 2$ and $f(b) = 3$.
   (b) B has two elements and A has four. Example:
       Let $f: \{a, b, c, d\} \to \{X, Y\}$ be defined by $a \to X$, $b \to X$, $c \to X$, $d \to Y$.

2. Let $f: \mathbb{R} \to \mathbb{R}$ be defined by $f(x) = x^2$. Find
   (a) $f([1, 2]) = [1, 4]$
   (b) $f(3) = 9$
   (c) $f^{-1}(4) = \{2, -2\}$
   (d) $f^{-1}([1, 2]) = [1, \sqrt{2}] \cup [-\sqrt{2}, -1]$
   (e) $f(f^{-1}([1, 2])) = [1, 2]$
   (f) $f^{-1}(f([1, 2])) = [1, 2] \cup [-2, -1]$

Section 2.2

1. Let $h: \mathbb{R} \to \mathbb{R}$ be defined by $h(z) = z^3 + z$ and $k: \mathbb{R} \to \mathbb{R}$ by $k(z) = z^2 - 2$. Define:
   (a) $h \circ k$: $h \circ k(x) = (x^2 - 2)^3 + (x^2 - 2)$
   (b) $k \circ h$: $k \circ h(x) = (x^3 + x)^2 - 2$

2. Let $f: \{1, 2, 3, 4, 5\} \to \{a, b, c, d, e\}$ be defined by $1 \to a$, $2 \to a$, $3 \to b$, $4 \to b$, $5 \to c$. Let $g: \{a, b, c, d, e\} \to \{1, 2, 3, 4, 5\}$ be defined by $a \to 5$, $b \to 4$, $c \to 4$, $d \to 3$, $e \to 2$. Define:
   (a) $f \circ g$: $a \to c, b \to b, c \to b, d \to b, e \to a$
   (b) $g \circ f$: $1 \to 5, 2 \to 5, 3 \to 4, 4 \to 4, 5 \to 4$

Section 2.3

1. Suppose $f: A \to B$
   1. Let $f(x) = f(y)$ for $x, y \in A$.
   2. ...
   3. ...
   4. $x = y$
   5. for all $x, y \in A$: if $f(x) = f(y)$, then $x = y$ (1—4; pr. $\forall \Rightarrow$)
   6. f is one-to-one.   (5; def.[1] 1-1)

Section 2.4

1. Let $f: A \to B$. Replace the * in Step 1 with a language statement.
   1. for all $y \in B$: $f(x) = y$ for some $x \in A$
   2. f is onto.   (1; def. onto)

Section 2.5

1. Let $f: A \to B$.
   (a) 1. g is a left inverse of f.
       2. $g \circ f = i_A$ (1; def. left inverse)
   (b) 1. g is a right inverse of f.
       2. $f \circ g = i_B$ (1; def. right inverse)
   (c) 1. g is an inverse of f.
       2. g is a left inverse of f.   (1; def. inverse)
       3. g is a right inverse of f.   (1; def. inverse)

Section 2.6

1. Let **E** denote the set of even integers $\{\ldots, -4, -2, 0, 2, 4, \ldots\}$. Prove that $f: \mathbf{Z} \to \mathbf{E}$ given by $f(x) = 2x$ is a bijection.

   Proof:

   Assume: $f: \mathbf{Z} \to \mathbf{E}$
                 $f(x) = 2x$ for all $x \in \mathbf{Z}$
   Show: (1) f one-to-one
                 (2) f onto

   Proof of (1):
   1. Let $f(x) = f(y)$ for $x, y \in \mathbf{Z}$.
   2. $2x = 2y$ (1; def. f)

3. $x = y$ (2; prop. **R**)
4. if $f(x) = f(y)$ for $x, y \in \mathbb{Z}$, then $x = y$ (1—3; pr. $\Rightarrow$ abbr.)
5. f is one-to-one.   (4; def. 1-1)

Proof of (2):

1. Let $y \in \mathbf{E}$.
2. $y = 2x$ for some $x \in \mathbb{Z}$ (1; def. **E**)
3. $y = f(x)$ (2, hyp.; def. f)
4. $y = f(x)$ for some $x \in \mathbb{Z}$ (2, 3; pr. $\exists$)
5. for all $y \in \mathbf{E}$: $y = f(x)$ for some $x \in \mathbb{Z}$ (1—4; pr. $\forall$)
6. f is onto.   (5; def. **onto**)

Section 2.8

1. 1. $(a, b) = (c, d)$
    2. $a = c$ (1; Thm. 2.8.1)
    3. $b = d$ (1; Thm. 2.8.1)
2. The following properties of $\mathbb{Z}$ are written in the usual infix notation. Write them using prefix notation.
    (a) for all $a, b, c \in \mathbb{Z}$: $a \cdot (b \cdot c) = (a \cdot b) \cdot c$
    for all $a, b, c \in \mathbb{Z}$: $\cdot (a, \cdot(b, c)) = \cdot(\cdot(a, b), c)$
    (b) for all $a, b, c \in \mathbb{Z}$: $a \cdot (b + c) = (a \cdot b) + (a \cdot c)$
    for all $a, b, c \in \mathbb{Z}$: $\cdot (a, +(b, c)) = +(\cdot(a, b), \cdot(a, c))$

Chapter 3

Section 3.2

1. State (1) which properties—reflexive (r), symmetric (s), and transitive (t)—hold for the following relations and (2) which of the relations are equivalence relations (e).
    (a) Congruence on the set of plane triangles (r, s, t, e).
    (b) Similarity on the set of plane triangles (r, s, t, e).
    (c) Parallelism of lines in the plane (s). If a line is defined to be parallel to itself, then also (r, t, e).
    (d) Perpendicularity of lines in the plane (s).
    (e) $=$ on $\mathbb{R}$ (r, s, t, e).
    (f) $<$ on $\mathbb{R}$ (t).
    (g) $\leq$ on $\mathbb{R}$ (r, t).
    (h) R defined by xRy for $x, y \in \mathbb{R}$ iff x and y agree to the tenth decimal place (r, s, t, e).
    (i) R defined by xRy for $x, y \in \mathbb{R}$ iff $|x - y| \leq 2$ (r, s).

Index

Abelian group, 205
Absolute value, 308
Accumulation point, 261
Aleph null, 271
And statement
　rule for proving, 40
　rule for using, 40
Antisymmetric property, 222
Arbitrary element, 16
Archimedean property, 104, 308
Axiom of Choice, 150, 153, 171
Axioms for a group, 205

Bernstein, Felix, 267
Bijection, 155
Binary operation, 169
Binomial Theorem, 182
Bound variable, 10
Bounded
　function, 260, 261
　sequence, 255

Cantor, Georg, 266
Cardinal numbers, 267
Cardinality
　aleph null, 271
　finite sets, 265
　infinite sets, 271
　same, 162
　smaller, 163
Cartesian product, 167
Cauchy sequence, 253
Cayley's Theorem, 299
Chain, 223
Closed
　group operation, 278
　subset, 261
Codomain, 109
Comparable, 222
Complement, 67, 72
Composition of functions, 121
Computation rule, 4
Conclusion, 4
Congruence modulo a subgroup, 290

Congruent, 190
Conjecture, 116
Connective, 32
Continuous function, 243
Continuum hypothesis, 276
Contradiction, proof by, 65
Contrapositive, 136, 194
Converge, 232
Corollary to proof, 153
Correspondence Theorem, 298
Coset, 287, 290
Countable
　definition, 272
　union, 273
Counterexample defined, 119

Decimal rational, 237
Decreasing sequence, 255
Definition rule, 9
　extended, 31
　negation version, 74
　no longer in language, 139
Delta-epsilon proofs, 243
Density of the rationals, 308
Difference, set, 67
Dihedral group, 283
Diverge, 232
Divides, 186, 223
Division algorithm, 184, 186
Division, defined, 308
Domain
　of a function, 109
　functions of a real variable, 238
Dual
　definition, 223
　proof, 224
Dummy variable, 1, 10

Element of a set, 1
Empty set, 2
　notation, 2
　proving empty, 104
Equal
　functions, 123, 169
　sets, 41

363

Index

Equipotent sets, 162
Equivalence
  class, 192
  relation, 188
  rule for proving, 60
  rule for using, 61
  of statements, 33
Equivalent statements, definition, 60
Euler phi function, 286
Extended definition rule, 31

Factor group, definition, 295
Factorial, 157
Fermat, Pierre de, 291
Field, 222
  complete, 306
  definition, 306
  ordered, 306
Finite set, definition, 160
First Isomorphism Theorem, 299
For all statement, 8, 106
  format for proving, 11
  format for using, 21
  negation, 95
  rule for proving, 10
  rule for using, 21
For all: if-then statement, 131
For some statement, 84
  negation, 95, 101
  rule for proving, 88, 90, 101
  rule for using, 86, 101
Formalism, 165
Free variable, 10, 137, 138, 240
Function, 109
  formal definition, 167, 169
Fundamental Homomorphism Theorem, 298
Fundamental Theorem of Arithmetic, 221

Generator
  of a group, 285
  for an ideal, 219
Godel, Kurt, 165
Greatest common divisor, 220
Greatest lower bound, 223
Group
  axioms, 205
  definition, 205, 277
Group table, 208

Hilbert, David, 161, 165
Homomorphism
  additive, 212
  group, 210, 297
  induced, 215
  multiplicative, 212
  natural, 297
  ring, 211
Hypothesis, 3

Ideal, 218
Identity
  element of a group, 206, 277
  function, 125
If-then statement
  definition, 53
  rule for proving, 54
  rule for using, 57
Iff
  chains, 81, 82
  definition, 78
  in definitions, 145
Image, 110
Implication
  definition, 53, 62
  rule for proving, 54
  rule for using, 57
Inclusion map, 271
Increasing sequence, 255
Index of a subgroup, 289
Indexed family of sets, 100
Induced homomorphism, 215
Induction, 174
  complete, 221
Inductive definition, 179
Infinite set, 159
  definition, 160
Inherited operation, 278
Injection, 155
Interior point, 263
Intermediate Value Theorem, 250
Intersection, definition, 39, 84, 100
Inverse
  function, 151
  in a group, 278
  left, 147
  right, 147
Isomorphic, 227
  groups, 209, 210
Isomorphism
  group, 210
  order, 226
  of Peano systems, 304
  ring, 211

Kernel, 214, 297

Lagrange's Theorem, 288
Language, 140
  formal, 3
Latin square, 211
Lattice, 225
Least common multiple, 224
Least element, 174
Least upper bound, 223
  axiom, 250
Left-hand notation, 282
Lemma, 148

Let
  choosing an arbitrary symbol, 85
  defining new symbols, 85
Limit
  of a function, 239, 241
  of a sequence, 232
Linear order, 223
Local variable, 1, 10
Lower bound, 223

Maximal element, 225
Member of a set, 1
Metalanguage, 3, 140
Minimal element, 225
Monotonic sequence, 255

Necessary condition, 293
Negation
  of convergence condition, 235
  for all, 13
  inference rules, 95
  symbolization, 35
Neighborhood, 246
Nondecreasing sequence, 255
Nonincreasing sequence, 255
Null sequence, 256

One-to-one function, 129
Onto function, 140
Open
  statement, 2
  subset of $\mathbb{R}$, 262
Or statement
  format for proving, 33
  format for using, 24, 35, 37
  rule for proving, 33, 37
  rule for using, 24, 35, 36
Order
  group, 279
  group element, 279
Ordered pair, 166

$\mathcal{P}$ or not $\mathcal{P}$ rule, 58
Paradox, Russell's, 269
Partial order, 222
Partially ordered set, 222
Partition, 196
Peano system, 303
Permutation, 156
Pick, contraction with *there exists*, 144
Pigeonhole Principle, 281
Poset, 222
Power set, 119
Predicate, 170
Preimage, 110
Prime
  ideal, 222
  number, 220

Proof organization, 2
Proper subset, 163

Quotient group, definition, 295

Range of a function, 110
Rational number, 190
Reasons
  legitimacy, 70
  tactical, 70
Reflection, 285
Reflexive property, 188, 222
Relation, 188
  equivalence, 188
Relatively prime, 220
Repeating decimal, 236
Respect multiplication, 213
Restriction of a function, 135
Right multiplication, 288
Right-hand notation, 282
Ring
  commutative, 211
  definition, 210
Rotation, 285
Routine method for proof steps, 27
Russell, Bertrand, 268
  paradox, 269

Same cardinality, 265
Schroeder, Ernst, 267
Schroeder-Bernstein Theorem, 164–165
Second Isomorphism Theorem, 299
Sequence, definition, 232
Set, 1
Set definition rule, 3
  negation version, 67
Smaller cardinality, 265
Smallest element, 174
Statement, 1
Subgroup, 278
  normal, 294
Subset
  definition, 8
  proper, 163
Substitution, inference rule, 47
Subtraction, defined, 308
Successor function, 303
Sufficient condition, 293
Surjection, 155
Symmetric
  group, 282
  property, 188
Symmetry, 43
  rule for using, 43

Tautology, 311
Theorems, rule for using, 49

Index

There exists, contraction with pick, 144
There exists statements, 84, 106
   negation, 95, 101
   rule for proving, 88, 90, 101
   rule for using, 86, 101
   several variables, 105
Total order, 223
Transitive property, 188, 222
Transitivity, 164
Triangle inequality, 308
Trichotomy, 225
Truth tables, 194

Undecidable, 276
Undefined symbols, 137
Union, definition, 24, 84, 100
Uniqueness
   format for proving, 130
   inference rule, 129
Universal set, 1
Upper bound, 223

Well defined, 133
   function, 215
   operation, 187
   operation, defined, 199

Well ordered, 174
Well-ordering axiom, 174

Zero-function, 212

Style Abbreviations

"Assume" implicit in cases, 52 (Section 1.6)
Contraction of steps
   extended definition, 31 (Section 1.4)
   for all: if-then, 131 (Section 2.3)
   there exists and pick, 144 (Section 2.4)
Defining property in language, 139 (Section 2.4)
Let, defining new variable
   as arbitrary element, 11 (Section 1.2), 85 (Section 1.10)
   in terms of previous variables, 85 (Section 1.10)
Paragraph outline for proof, 68 (Section 1.8)
Paragraph proof, 141, 144 (Section 2.4)
Variables
   free, 240 (Section 4.2)
   omit in hypotheses, 78 (Section 1.9)